AF894850

Springer Tracts in Mechanical Engineering

Board of editors

Seung-Bok Choi, Inha University, Incheon, South Korea
Haibin Duan, Beijing University of Aeronautics and Astronautics, Beijing, P.R. China
Yili Fu, Harbin Institute of Technology, Harbin, P.R. China
Jian-Qiao Sun, University of California, Merced, USA

About this Series

Springer Tracts in Mechanical Engineering (STME) publishes the latest developments in Mechanical Engineering - quickly, informally and with high quality. The intent is to cover all the main branches of mechanical engineering, both theoretical and applied, including:

- Engineering Design
- Machinery and Machine Elements
- Mechanical structures and stress analysis
- Automotive Engineering
- Engine Technology
- Aerospace Technology and Astronautics
- Nanotechnology and Microengineering
- Control, Robotics, Mechatronics
- MEMS
- Theoretical and Applied Mechanics
- Dynamical Systems, Control
- Fluids mechanics
- Engineering Thermodynamics, Heat and Mass Transfer
- Manufacturing
- Precision engineering, Instrumentation, Measurement
- Materials Engineering
- Tribology and surface technology

Within the scopes of the series are monographs, professional books or graduate textbooks, edited volumes as well as outstanding Ph.D. theses and books purposely devoted to support education in mechanical engineering at graduate and post-graduate levels.

More information about this series at http://www.springer.com/series/11693

Ruxu Du · Zheng Li · Kamal Youcef-Toumi
Pablo Valdivia y Alvarado
Editors

Robot Fish

Bio-inspired Fishlike Underwater Robots

Springer

Editors
Ruxu Du
The Chinese University of Hong Kong
Hong Kong
China

Kamal Youcef-Toumi
Massachusetts Institute of Technology
Cambridge, MA
USA

Zheng Li
The Chinese University of Hong Kong
Hong Kong
China

Pablo Valdivia y Alvarado
Massachusetts Institute of Technology
Cambridge, MA
USA

ISSN 2195-9862 ISSN 2195-9870 (electronic)
Springer Tracts in Mechanical Engineering
ISBN 978-3-662-46869-2 ISBN 978-3-662-46870-8 (eBook)
DOI 10.1007/978-3-662-46870-8

Library of Congress Control Number: 2015936287

Springer Heidelberg New York Dordrecht London
© Springer-Verlag Berlin Heidelberg 2015
This work is subject to copyright. All rights are reserved by the Publisher, whether the whole or part of the material is concerned, specifically the rights of translation, reprinting, reuse of illustrations, recitation, broadcasting, reproduction on microfilms or in any other physical way, and transmission or information storage and retrieval, electronic adaptation, computer software, or by similar or dissimilar methodology now known or hereafter developed.
The use of general descriptive names, registered names, trademarks, service marks, etc. in this publication does not imply, even in the absence of a specific statement, that such names are exempt from the relevant protective laws and regulations and therefore free for general use.
The publisher, the authors and the editors are safe to assume that the advice and information in this book are believed to be true and accurate at the date of publication. Neither the publisher nor the authors or the editors give a warranty, express or implied, with respect to the material contained herein or for any errors or omissions that may have been made.

Printed on acid-free paper

Springer-Verlag GmbH Berlin Heidelberg is part of Springer Science+Business Media
(www.springer.com)

Editorial

How fish swim is a wonder of nature—they are elegant, fast, quiet, and move with high efficiency. It is known that the propulsion efficiency of fish is about 90 %. By comparison, the efficiency of the currently vastly adopted rotary propeller can hardly reach even half of it. In fact, the performance of a typical ship, in terms of efficiency, speed, acceleration, turning, noise, etc., is far worse than that of a fish. The outstanding performance of fish swimming is the result of millions of years of evolution and is attributed to a variety of factors including a streamlined body, mucous skin, perfect coordination control, etc. Though the key reason is its propulsion method. This motivates scientists and engineers around the world to study how fish swim and accordingly, to design and build robot fish to mimic fish swimming. Particularly, in the past decade or so, like many other science and technology subjects, the research and development on robot fish has experienced an exponential growth. Therefore, it has become necessary to have a book that summarizes the current state-of-the-art technologies, analyzes the roadblocks, and gives the future directions. The objectives of this book are threefold: First, it summarizes the state-of-the-art technologies in designing, building, and controlling robot fish. Second, it analyzes the major obstacles for future development. These obstacles include scientific difficulties, such as materials, two-phase flow, and underwater sensing, as well as engineering challenges ranging from water leakage to autonomous control. Third, it points out the future research and development directions, which include efficient power generation, efficient power transmission, efficient propulsion, sensing, intelligent control, autonomous swimming, etc.

The book contains a total of 12 chapters. The first chapter, "Introduction" gives an overview of the research and development of robot fishes. "Fish Locomotion: Biology and Robotics of Body and Fin-Based Movements" presents a systematic study on the locomotion of fish and robot fish. "Wire-driven Robot Fish" presents a unique robot fish design using the wire-driven mechanism, while "Design and Control of a Multi-Joint Robot Fish" and "Research on the Robotic Fish Propelled by Oscillating Pectoral Fins" give two classical designs, one flaps the caudal fin and the other flaps the pectoral fin. "Soft-Body Robot Fish" presents a design based on soft body, by which the robot fish structure and control can be simplified. "*iSplash*:

Realizing Fast Carangiform Swimming to Outperform a Real Fish" discusses how to make robot fish swimming as fast as real fish. "IPMC-Actuated Robotic Fish" and "Macro-Fiber Composite Actuated Piezoelectric Robotic Fish" present two robot fish that are driven by smart materials (specifically, Ionic polymer-metal composites (IPMC) and Macro-fiber composite). "A Multifunctional Underwater Microrobot for Mother-Son Underwater Robot Systems" and "Fish-Robot Interactions: Robot Fish in Animal Behavioral Studies" deal with intelligent control, which include robot fish swimming in groups, cooperating to achieve specific tasks, and interact with real fish.

The robot fish research is still in its infancy. However, we believe that 10 years from now, robot fish will be able swim with real fish in open water. It can carry out many tasks such as underwater exploration, pollution monitoring, and water transportation. This book will help to achieve this goal.

We wish that the readers find the book useful and inspiring.

Ruxu Du
Zheng Li
Kamal Youcef-Toumi
Pablo Valdivia Y Alvarado

Acknowledgments

Robot fish is a new area that requires knowledge from various fields including physics, biology, mechanical engineering, electronics engineering, materials engineering, and computer science. Moreover, it is a new area, so few can cover the entire area. The authors wish to thank all the contributors; without them this book would not have been possible. The authors also wish to thank Ms. Li Shen, Ms. Jessie Guo and Mr. Prasanna Kumar Narayanasamy of Springer, who helped to edit the book.

Additionally, the first editor, Ruxu Du, wishes to thank his wife, Hui Yang, who gave all possible encouragement and support throughout the years. The second editor, Zheng Li, wishes to thank his family, especially his wife Hui Peng, for the everlasting encouragement and support in his life.

Contents

Introduction

Ruxu Du, Zheng Li, Kamal Youcef-Toumi and Pablo Valdivia y Alvarado

Abstract This introductory chapter includes four sections. The first section presents various methods of propulsion in water and brings in interesting facts about fish swimming. The second section briefly introduces how fish and other aquatic animal swims. The third section reviews the current status of robot fish. Finally, a methodology for the design of biomimetic robot fish is presented in the fourth section. The chapter also serves an introduction to the book.

1 Propulsion in Water from Ancient to Modern

Water is a magical fluid. Setting aside its many characteristics that support life on earth, the buoyance of water allows animals to swim and ships to sail. For instance, the world of shipping depends greatly on water from ancient to modern. Currently, nearly 90 % of world shipping is on water. Shipping consumes approximately 17.5 Exa-Joule (EJ) per year. Assuming that improved propulsion efficiency saves just 1 %, it will be sufficient to power entire Hong Kong for a year. Moreover, the ocean is the frontier of the modern world; with ever-increasing population, we will inevitably look for food, energy, and living space in the ocean. Clearly, propulsion in water is extremely important and worthwhile to investigate, invest, and improve.

R. Du (✉) · Z. Li
The Chinese University of Hong Kong, Hong Kong, China
e-mail: rdu@mae.cuhk.edu.hk

Z. Li
e-mail: lizheng@cuhk.edu.hk

K. Youcef-Toumi · P.V. y Alvarado
Massachusetts Institute of Technology, Cambridge, MA, USA
e-mail: youcef@mit.edu

P.V. y Alvarado
e-mail: pablov@mit.edu

© Springer-Verlag Berlin Heidelberg 2015

R. Du et al. (eds.), *Robot Fish*, Springer Tracts in Mechanical Engineering,
DOI 10.1007/978-3-662-46870-8_1

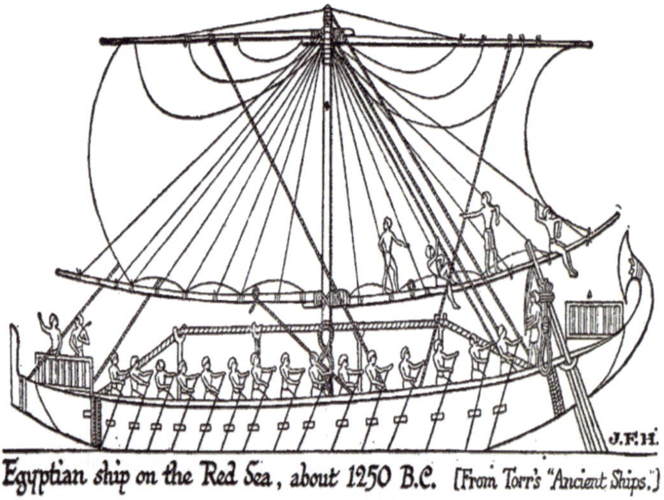

Fig. 1 Ancient Egyptian ship (adopted from Wikipedia) [30]

It is not clear when our ancestors built the first ship to navigate in water. However, it is known that reeds, canoes, and rafts are at least 7,000 years old. How to propel ships? Our ancestors were probably inspired by fish swimming in developing propulsion methods. The first method is rowing, which is imitating the moving of the pectoral fin of the fish. A small canoe can be paddled by one person, while a large ship may have dozens of pedals as shown in Fig. 1.

The second method is sculling. As shown in Fig. 2, sculling is a technique mastered by the ancient Chinese. It was perhaps inspired by the moving of body

Fig. 2 Chinese "sampan" propelled by "yáolŭ" via single-oar sculling (adopted from Wikipedia) [31]

Fig. 3 The large screw propeller of a modern ship (adopted from Wikipedia) [1]

and caudal fin of fish. When the first author was a child in the 1960s, sculling ships were still very popular in the Pearl River delta.

With the advance of civilization, our ancestors gradually learned to use the wind sail and then the screw propeller. In the modern era, the theory and practice of screw propeller [1] is well developed around the world. It is simple to make, simple to control, and efficient. Nowadays, few large ships are not driven by the screw propeller (Fig. 3).

2 How Do Fish Swim

Compared to fish swim, however, man-made propulsion methods are far inferior. It is known that the propulsion efficiency of fish exceeds 90 % [2]; the speed of a sailfish can reach 110 km/h, while the maximum acceleration of a pike is as high as 249 m/s^2, which is over 25 g. Also, fish swim silently and can turn around sharply. The outstanding performance of fish is attributed to multiple factors, such as streamlined body, mucous body surface, etc. Among all these factors, the most important is their propulsion method.

Then, how does fish propel? In the past century, science and technology started an explosive growth. A number of pioneers started to examine the propulsion of fish and other aquatic animals. In the 1960s, Jacques Yves Cousteau [3] (Fig. 4), a world-renowned explorer, conservationist, filmmaker, innovator, scientist, author, and photographer directed an award-winning film, "The Silent World," which inspired a generation of people. Subsequently, he directed a number of films and published a series of books. In his book entitled "The Ocean World of Jacques Cousteau, the Art of Motion," he studied the swimming of fish and other aquatic animals. First, he pointed out that the shape of the fish determines how does the fish swim. Fish shapes can be roughly divided into the following categories: the

Fig. 4 Jacques Yves Cousteau (1910–1997), photo adopted from Wikipedia [3]

laterally fattened (e.g., angelfish); the fusiform and laterally fattened (e.g., soldier fish); the fusiform (e.g., cod), the attenuated and cylindrical (e.g., trumpet fish), the attenuated (e.g., the moray), the dorsoventrally compressed (e.g., stingray), and the combining shapes (e.g., the jack). Next, he pointed out that during swimming, fish are subject to various forces causing it to roll, pitch, and yaw, as shown in Fig. 5. To achieve highly effective swim, fish with different shapes evolve different fins (or fish with different fins evolve to different shapes) and hence, different propulsion methods. Additionally, fins are also used to start, lift, stabilize, hover, turn, and stop. Nature is full of wonder. Some seemingly impossible designs, such as the ocean sunfish, (also called *Mola mola*), oarfish, and seahorse can swim just fine. Moreover, the method of swimming defines the lifestyle of the fish. Most fast swimmers, big or small, adopt tail flapping for propulsion.

At about the same time, British scientist Sir Michael James Lighthill (Fig. 6) studied fish swimming using fluid mechanics. From a history point of view, the

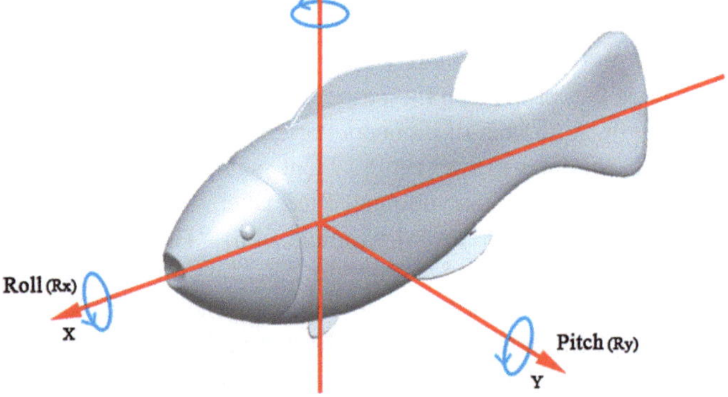

Fig. 5 Illustration of fish motion during swimming

Fig. 6 Michael James Lighthill (1924–1998), photo adopted from Wikipedia [32]

earliest scientific study of fish swimming could be dated back to the 1930s [4]. Though, it was Lighthill who laid the cornerstone [5–8].

For fish and other aquatic animals, there are generally two types of swimming methods: one using Body and/or Caudal Fin (BCF) and the other using the Medium and/or Paired Fin (MPF). There is also jet propulsion (e.g., jellyfish and scallop) as well as walkers and crawlers (e.g., shrimp and crab), though these are not mainstream. Moreover, BCF (i.e., tail flapping) is used by approximately 85 % of the fish species, including many fast swimmers such as sailfish, tuna, and pike. Therefore, it is most studied form of swimming. Within BCF, there are roughly two kinds of motion modes: the oscillatory motion, or the "C" mode, (e.g., carp) and the undulatory motion, or the "S" mode (e.g., eel).

Lighthill proposed that BCF flapping can be described by the traveling wave model below:

$$y(x,t) = \left[c_1 x + c_2 x^2\right][\sin(kx + \omega t)] \tag{1}$$

where y and x are sideward and forward displacements respectively (refer to Fig. 5), t is the time, c_1 and c_2 are the linear and quadratic wave amplitudes, k is the body wave number and ω is the wave frequency. Note that the increase of k changes the swim from oscillatory swimming to undulatory swimming. As an example, Fig. 7 shows a couple of computer simulation results. Figure 7a is oscillatory swimming, in which the model coefficients are $c_1 = 0.1$, $c_2 = 0.2$, $k = 0.5$, and $\omega = \pi$. In the figure, the five curves represent the fish body curves at different time instances. Additionally, it is seen that the head of the fish is kept straightforward (i.e., $y(0, t) = 0$) and the tail of the fish has the largest displacement. Figure 7b is undulatory swimming, in which the model coefficients are $c_1 = 0.02$, $c_2 = 0.0835$, $k = 0.6$, and $\omega = 3.86$. In this case, each fish body curve can be viewed as a number of circular arcs connected serially. The traveling wave model has been validated by many scientific experiments.

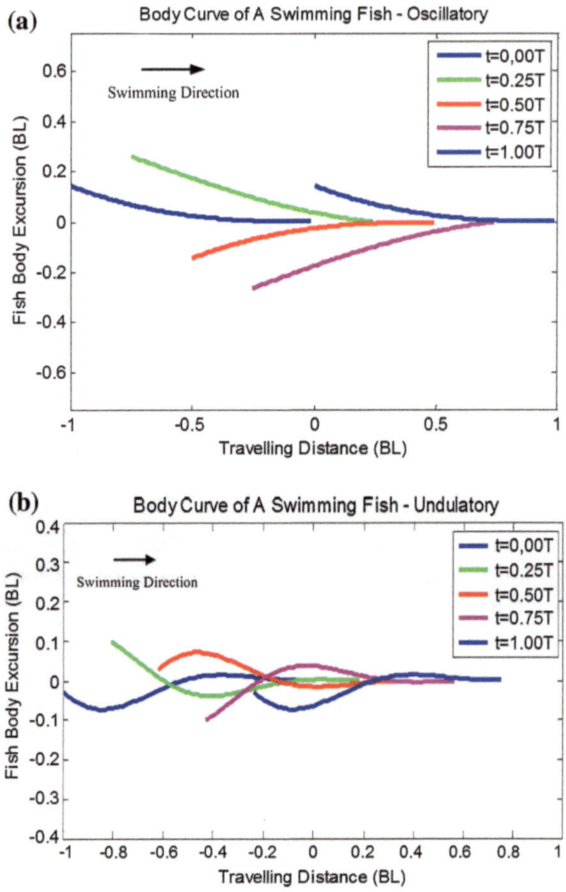

Fig. 7 Computer simulation of the traveling wave model: **a** An example of oscillatory swimming; and **b** an example of undulatory swimming

Following his footsteps, many people contributed to modeling fish swimming. For example, Alvarado and Youcef-Toumi [9] suggested a wave model that counts the "recoil" of the fish (i.e., when the fish's tail flaps, its head also turns):

$$y(x,t) = \left[c_0 + c_1 x + c_2 x^2\right]\left[\sin(kx + \omega t)\right] \tag{2}$$

where c_0 is a constant.

A more detailed classification of fish swimming includes anguilliform, subcarangiform, carangiform, and thunniform in which anguilliform is considered as S mode while the others are considered as C mode, as shown in Fig. 8. To mathematically differentiate these swimming forms, we can use the characteristic function defined below:

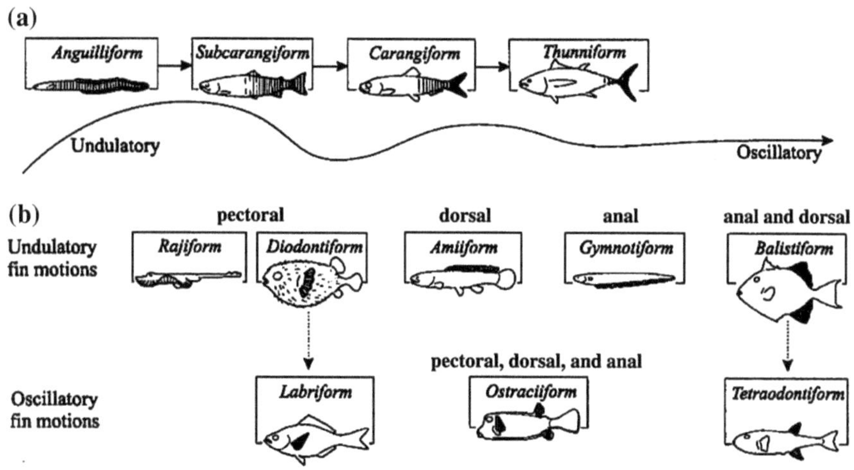

Fig. 8 The classification of types of fish swim [33, 34], the *dark parts* indicate where the flapping occurs [35]. **a** BCF; **b** MPF

$$\Phi(\lambda, x) = \sin\left(\frac{2\pi}{\lambda}x\right), \quad 0 \leq x \leq 1, \lambda \geq 0 \tag{3}$$

where $\lambda = 2\pi/k$ (i.e., $k = 2\pi / \lambda$). The characteristic function describes the traveling wave pattern with respect to the Body Length (BL), x ($0 \leq x \leq 1$) and the wave length, λ.

According to the characteristic function, the swimming forms are differentiated as follows:

(a) For Thunniform swimming, $2\pi/\lambda \leq \pi/2$, or $\lambda \geq 4$;
(b) For Carangiform swimming, $\pi/2 < 2\pi/\lambda \leq 3\pi/2$, or $4 > \lambda \geq 4/3$;
(c) For Subcarangiform swimming, $3\pi/2 < 2\pi/\lambda \leq 5\pi/2$, or $4/3 > \lambda \geq 4/5$; and
(d) For Anguilliform swimming, $2\pi/\lambda > 5\pi/2$, or $\lambda < 4/5$.

Figure 9 shows the characteristic curve of the four swimming forms. In thunniform swimming, waving is confined to a quarter of the wavelength. In carangiform swimming, waving is between 1/4 and 3/4 of the wavelength. In subcarangiform swimming, waving is between 3/4 and 5/4 of the wavelength. In anguilliform swimming, waving is over 5/4 of the wavelength. This categorization is compatible to Wardle's observations [10]. In his findings, the wavelength of a scup (carangiform swimming) is around 1.5 BL ($\lambda = 1.5$); the wavelength of a saithe (subcarangiform swimming) is 1 BL ($\lambda = 1.0$), and the wavelength of an eel (anguilliform swimming) is 0.5 BL ($\lambda = 0.5$).

The traveling wave is a function of time. Figure 10 shows a set of computer simulation results. From the figure, it is seen that the head of the fish is kept still while the body flaps following the characteristic function.

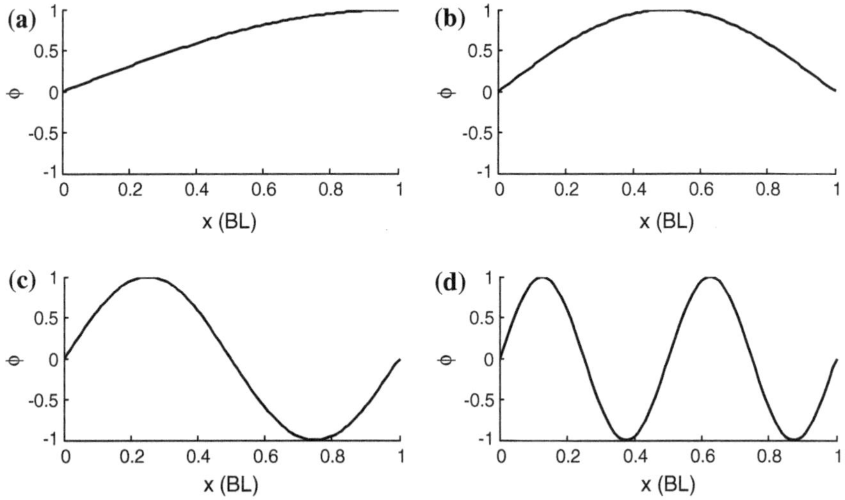

Fig. 9 The characteristic function of fish swimming forms: **a** Thunniform; **b** Carangiform; **c** Subcarangiform; and **d** Anguilliform

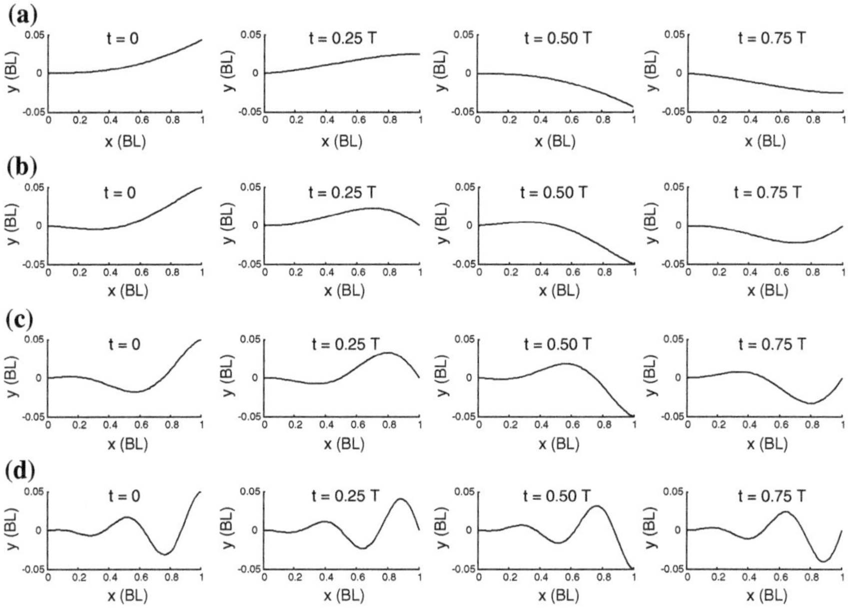

Fig. 10 Fish swimming curves based on the traveling wave model: **a** Thunniform swimming; **b** Carangiform swimming; **c** Subcarangiform swimming; and **d** Anguilliform swimming

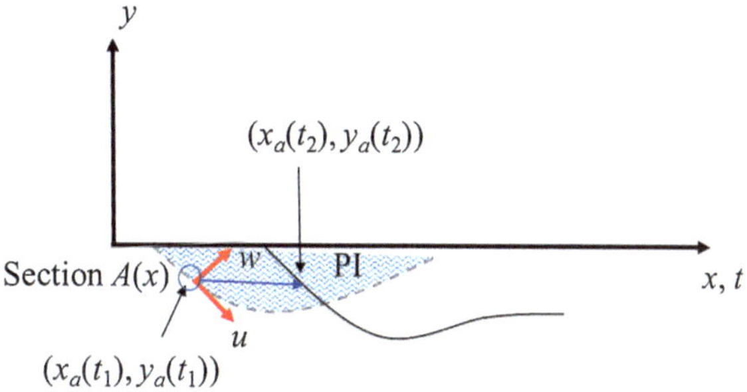

Fig. 11 Illustration of the traveling wave lines of two successive instants

Lighthill's most significant work is his Elongated Body Theory (EBT). In EBT, as illustrated in Fig. 11, it is assumed that: (1) The water momentum near a section of a fish ($A(x)$) is in the direction perpendicular to the body of the fish (w). It has magnitude equal to the virtual mass per unit length of the fish (m) times the perpendicular component of fish velocity in that direction (u); (2) The thrust can be obtained by considering rate of change of momentum within the volume enclosing the fish whose boundary at each instant includes a flat surface, PI, (the shaded area) perpendicular to the caudal fin through its posterior end; and (3) In the momentum balance it is necessary to take into account the transfer of momentum across the PI ($mw^2/2$) not only by convection but also by the action of the resultant of the pressures generated by the motions within the plane PI. Note that in the figure, the dashed line represents the traveling wave of the fish in time t_1, and the solid line represents the travel wave of the fish in time t_2.

According to EBT, the instantaneous push per unit length of the fish is:

$$L(x,t) = -\rho\left(\frac{\partial}{\partial t} + U\frac{\partial}{\partial x}\right)\{V(x,t)A(x)\} \tag{4}$$

where ρ is the density of the flow; U is the passing flow velocity; $V(x, t) = (u(x, t), w(x, t))$ is the velocity of the fish relative to the flow; $A(x)$ is the cross-section area of the fish body mentioned in the previous paragraph, and x varies from 0 to L corresponding to the length of the fish. Note that the mass of the fish is not in the equation because it is the same as the water approximately equal to 1. Thus, the rate of work done by the fish can be represented as follows:

$$W = \frac{\partial}{\partial t}\left\{\rho\int_0^L \frac{\partial y}{\partial t} VA(x)\mathrm{d}x - 0.5\rho\int_0^L V^2 A(x)\mathrm{d}x\right\} + \rho U\left[\frac{\partial y}{\partial t} VA(x)\right]_0^L \tag{5}$$

where $\partial y/\partial t$ is the traversing velocity of the fish. Over a long period of time the mean work done by the fish is:

$$\bar{W} = \rho U A(l)\left\{\overline{\frac{\partial y}{\partial t}\left(\frac{\partial y}{\partial t} + U\frac{\partial y}{\partial x}\right)}\right\}_{x=L} \tag{6}$$

where $\partial y/\partial t$ is the slope of the fish body. The rate of work available for thrust is obtained by subtracting the rate of kinetic energy of lateral fluid shedding $\left(0.5\rho\bar{V}^2 A\right)U$. As a result, the mean thrust is:

$$\bar{T} = \left[\frac{1}{2}\rho A(l)\left(\overline{\left(\frac{\partial y(x,t)}{\partial t}\right)^2} - U^2\overline{\left(\frac{\partial y(x,t)}{\partial x}\right)^2}\right)\right]\Bigg|_{x=L} \tag{7}$$

Moreover, the drag force of the fish is:

$$F_D = \frac{1}{2}C_D\rho U^2 S \tag{8}$$

where C_D is the drag coefficient. For long cylinders, C_D would be equal to 0.82; for cone shape, it is 0.5 [6, 7]. The cruising speed of the fish is achieved when the mean thrust is equal to the mean drag force. Thus, we have the cruising speed as:

$$U = \left[\sqrt{\frac{m\cdot\overline{\left(\frac{\partial y(x,t)}{\partial t}\right)^2}}{C_D\rho S + m\cdot\overline{\left(\frac{\partial y(x,t)}{\partial x}\right)^2}}}\right]\Bigg|_{x=L} \tag{9}$$

Here $m = \rho A(L)$ is the virtual mass at the fish tail distal tip, which is constant for a given fish. From this representation, the variables affecting the cruising speed include the tail distal tip lateral velocity and the tail distal tip slope. Once these two terms are known, the mean swimming speed and the thrust force can be obtained. The mean power of the fish is the product of cruise speed and mean thrust:

$$\overline{P} = \rho U A(l)\left(\overline{\frac{\partial y(x,t)}{\partial t}\left(\frac{\partial y(x,t)}{\partial t} + U\frac{\partial y(x,t)}{\partial x}\right)}\right)\Bigg|_{x=L} \tag{10}$$

Finally, the propulsion efficiency (Froude Efficiency) of the fish is the rate of work done and the total work. It is obtained as:

$$\eta = \frac{U\bar{T}}{\bar{P}} \times 100\,\% \tag{11}$$

An improved efficiency model for the EBT is shown in Eq. (12). It is also noted as the improved Froude efficiency model [7].

$$\begin{cases} \eta = \frac{1}{2}(1+\beta) - \frac{1}{2}\alpha^2\frac{\beta^2}{1+\beta} \\ \alpha = \frac{\lambda}{2\pi}\frac{h'(\mathrm{L})}{h(\mathrm{L})} = \frac{1}{2\pi}\left(\frac{\lambda}{L}\right)\left(\frac{\Delta L}{L}\right)^{-1}\left(1-\frac{h_t}{h_m}\right) \\ \beta = U/V \end{cases} \tag{12}$$

(a) **(b)**

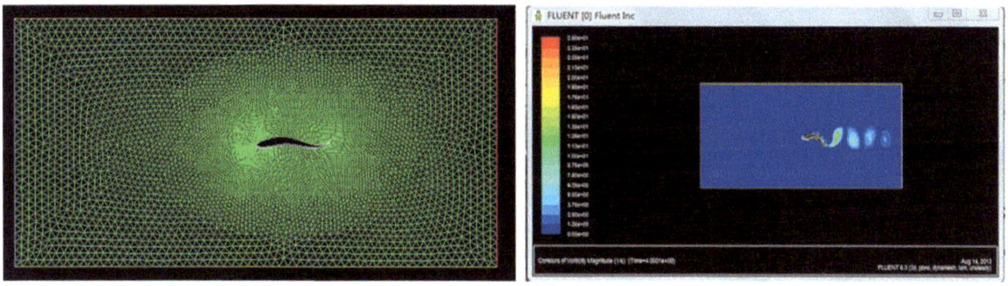

Fig. 12 Modeling of fish swim using CFD software Fluent®. **a** The mesh in CFD modeling; **b** A graphic simulation result

where h_m is the amplitude at the tail end $x = L$; and h_t is amplitude at $x = L + \Delta L$.

Following in the footsteps of Lighthill, many analytic studies were carried out and the results are documented in the monographs by Videler [11] and Romanenko [12], as well as the review papers [2, 13].

The classical study is unfortunately stained by the complex mathematics, oversimplified model and boundary conditions, as well as low accuracy approximation. In the past two decades, with the rapid advance of computer technology, Computational Fluid Dynamics (CFD) is increasingly used to model fish swimming [14–16]. We can use commercial software systems, such as FLEUNT®, to model the fish swim. Note that the modeling is not straightforward. First, it needs to develop the model of the fish and mesh the surrounding fluid (Fig. 12a). Second, it needs to setup the motion of the fish (we can use the aforementioned travel wave model). Third, it requires to setup the moving boundary for each steps and choose a solving algorithm. The simulation may take from several minutes to a number of hours, depending on the mesh and the motion. Finally, the simulation can be visualized in a number of ways, such as figure (Fig. 12b), graph, and/or animation. With proper modeling, we can model various fish swim modes and conditions (such as cruising, accelerating, turning, diving, stopping, etc.). Figure 12 shows a set of simulation examples and Table 1 summarizes the simulation results. In the table, the first three rows are the model parameters (Italicized values). The flapping frequency is set as 2 Hz and the flapping amplitude is 0.5 BL. The last two rows show the speed and the efficiency of the fish swim (Bolded values) (Fig. 13).

Table 1 Samples of CFD simulation of fish swimming

	Anguiliform	Subcarangiform	Carangiform	Thunniform
c_1	*1.0*	*0.7*	*0.4*	*0.1*
c_2	*0.0*	*0.3*	*0.6*	*0.9*
λ	*0.5*	*0.75*	*1.0*	*1.25*
V	**1.02**	**1.35**	**1.43**	**1.36**
η	**0.74**	**0.69**	**0.60**	**0.50**

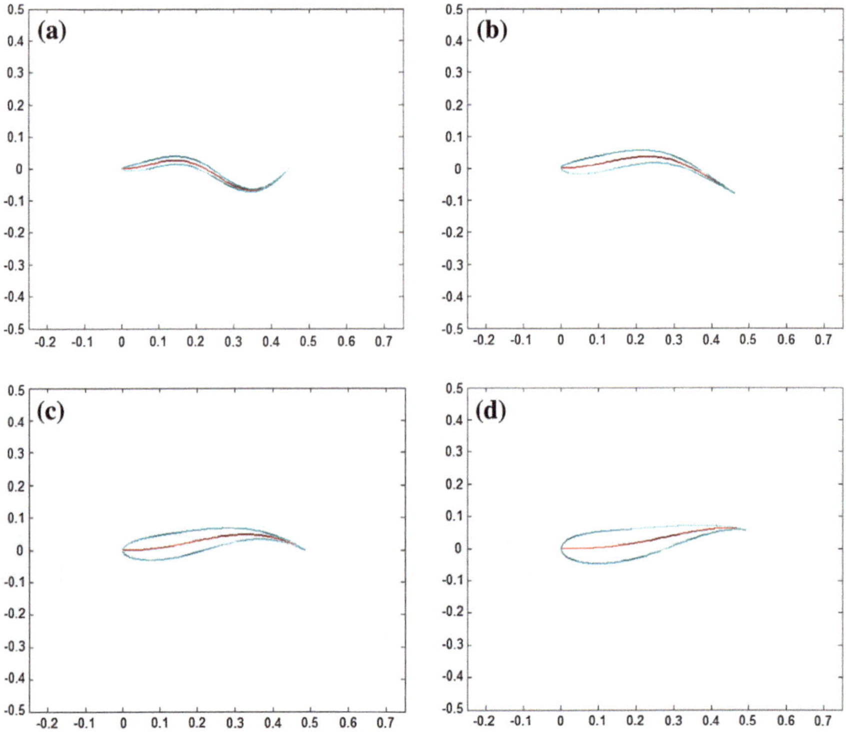

Fig. 13 Several examples of CFD models of fish swim. **a** Anguiliform; **b** Subcarangiform; **c** Carangiform; **d** Thunniform

It shall be pointed out that our CFD study is rather preliminary. More detailed studied have been carried out as shown in [15, 16].

In comparison to BCF swim, MPF swim is less understood. As shown in Fig. 8, MPF swim can be further divided into two categories: undulatory fin motion and oscillatory fin motion. The former includes pectoral fin motion, dorsal fin motion, anal fin motion and a combination of them. The latter includes labriform motion, ostraciiform motion, and tetraodontiform motion. The MPF swim is equally effective in nature, but is less studied due in part to its complexity. Similar to the BCF swim, the flapping of the pectoral fin can be considered as a traveling wave, but in 3D. No simple model is available in the literature, though experimental data is available. Figure 14 shows the model of a ray fish. Figure 15 shows a set of experiment results from a cow-nosed ray. These data are borrowed from Chap. 5 of this book. In the figure, each curve represents the edge of the pectoral fin in the (x, z) plane at a time instant. Note that the downstroke and the upstroke are different. In order to find the velocity, thrust force, and efficiency, CFD modeling is perhaps the only way besides experiments.

As shown in Chap. 2 of this book, in nature fish swimming is achieved by a combination of multiple fins and each fin is controlled by a multiple group of

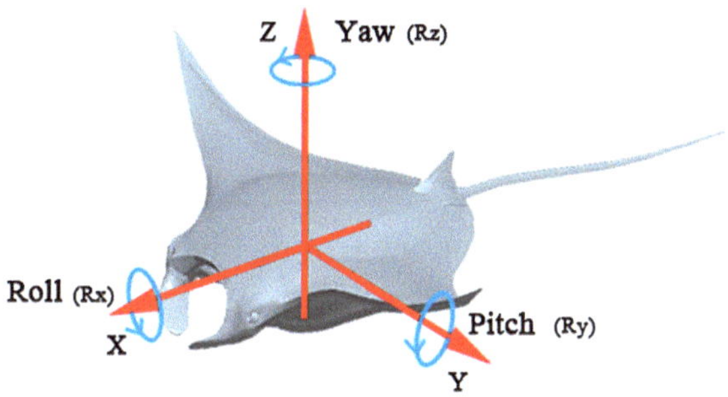

Fig. 14 The model of a ray fish swimming

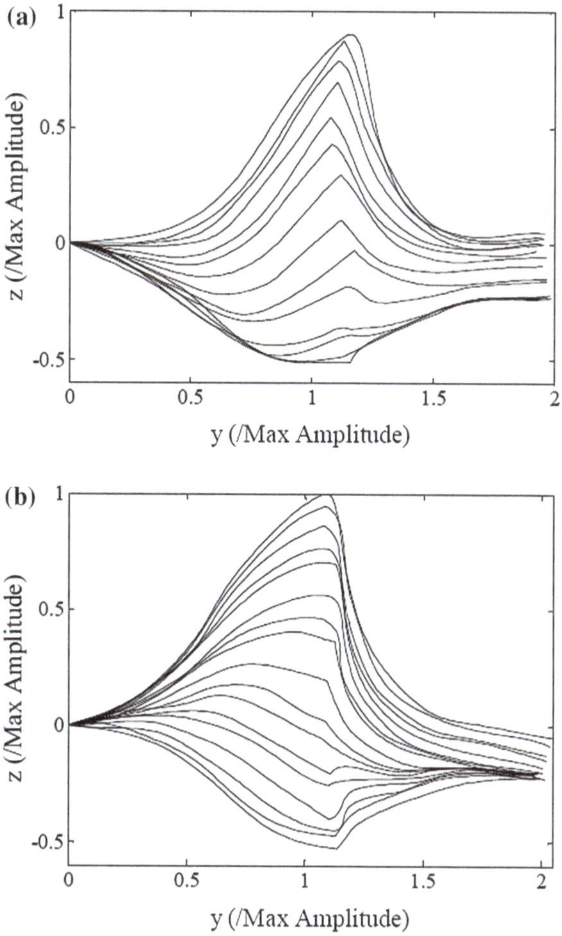

Fig. 15 The flapping motion of a cow-nosed ray (adopted from Chap. 5 of this book). **a** Downstroke; **b** Upstroke

muscles. As a result, the aforementioned models, including complex CFD models, are but much simplified models. Though, these models help us to understand the basics of fish swimming and to build a robot fish that can swim efficiently.

3 A Brief Review of Robot Fish

Inspired by fish swimming, many engineers and scientists have tried to design and build flapping-based robot fish. Since the first robot fish reported from MIT in 1989 [17], presently there are at least 400 papers published and some 40 different robot fish built (excluding some 30 different robot fish toys that can only flap and drift in water). This book presents a number of most representative robot fish built around the world. It shall be pointed out though, the research on robot fish is in its infancy and much work is still needed.

In general, the main efforts on robot fish research can be divided into two categories: design and prototyping, as well as control and navigation.

According to the literature survey, most robot fish are of BCF type. The designs of these robot fish can be further divided into three kinds: the Single Joint (SJ) design, the Multi-Joint (MJ) design, and the smart material-based design as illustrated in Fig. 16. The first design uses direct drive or conventional mechanisms, such as crank and four-bar mechanism, to transform the rotation motion of a motor to the flapping motion. This design is commonly adopted by robot fish toys. It is simple, easy to make, easy to control, and can generate a large flapping force. However, its flapping motion is stiff and the propulsion efficiency is low. The use of passive flexible fins can improve the efficiency but scarifies the thrust. The second kind is to use a number of serially linked motors (typically 3). This is the most popular method and a couple of typical examples are detailed in Chaps. 4 and 7 of this book. In this design, the flapping is approximated by a polyline with a limited number of nodes (same as the number of motors) [18]. Compared to the single joint design, it more closely resembles the fish's flapping (in both C mode and S mode) and hence, is more efficient. However, it is difficult to make and difficult to control. Particularly, it requires complex synchronization, without which the propulsion efficiency could be jerpodized. The third method is to use smart materials, such as Ionic Polymer Metal Composite (IPMC) (Chap. 8 of this book), Piezoelectric Material (PZT) (Chap. 9), and Shape Memory Alloy (SMA) [19]. With proper control, these

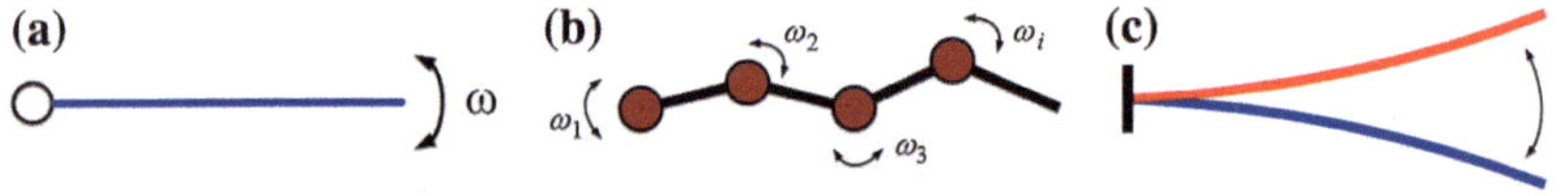

Fig. 16 Three kinds of BCF robot fish design. **a** Single joint design with rigid or flexible tail, driven by an motor; **b** Multi-joint design, each joint is driven by a motor; **c** Smart material design, based on the deformation of the material

materials could bend to circular arcs generating flapping motion. A major limitation of this method is its inefficiency in converting electric power into mechanical motion is low (typically only 20–30 %); as a result, its overall efficiency is not very high.

There are also MPF type robot fish. An example is presented in Chap. 5 of this book. Because the pectoral fin must move in 3D, at least two or three actuators are needed, plus one for turning.

Controlling and navigating of robot fish is the other main topic of research and development. In general, there are three levels of control and navigation: controlling the flapping of robot fish, interacting with the environment, and navigating through a specific path. The first level is to control the flapping of robot fish. The objective is achieve high efficiency and other desirable performances, such as speed, acceleration, turning radius, etc. The mode of control is usually semi-automatic, i.e., a human operator would steer the robot fish in open water (or in controlled environment for experiment testing). This level is relatively simple. If the robot fish is driven by multiple actuators, then the synchronization of the actuators is the key.

The second level is the control for autonomous swim. When examining fish swim in nature, we can see that fish constantly adapt to the changing environment, such as the water flow, light, obstacles, predator and prey, etc. In many cases, fish also swim in a cooperative manner. At this level, for robot fish, the main objective of control is autonomous swim. This involves sensing, information processing, and control. It becomes very complex if the robot fish needs to accelerate, dive, turn, fast-start, or to cooperate. These issues are discussed in Chaps. 10, 11, and 12 of the book.

The third level is navigation in long distance following a specified path. It involves technological issues such as energy supply, positioning, and water resistance, as well as environment issues such as tide, turbulence, and shallow water. According to the literature, no robot fish is able to reach this level. Fortunately, the Autonomous Underwater Vehicle (AUV) has proved its feasibility [20].

In December 2014, the US Navy tested a robot fish [21]. Based on the news announcement, the robot fish is controlled using a joystick. This indicates that it is still in the first level of control—a long way to go to catch a real fish.

In the rest of this chapter, we will focus on the design and prototyping of robot fish.

4 A Methodology for Robot Fish Design

How to design a robot fish that can perform as good as a fish? An effective methodology is to follow the design of a fish.

Figure 17 shows the brief anatomy of a fish [22]. As shown in Fig. 17a, the fish skeleton is composed of a number of short vertebrae; two successive vertebrae form an intervertebral joint, which allows only a small rotation. Note that the lengths of the vertebrae are roughly the same. As shown in Fig. 17b, the driving muscle arranges from the head to tail.

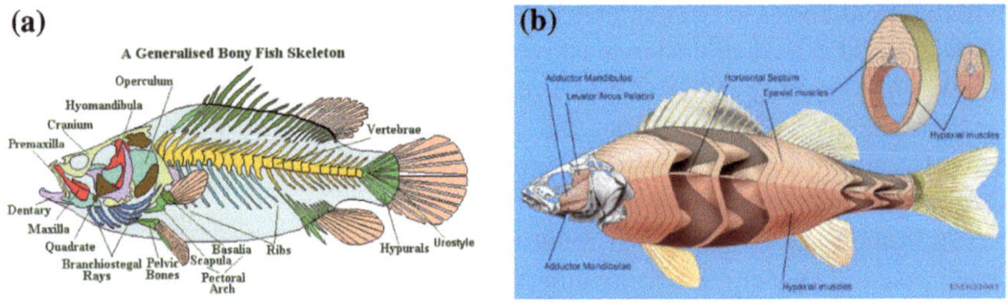

Fig. 17 The anatomy of a fish [22, 36]. **a** The skeleton; **b** The muscle arrangement

The flapping of fish is due to muscle contraction. As illustrated in Fig. 18 [23], the muscle contraction is transmitted to the bending of the backbone by the tendons. There are two types of tendons: the Anterior Oblique Tendon (AOT) connected to the muscle and the Posterior Oblique Tendon (POT) attached to the vertebrae. POT and AOT converge at the main horizontal septum. A band of stout collagen fibers form a thick lateral portion of the AOP near the main horizontal septum forming a sling, through which the POT passes (Fig. 18a). The contraction of muscle pulls the POT and draws it through the slings (Fig. 18b). A moment is therefore exerted on the backbone causing it to bend. The angular rotation of each joint is about the same and hence, can be assumed equal. Thus, the total bending angle due to a unit of muscle contraction can be determined through the geometric relationship. Additionally, the ratio between the muscle contraction and the lateral displacement of the vertebra is determined by the length of the POT, the angle of

Fig. 18 Illustration of the biomechanical model of fish backbone bending. **a** Fish backbone musculoskeletal structure; **b** Backbone bending by muscle contraction

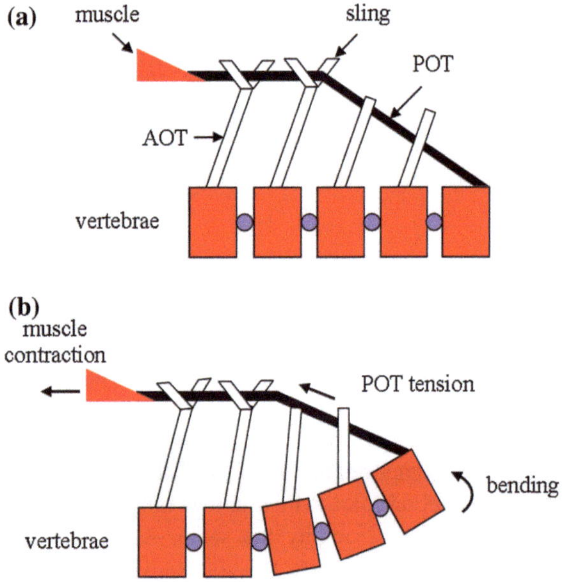

the POT to the backbone, and the lateral distance from the backbone to the AOT-POT sling. The larger the angle of POT to the backbone, the smaller the ratio. For example, for tuna the ratio is around 3.4, and for mackerel the ratio is around 2.1 [24]. The lower ratio of mackerel is suitable for powerful flapping while the higher ratio of tuna is better for energy storage.

For robot fish, just like real fish, delivering the power to generate the desirable flapping is key. In practice, it is not feasible to use a large number of actuators (equivalent to the muscle groups of the fish) because of the complexity and power consumption. A classical design is shown in Chap. 4 of this book. It uses four motors connected in series to generate flapping. The other design is shown in Chap. 7 of the book. It uses one large motor driving four joints. This allows the robot fish, named iSplash, to flap at high frequency (20 Hz) generating a fast speed of 3.6 m/s. Perhaps the most intricate design is the soft body robot fish presented in Chap. 6 of the book. It has just one motor and utilizes the material properties of the soft body to generate desirable flapping.

Another interesting design is the use of biomimetic wire-driven mechanism (BWDM) as presented in Chap. 3 of this book. Let us use BWDM to show how to generate desirable flapping curve, as it nicely resembles the anatomy of fish. Briefly, there are two types of wire-driven robot fish [25–29], one with serpentine backbone (like a bony fish) and the other with continuum backbone (like an earthworm), as shown in Fig. 19. For serpentine BWDM, the vertebrae are the same as the fish and the wires act as muscles. The wires are a pair. When one wire is pulled, the other is

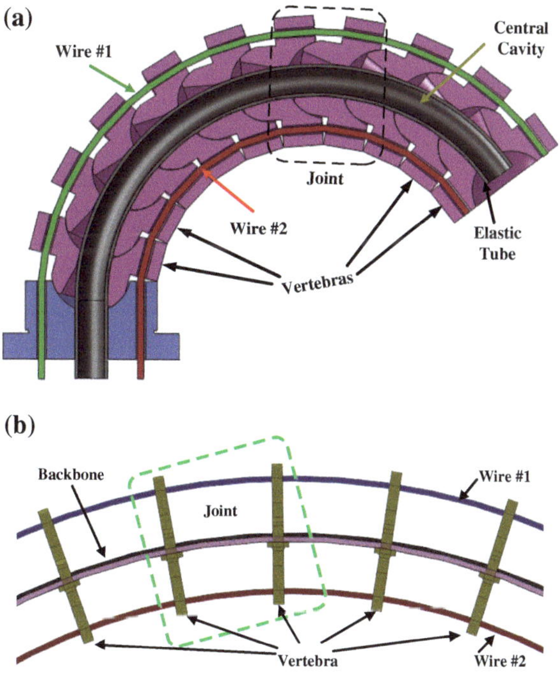

Fig. 19 The structure of BWDM. **a** Serpentine backbone; **b** Continuum backbone

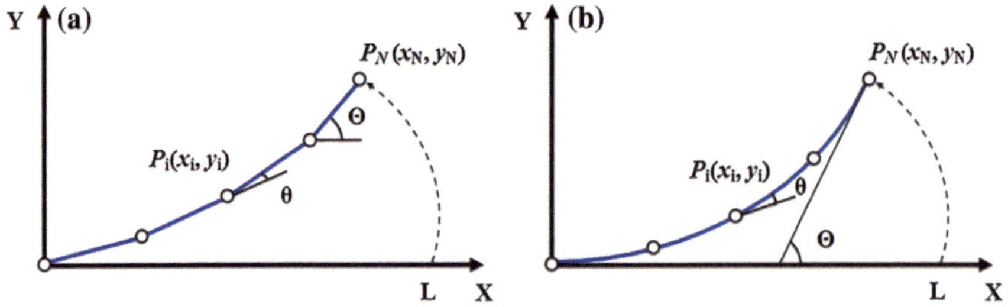

Fig. 20 The model of wire-driven robot fish. **a** Serpentine BWDM; **b** Continuum BWDM

extended causing bending. For continuum BWDM, the wires are the same but the vertebrae are formed by simple eyelets. Figure 20 shows the model of the BWDM. For serpentine BWDM (Fig. 20a), bending can be modeled by a series of linear line segments. During bending, each line segment bends at the same angle, θ, and the total bend angle is therefore $\Theta = N \times \theta$, where N is the number of vertebrae. For continuum BWDM (Fig. 20b), the bending is simply an arc, and the bending angle is similar to the serpentine BWDM. Also, a robot fish may have several such BWDM segments in order to generate anguilliform swimming.

As discussed in Sect. 2, fish flapping can be described by the traveling wave equation (Eq. 1). On the other hand, one BWDM segment can bend into a circular arc. As a result, with a few segments, the wire-driven robot fish can fit fish flapping very well as illustrated in Fig. 21. Specifically, for oscillatory swimming, one segment is enough; for undulatory swimming, multiple segments are needed and

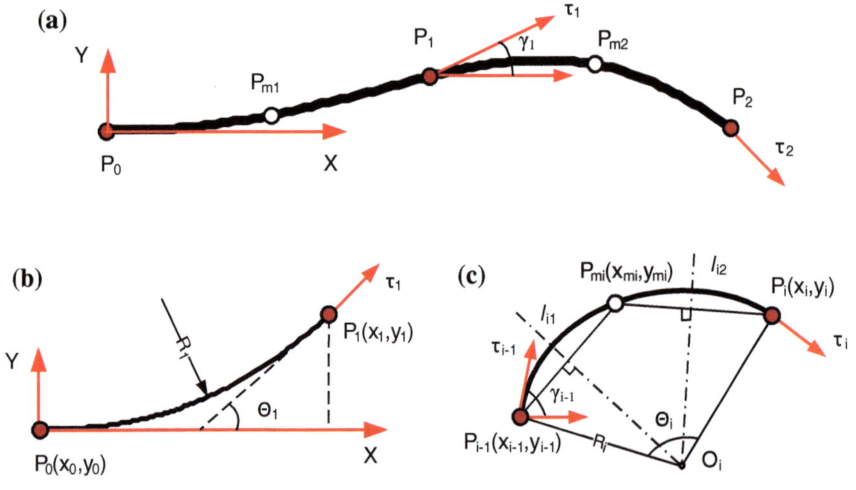

Fig. 21 Fitting the traveling wave with BWDM: **a** a typical traveling wave; **b** fitting of the first segment; **c** fitting of the subsequent segment

Table 2 Different forms of swimming fitting by BWDM

Range of λ	Swimming form	Number of segments
[4, +∞)	Thunniform	1
[4/3, 4)	Carangiform	2
[4/5, 4/3)	Subcarangiform	3
(0, 4/5)	Anguilliform	[0.5 + 2/λ]

the number of segments is M = [0.5 + 2/λ], where [*] denotes the integral part of "*". Table 2 shows the required number of segments for different forms of swim.

To evaluate how well the robot fish emulate real fish, a curve fitting index, ξ, is introduced as shown in Eq. (13). This index measures the discrepancy between the designed robot fish flapping and the targeted traveling wave in a full flapping cycle with flapping amplitude, r. For ease of computation, one can also use the discrete form as shown in Eq. (14). In this equation, the flapping cycle is divided into S intervals.

$$\xi = \frac{1}{T \cdot L}\int_0^T \int_0^L \left|\frac{y(x,t) - y_f(x,t)}{r}\right| \mathrm{d}x\mathrm{d}t \tag{13}$$

$$\xi \approx \frac{1}{S \cdot N \cdot M} \cdot \sum_{n=1}^{S}\sum_{i=1}^{M}\sum_{j=1}^{N_i} \left|\frac{y(x_{i,j}, t_n) - y_f(x_{i,j}, t_n)}{r}\right| \tag{14}$$

Now, let us examine how well the wire-driven robot fish fits the targeted traveling wave form. First, for the Thunniform swim, it is the oscillatory swim and can be achieved by a single segment BWDM. Assume at a time instance, t, the traveling wave has the anterior position $P_0(x_0, y_0)$ and the posterior position $P_1(x_1, y_1)$. For the wire-driven robot fish that has N identical vertebrae and total length L = 1 BL, let both ends of the wire-driven robot fish be coincident with P_0 and P_1, then the bending angle of the wire-driven robot fish is:

$$\Theta = \frac{2N}{N+1}\arctan\left(\frac{y_1}{x_1}\right) = \frac{2N}{N+1}\arctan\left(r \cdot \sin(\frac{2\pi}{\lambda} + 2\pi \cdot t)\right) \tag{15}$$

Moreover, for each vertebra, the joint positions can be found using Eq. (16). Figure 21a shows the simulation result. In the figure, the traveling wave curve is in black. The wire-driven robot fish has 10 vertebrae and its flapping curve is in red. For comparison, the flapping curve of an SJ robot fish is also shown in blue. From the figure, it is seen that at time t = 0.25 T and t = 0.75 T, both the SJ robot fish and the wire-driven robot fish match the travel wave curve well; however, at t = 0 and t = 0.5 T, the wire-driven robot fish matches the traveling wave curve much better than that of the SJ robot fish. This implies that the wire-driven robot fish can outperform the SJ robot fish. Quantitatively, using the index defined in Eq. (14), for the wire-driven robot fish the average fitting error is 0.0157r, while that of the SJ robot fish it is 0.093r. In other words, the fitting error of the SJ robot fish is 5.94 times larger than that of the wire-driven robot fish.

$$\begin{bmatrix} x_j \\ y_j \end{bmatrix} = \begin{bmatrix} \frac{L}{N} \cdot \frac{\sin[(j\cdot\Theta)/(2N)]\cdot\cos[(j+1)\cdot\Theta/(2N)]}{\sin[\Theta/(2N)]} \\ \frac{L}{N} \cdot \frac{\sin[(j\cdot\Theta)/(2N)]\cdot\sin[(j+1)\cdot\Theta/(2N)]}{\sin[\Theta/(2N)]} \end{bmatrix} \quad 0 \le j \le N \tag{16}$$

For undulatory swim, including carangiform swim, subcarangiform swim, and anguilliform swim, multiple BWDM segments are needed as shown in Table 2. For the first segment, the length is $\lambda/4$, for the middle segment the length is $\lambda/2$, and for the last segment, it is the rest. Moreover, for the first segment, the curve fitting is the same as in the thunniform swim as shown in Fig. 21b. For the remaining segments, the curve fitting is shown in Fig. 21c. Specifically, for the ith segment, the two endpoints are P_{i-1} and P_i, and the middle point is P_{mi}. The orientation of the starting point of the section is γ_{i-1}. Each segment bends to a circular arc with both ends coincident with P_{i-1} and P_i, and the middle point coincident with P_{mi}. The bending angle and orientation can be calculated using the following procedure:

Step 1 Find the perpendicular bisector for $P_{i\text{-}1}P_{mi}$ and $P_{mi}P_i$ respectively:

$$\begin{cases} l_{i1}: & \left(y - \frac{y_{i-1}+y_{mi}}{2}\right) + \frac{x_{i-1}-x_{mi}}{y_{i-1}-y_{mi}}\left(x - \frac{x_{i-1}+x_{mi}}{2}\right) = 0 \\ l_{i2}: & \left(y - \frac{y_{mi}+y_i}{2}\right) + \frac{x_{mi}-x_i}{y_{mi}-y_i}\left(x - \frac{x_{mi}+x_i}{2}\right) = 0 \end{cases} \tag{17}$$

Step 2 Find the center of the circular arc, $O_i(x_{oi}, y_{oi})$, which is the intersection of l_{i1} and l_{i2}

Step 3 Find the bending angle:

$$\Theta_i = 2\left\{\arctan\left(\frac{y_i - y_{mi}}{x_i - x_{mi}}\right) - \arctan\left(\frac{y_{mi} - y_{i-1}}{x_{mi} - x_{i-1}}\right)\right\} \tag{18}$$

Step 4 Find the orientation:

$$\gamma_{i-1} = \arctan\left(-\frac{x_{oi} - x_{i-1}}{y_{oi} - y_{i-1}}\right) \tag{19}$$

Subsequently, the joint positions are:

$$\begin{bmatrix} x_{i,j} \\ y_{i,j} \\ 1 \end{bmatrix} = \begin{bmatrix} \cos(\gamma_{i-1}) & -\sin(\gamma_{i-1}) & x_{i-1,N_{i-1}} \\ \sin(\gamma_{i-1}) & \cos(\gamma_{i-1}) & y_{i-1,N_{i-1}} \\ 0 & 0 & 1 \end{bmatrix} \begin{bmatrix} x_j \\ y_j \\ 1 \end{bmatrix} \quad 0 \le i \le M, 0 \le j \le N_i \tag{20}$$

where M is the total number of segment; N_i is the number of vertebrae in the ith segment; x_j, and y_j are as per Eq. (16), in which the bending angle is $\Theta = \Theta_i$, and the vertebra length is $l = l_{ij}$. In addition, the joint rotation in the ith segment is $\theta_i = \Theta_i/N_i$.

The distal end joint orientation of the ith segment is $\gamma_{i-1} + \Theta_i$ and the orientation of the $(i + 1)$th segment is γ_i. The phase lag between the two successive segments is then:

$$\phi_i = \gamma_i - \gamma_{i-1} - \Theta_i \tag{21}$$

Finally, for the $(i + 1)$th segment, the joint rotation is:

$$\theta_{i+1} = \varphi_i + \Theta_{i+1}/N_{i+1} \tag{22}$$

A set of computer simulation results are shown in Fig. 22, in which the traveling wave is shown in black and the wire-driven robot fish flapping curve is shown in red. For comparison, the MJ robot fish flapping curve is shown in blue. Figure 22b is the case of carangiform swim. In this case, two segments are used for the wire-driven robot fish and two joints are used for the MJ robot fish. From the figure, it is seen that the wire-driven robot fish can match the traveling wave well in the entire flapping cycle. In contrast, the MJ robot fish matches the traveling wave only in some time instances. For example, when t = 0.25 T and t = 0.75 T, the flapping curves of the MJ robot fish are greatly different from the traveling wave. Using the curve fitting index, the average fitting error of the wire-driven robot fish is 0.0107r, and the average fitting error of the MJ robot fish is 0.0714r, which is 6.66 times larger.

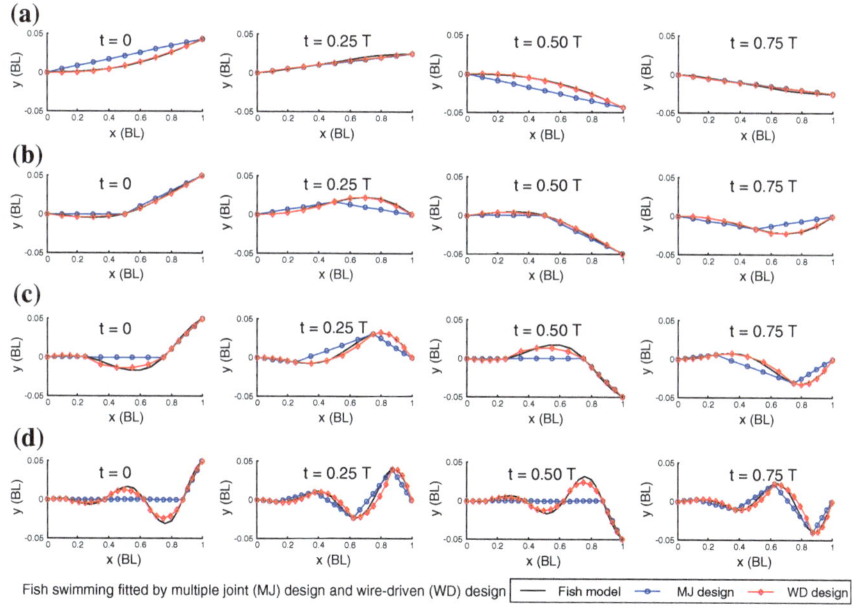

Fig. 22 Comparison among the traveling wave, the flapping of the multi-joint robot fish, and the wire-driven robot fish. **a** Thunniform swimming; **b** Carangiform swimming; **c** Sub-carangiform swimming; **d** Anguilliform swimming

Figure 22c shows the case of subcarangiform swim. In this case, three segments are used for the wire-driven robot fish and three joints are used the MJ robot fish. The fitting error of the wire-driven robot fish is $0.0587r$, and the fitting error of the MJ robot fish is $0.1331r$, or 2.2 times larger.

Figure 22d shows the case of anguilliform swim. In this case, five segments are needed for the wire-driven robot fish and five joints are used for the MJ robot fish. The fitting error of the wire-driven robot fish is $0.0568r$ and the fitting error of the MJ design is $0.1567r$, or 2.3 times larger.

It is shall be pointed out that, for the MJ robot fish, the length of the segments are sensitive. For example, to match the traveling wave curve at $t = 0.25$ T, the length of each segment should be half of the wavelength, λ. However, this will result in the largest fitting error at some other time instances, for example, $t = 0.50$ T , at which time, the first $(M - 1)$ segments are lined together. In order to solve this problem, more segments are therefore required. For the wire-driven robot fish, there is no such problem as it can fit the traveling curve well during the entire flapping cycle.

In summary, to design a robot fish one shall consider the following:

(a) The type of robot fish: SJ, MJ, smart material-based or BWDM; and the driving mechanism of the robot fish: direct drive, crank, etc.;
(b) The number of segments: M (which will determine where the robot fish will do oscillating swim or undulating swim). For multiple segments, the length of the first segment shall be $\lambda/4$, the length of the second segment shall be $\lambda/2$, etc.
(c) The geometry of each segment (which will determine how well the flapping of the robot fish matches the travel wave equation). Specifically, the following optimization can be used to determine the geometry of each segment:

$$\min \sum_{n=1}^{S} \sum_{i=1}^{M} \sum_{j=1}^{N_i} \left| \frac{y(x_{ij}, t_n) - y_f(x_{ij}, t_n)}{r} \right| \tag{23}$$

It should be mentioned that the presented design methodology focuses on the matching of the traveling wave. Many practical issues are not considered such as the soft body of the robot fish, the dynamics of flapping, the fluid dynamics of the water, etc. Therefore, much more research is still needed, especially for high speed swim, turning, and diving.

5 Discussions and Conclusions

Inspired by fish swim, engineers and scientists around the world have been trying to design and build flapping-based underwater robot, i.e., robot fish, for more than two decades. This chapter first gives a brief review of the theory of fish swim. Next, it examines the existing designs of robot fish. Then, a design methodology is presented: the robot fish shall emulate the flapping of real fish: the better the emulation, the better the performance. In practice, the flapping robot fish can be realized using various methods, including single joint, multiple joints, smart-materials and soft body. Additionally, the joints and/or the soft body can be driven directly, or through various mechanisms (such as cranks), as well as through BWDM. The design parameters can be optimized by comparing to the flapping of real fish (i.e., the travel wave equation).

Lastly, it should be pointed out that the study on robot fish is still in its infancy. Many problems are yet to be resolved, such as the soft body of the robot fish, the dynamics of flapping, the fluid dynamics of the water, etc. The control of the robot fish is another set of problems, which include sensing the position and orientation, detecting obstacles, receiving supervision commands, etc. Navigation in the long distance is a long-term goal and hopefully, it can be achieved in a decade or so.

Acknowledgment The authors wish to thank the following people who have helped in their study of robot fish: Mr. Baofeng Liao, Mr. Yong Zhong, Miss Candy Wai Pik Lau, and Mr. Allan Mok.

References

1. http://en.wikipedia.org/wiki/Propeller
2. Triantafyllou MS, George S (1995) An efficient swimming machine. Sci Am 271(3):64–70
3. http://en.wikipedia.org/wiki/Jacques_Yves_Cousteau
4. Gray J (1933) Studies in animal locomotion. J Exp Biol 10:88–103
5. Lighthill MJ (1960) Note on the swimming of slender fish. J Fluid Mech 9(2):305–317
6. Lighthill MJ (1970) Aquatic animal propulsion of high hydro mechanical efficiency. J Fluid Mech 44(2):265–301
7. Lighthill MJ (1971) Large-amplitude elongated-body theory of fish locomotion. In: Proceedings of 1971 royal society annual meeting, London, England, pp 125–138
8. Lighthill MJ (1997) Hydromechanics of aquatic animal propulsion. In: Collected papers of sir James Lighthill, vol 4. Oxford University Press, Oxford
9. Alvarado PV, Youcef-Toumi K (2006) Design of machines with compliant bodies for biomimetic locomotion in liquid environments. Trans ASME J Dyn Syst Meas Control 128 (1):3–13
10. Westneat MW, Wainwright SA (2001) Mechanical design for swimming: muscle, tendon, and bone. Fish Physiol 19:271–311
11. Videler JJ (1993) Fish swimming. Chapman & Hall, London
12. Romanenko EV (2002) Fish and dolphin swimming, Sofia—Moscow
13. Sfakiotakis M, Lane DM, Davies JBC (1999) Review of fish swimming models for aquatic locomotion. IEEE J Ocean Eng 24(2):237–252
14. Lauder GV, Tytell ED (2005) Hydrodynamics of undulatory propulsion. Fish Physiol 23: 425–468
15. Lamas MI, Rodriguez JD, Rodriguez CG (2012) CFD analysis of biologically-inspired marine propulsors. Brodogradnja 63:125–133
16. Adkins D, Yan YY (2006) CFD simulation of fish-like body moving in viscous liquid. J Bionic Eng 3(3):147–153
17. http://www.sciencemuseum.org.uk/images/ManualSSPL/10328064.aspx (MIT Robot Tuna)
18. http://dces.essex.ac.uk/staff/hhu/jliua/index.htm (Essex University Robot Fish)

19. Wang ZL et al (2008) A micro-robot fish with embedded SMA wire actuated flexible biomimetic fin. Sens Actuators A Phys 144:354–360
20. http://en.wikipedia.org/wiki/Autonomous_underwater_vehicle
21. https://www.youtube.com/watch?v=tupVgBxFCBA
22. http://www.earthlife.net/fish/skeleton.html
23. Alexander RM (1988) Elastic mechanisms in animal movement. Cambridge University Press, Cambridge
24. Westneat MW, Wainwright SA (2001) Mechanical design for swimming: muscle, tendon, and bone. Fish Physiol 19:271–311
25. Li Z, Du R (2013) Design and analysis of a bio-inspired wire-driven multi-section flexible robot. Int J Adv Robot Syst 10:1–11
26. Li Z, Du R (2012) Design and analysis of a biomimetic wire-driven flapping propeller. In: The 4th IEEE RAS & EMBS international conference on biomedical robotics and biomechatronics (BioRob 2012), Roma, Italy, pp 276–281
27. Li Z, Gao W, Du R, Liao B (2012) Design and analysis of a wrie-driven robot tadpole. In: International mechanical engineering congress and exposition (IMECE 2012), Houston, Texas
28. Li Z, Du R, Zhang Y, Li H (2013) Robot fish with novel wire-driven continuum flapping propulsor. Appl Mech Mater 300:510–514
29. Liao BF, Li Z, Du R (2012) Robot tadpole with a novel biomimetic wire-driven propulsor. In: The IEEE international conference on robotics and biomimetics (ROBIO 2012), Guangzhou, China
30. http://en.wikipedia.org/wiki/Ancient_Egyptian_technology
31. http://en.wikipedia.org/wiki/Sculling
32. http://en.wikipedia.org/wiki/Michael_James_Lighthill
33. Lindsey CC. (1978) Form, function and locomotory habits in fish. Fish Physiol 7
34. Colgate JE, Kevin ML (2004) Mechanics and control of swimming: a review. Ocean Eng IEEE J 29(3):660–673
35. Colgate JE, Lynch KM (Jul 2004) Mechanics and control of swimming: A review. IEEE J Oceanic Eng 29:660–673
36. http://animaldiversity.ummz.umich.edu/collections/contributors/Grzimek_fish/structure_function/v04_id131_con_axialmu/

Fish Locomotion: Biology and Robotics of Body and Fin-Based Movements

George V. Lauder and James L. Tangorra

Abstract The study of fish locomotion provides a rich source of inspiration for the design of robotic devices. Fish exhibit an array of complex locomotor designs that involve both diversity of structures used to generate locomotor forces, and versatile behaviors to engage with the aquatic environment. The functional design of fish includes both a flexible body exhibiting undulatory motion as well as numerous control surfaces that enable fish to vector forces and execute rapid maneuvers in roll, pitch, and yaw directions. Patterns of body undulation have often been misunderstood, and fish with propulsive mechanics as diverse as tuna and eels can display similar patterns of body bending during swimming. Many of the often-cited classical locomotor categories are based on a misunderstanding of body and fin kinematics. Fish fins can exhibit remarkably complex conformational changes during propulsion, and do not function as flat plates but have individual mobile fin rays actuated by muscles at the fin base. Fin motion and surface bending in most fish is actively controlled. Even during steady horizontal locomotion, median fins such as the dorsal and anal fins function to balance torques and can contribute to thrust. Locomotion using body undulation is not achieved independently from fin motion, and the vast majority of fish locomotor patterns utilize both the body and fins. Robotic systems derived from fish templates can range from simple flexible plastic panels to more complex models of whole body and fin design. Experimental test platforms that represent individual fins or specific components of fish locomotor design allow for detailed testing of hydrodynamic and mechanical function. Actuating and controlling complex fish robotic systems involving both the body and multiple individual fins are a major challenge for the future.

Keywords Swimming · Hydrodynamics · Fish · Locomotion · Robotics

G.V. Lauder (✉)
Museum of Comparative Zoology, Harvard University, 26 Oxford Street, Cambridge
MA 02138, USA
e-mail: glauder@oeb.harvard.edu

J.L. Tangorra
Department of Mechanical Engineering, Drexel University, Philadelphia, PA 19104, USA
e-mail: tangorra@coe.drexel.edu

© Springer-Verlag Berlin Heidelberg 2015

R. Du et al. (eds.), *Robot Fish*, Springer Tracts in Mechanical Engineering,
DOI 10.1007/978-3-662-46870-8_2

1 Introduction

The study of fish locomotion provides a rich source of inspiration for the design of robotic devices. Fish exhibit an array of complex locomotor designs that involve both diversity of structures used to generate locomotor forces, and versatile behaviors to engage with the aquatic environment which can range from calm stratified lakes to turbulent high-velocity rivers. The functional design of fish includes both a flexible body exhibiting undulatory motion as well as numerous control surfaces that enable fish to vector forces and execute rapid maneuvers in roll, pitch, and yaw directions. Designers of robotic systems that must deal with the unpredictable aquatic environment can turn to the mechanics and function of fish as a source of ideas for how to approach the construction of new mechanical devices that use flexible control surfaces and patterns of wave-like motion to produce thrust and maneuvering forces.

In this chapter we provide a general background and an overview of the mechanics of fish locomotion, and then explore some of the mechanical systems that we have constructed to better understand the function of both the body and fins of fish. Research on fish robotics has the potential not only to demonstrate novel designs for autonomous underwater vehicles that are not based on propellers, but also to use robotic systems and their control to better understand the biology of fish swimming. In our view, the potential for using mechanical systems to test biological hypotheses about how and why fish are designed the way they are is one of the most intriguing areas for future work.

Within the past 15 years, there have been many reviews and overviews of the topic of fish swimming and the increase in publication rate on this topic reflects the growing interest in understanding the functional design of fish and the implications for robotic system construction. In this chapter, we will not duplicate material from these previous articles, but will instead focus on summarizing key features of fish functional design; note several misunderstandings in some of the current literature on the mechanics of fish swimming; address specific fish systems that are of special relevance for the construction of fish-like mechanical devices; and finally illustrate some of the mechanical systems that we have developed ranging from simple to complex representations of fish functional design.

Readers interested in the literature review of one or more aspects of fish locomotor function can examine classical summaries such as [4, 47–49, 76, 131] as well as the more contemporary books [12, 27, 75, 86, 110, 127, 134, 139] and articles [36, 39–41, 60, 62, 73, 87, 106, 109, 122, 123, 130, 132, 133].

2 Overview of Fish Swimming Dynamics

2.1 General Background

Perhaps the most common image of fish as swimming objects is the often-published horizontal section through a fish body that resembles a NACA airfoil in shape. This view emphasizes the overall streamlined nature of fish bodies and allows for comparisons between fish and objects that are designed to minimize drag or to generate lift. But fish shapes are complex in three dimensions [125, 126] and have multiple projections (fins) and changes in geometry along the body length that are not well represented in a horizontal NACA-like section. Figure 1 illustrates some of the complexity in fish shape with transverse sections obtained using a micro-CT scanner of a bluegill sunfish *Lepomis macrochirus*. Toward the front of the body, fish often are thickest due to the head and gill region housing the brain, feeding and respiratory systems. This portion of the body is not symmetrical in the transverse plane, and the top is often more pointed while the bottom (ventral) region is rounded. Moving back along the body reveals increased dorsal–ventral symmetry and the median fins which project into the water above and below the fish body. The surfaces of these fins are not smooth, and instead have a roughened exterior due to the fin rays that support the thin fin membrane. Near the base of the tail (the region known as the caudal peduncle: sections d and e in Fig. 1) the body becomes more symmetrical with sharp edges that cause flow separation as the body and tail move from side to side during swimming [124]. The consequences of body asymmetry for the generation and control of locomotor torques has not been addressed, and remains an interesting area for future work.

2.2 Fish Locomotion Using Body Deformation

Although the notion of a gait is most common in studies of terrestrial locomotion (as when a horse moves from a walk to a trot to a gallop as speed increases), many fish have gaits too. As swimming speed increases, fish can change from a primarily pectoral fin-based swimming style in which the body remains relatively still, to body undulation with fin motion, to exclusively steady wave-like body motions for thrust generation. If speed increases still further, unsteady locomotion can occur with burst-and-glide swimming with intermittent high-frequency beating of the tail. These gait transitions are significant because they reflect the use of different muscular systems and a shift from steady swimming to high-speed unsteady locomotion caused by a change in muscle fiber activation from slower red fibers to the faster-contracting white fibers [14, 30, 108].

Figure 2 shows patterns of body bending in several species of swimming fish. In most species, the head oscillates relatively little until swimming speeds increase beyond two body lengths per second. Species as diverse as eels, sunfish, and clown

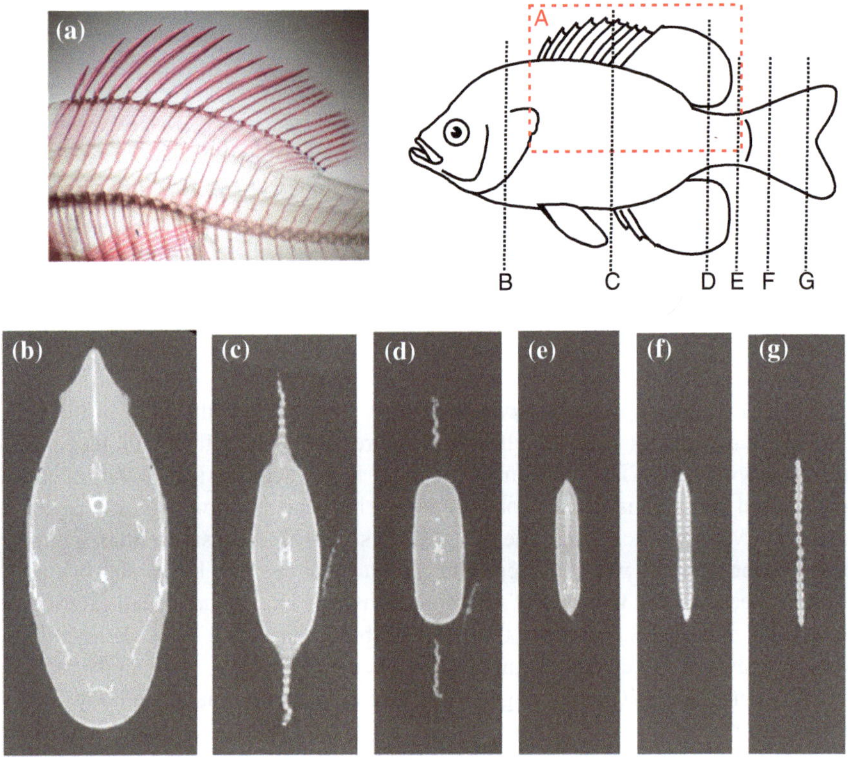

Fig. 1 Fish vary greatly in cross-sectional shape along the body. This figure illustrates shape cross-sections from micro-CT scans at several points (**b**–**g**) along the body of a 15 cm long bluegill sunfish, *Lepomis macrochirus*. Panel **a** shows the skeleton of this species (bones are stained *red* and the tissue has been enzymatically rendered transparent) to illustrate how fins such as the dorsal fin are supported by the axial skeleton. Panel **b** provides a guide to the location of the images below. Body shape changes from rounded anteriorly to the tail region where the *upper* and *lower* body surfaces have sharper edges. Fin surfaces are bumpy (see sections **c** and **g**) with projecting fin rays. All micro-CT images are shown at the same scale. From Lauder [61]

knifefish (Fig. 2) show remarkably similar patterns of body bending, and even quantitative analyses of tuna locomotion [29, 70] show similar patterns of head oscillation to other species. This result is not generally appreciated in the fish mechanics literature which suggests that tuna show reduced head oscillation amplitudes relative to other fish: the published quantitative analyses do not support this conclusion, although considerably more research needs to be done on tuna kinematics under controlled conditions.

Figure 3a shows an analysis of the amplitude of body bending as speed changes in largemouth bass [56]. At low speeds there is virtually no amplitude along the anterior third of the body, and even as speed increases considerably head oscillation does not increase greatly. There is a relatively sharp transition two-thirds of the way along the body where lateral body amplitudes begin to increase rapidly, reaching

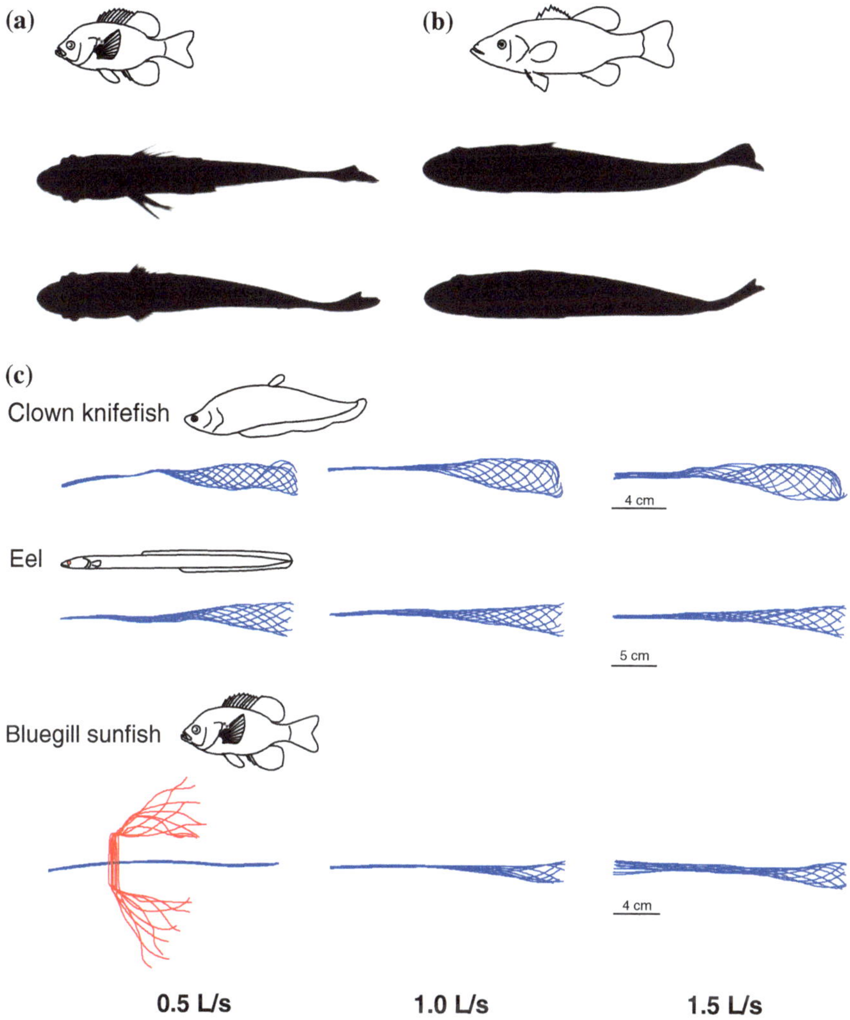

Fig. 2 Patterns of body motion in fish swimming with undulatory motion. During swimming, a traveling wave of bending passes down along the body. Panels **a** and **b** show body outlines at two different times in bluegill sunfish (*L. macrochirus*) and largemouth bass (*Micropterus salmoides*). Panel **c** shows how the pattern of body bending changes as fish increase swimming speed from slow (0.5 L/s) to moderate (1.5 L/s) speeds. At 0.5 L/s, bluegill sunfish use labriform locomotion and generate thrust only with their pectoral fie. Note the relatively low amplitude sideways (yaw) head motion and increasing amplitude along the body in most species. Panel **c** modified from Xiong and Lauder [140]

a maximum at the tail tip. Furthermore, graphs of amplitude versus length often have two inflection points (e.g., Fig. 3a), and are more complex than the relatively simple functions often used to program undulatory robotic devices.

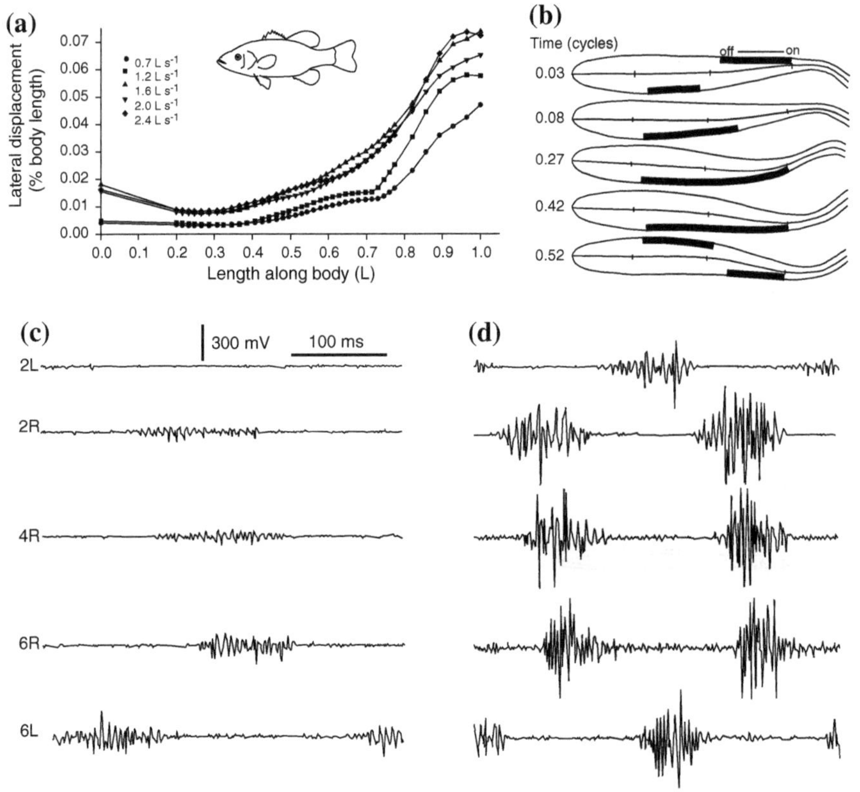

Fig. 3 Pattern of body bending during undulatory locomotion (**a**) and body muscle activity used to drive the bending wave (**b** and **c**). Body oscillation amplitude in swimming fishes increases with speed, and is minimal at a point approximately 30 % down the body from the tip of the snout. Waves of muscle activity (**b**, *black bars*) pass along the body to generate this wave, and muscle activity in the *red* fibers that power slow to moderate swimming increases as swimming speed increases (**c**, **d**). Labels in **c** refer to the positions down the body: positions 2, 4 and 6 correspond to 43, 57, and 72 % down the body. *R* and *L* refer to the Right and Left sides of the body. Modified from Jayne and Lauder [55, 56]

Fish kinematics in a wide diversity of species match this general pattern of reduced head oscillation with a wave-like pattern of body bending (Fig. 2). While fish species certainly differ in the number of waves and in specific details of the amplitude envelope [45], the differences in this two-dimensional view are minor, and recent quantitative comparisons of the midline kinematics of a diversity of species does not lend much support to the often-cited qualitative descriptions of fish swimming modes and simple graphical summaries that inaccurately depict patterns of body bending [76, 105].

Fish body undulations are caused by a wave-like pattern of muscle activation that sweeps from the head toward the tail (Fig. 3). During steady sustainable swimming, fish use almost exclusively red muscle fibers that have higher aerobic capability even though these fibers usually form only a very small percentage of total fish muscle mass, often less than 5 %. The bulk of fish muscle mass is composed of larger white fibers that are used for rapid unsteady motions and escape responses [53, 54, 107, 108, 115]. Thus, the bulk of fish bodies are passive during slower, routine, steady swimming behaviors, and body wave motion is controlled actively only by a relatively small proportion of muscle fibers. Figure 3b illustrates that muscle electrical activity alternates between the left and right sides of a swimming fish, but due to the movement of the wave of electrical activity toward the tail and the relatively large extent of activation on one side, there is considerable co-activation of red muscle fibers on the right and left sides of the body at any one time. This may stiffen the body actively and reduce the amplitude of motion which could be tuned to particular swimming speeds based on the intensity of muscle activation. Figure 3c documents that the intensity of muscle activity increases with swimming speed as does the velocity of the muscle activation wave and frequency, while the overall pattern of red fiber activity remains similar until speeds increase to the point where white muscle fibers are activated. Fish certainly possess the capability of actively controlling body stiffness with their locomotor musculature, but the extent to which this alters locomotor function is largely unknown [81, 82, 88].

One area of fish swimming that has received relatively little attention but that has great promise for facilitating the comparison of fish swimming mechanics to that of robotic devices is the study of center of mass (COM) motion [135]. In contrast to biomechanical investigations of walking and flying animals, almost nothing is known about motion of the center of mass in swimming fish. One recent study of ours quantified the three-dimensional center of mass motion in several species of fish using different swimming modes [140], but much more remains to be done to better understand the pattern and magnitudes of fish center of mass movement during steady swimming behaviors. We propose that comparison of COM motion in live fish to that of robotic devices provides a useful performance metric by which autonomous robotic fish-like mechanical systems can be compared to the swimming performance of live fish. High COM oscillation amplitude during swimming would be expected to reflect high costs of transport, and autonomous robotic designs could be evaluated and compared based on the amplitude of COM oscillation in the three movement planes.

2.3 *Fish Locomotion Using Fin Movements*

Fish fins also play an important, and in many species a dominant, role in propulsion. Pectoral fins are commonly used for slower speed swimming and to generate thrust for steady swimming, and off-axis forces for maneuvering (e.g., [31, 34, 44, 128, 129, 137, 138]. Pectoral fin propulsion may involve complex conformational and wave-like motions (Fig. 4) [16, 28, 34, 68, 100], and both experimental and computational fluid dynamics of pectoral propulsion have shown that thrust can be

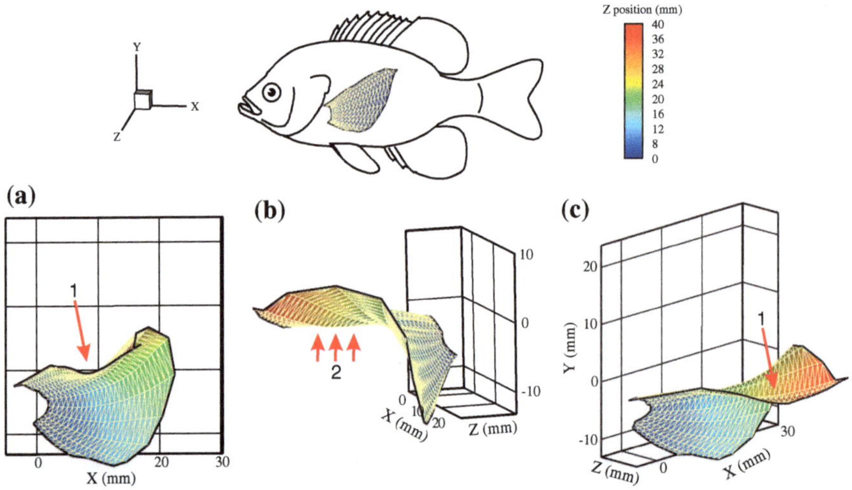

Fig. 4 Kinematics of the pectoral fins during locomotion in bluegill sunfish (*L. macrochirus*). Pectoral fins are often used by fish for slow to moderate speed locomotion, and can undergo complex conformational changes as the fin beats. This figure shows three views of a single time during the fin beat cycle; the three fin images are color coded by z-position, which indicates the distance of the fin element from the body. Note the bent and *cupped* shape of the fin at this time. *Arrow* #1 points to the wave of bending that travels from root to tip of the fin, while *arrows* labeled #2 point to the region of the fin surface that generates thrust during the outstroke. From Lauder [61]

generated on both the instroke and outstroke of fin motion, and that fin conformational changes may act to reduce vertical center of mass oscillations.

Fins such as median dorsal and anal fins (Fig. 5) are also important players in controlling fish body motion during undulatory swimming and these fins can also contribute thrust as they are actively moved by basal fin musculature [33, 35, 113, 114]. Dorsal and anal fins act to balance body torques during steady swimming (Fig. 5b) and to generate rolling moments during maneuvers. In addition, wakes from the dorsal and anal fins (and even the pelvic fins) can greatly alter incident flow on the tail [1, 112, 114]. Figure 5b shows the path of the caudal fin of a trout through the vortex wake shed by the dorsal fin and demonstrates that the tail of swimming fish does not move through undisturbed free-stream fluid, but instead encounters vortices that greatly alter the angle of attack of incident flow.

The fin rays of bony fish (but not sharks or rays) possess an interesting mechanical design that involves bilaminar jointed bony half-rays connected by collagen fibers and elastic tissue. The mechanical properties of these bilaminar fin rays have recently been studied in some detail [2, 42, 69, 116, 117] and represent a key evolutionary innovation in fish functional design that allows active curvature control of the fin propulsive surface and permits fish to resist fluid loading on their

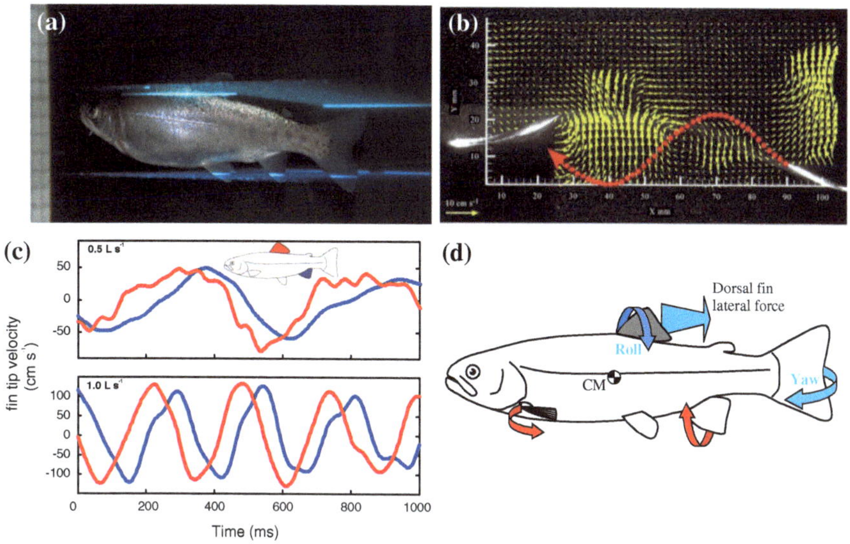

Fig. 5 Median fins such as the dorsal and anal fins play important roles during undulatory locomotion. **a** Trout (*Salvelinus fontinalis*) swimming in two horizontal laser light sheets to permit simultaneous imaging of flows generated by the dorsal and anal fins and the tail. **b** The dorsal fin (to the *left*) generates a clear vortex wake (*yellow* vectors) that the tail must pass through (*red dots* show the path that the tail will take through the dorsal vortex wake) during swimming rainbow trout (*Oncorhynchus mykiss*); free-stream flow has been subtracted. **c** Velocity through time of the dorsal and anal fins during locomotion at two speeds. These fins contribute significantly to both generating thrust and modulating body torques. **d** Schematic summary of the function of different median fins and their roles in controlling body position. Modified from Drucker and Lauder [35] and Standen and Lauder [114]

control surfaces [63]. Fin rays of sharks and rays are simpler collagenous rods that are not capable of active curvature control.

2.4 Multi-Fin and Body Locomotion Together: The Norm and not the Exception

Although in this chapter we have for the most part treated locomotion using body deformations separately from fin-based motions, in most fish species patterns of body deformation during undulatory propulsion occurs in concert with the active movements of fins. Two of the more common qualitative descriptive terms for fish swimming are "body and caudal fin" (BCF) and "median and paired fin" (MPF) locomotion. But dividing fish swimming into these two categories is entirely artificial and obscures the important fact that most species use the body *and* fins at the same time during swimming. And such coordinated use of the body and fins is critical for maintaining control of body position and for effecting changes in body orientation.

As examples of the contribution of the median and paired fins, first consider swimming bluegill sunfish as speed increases [32]. At slow speeds below 1.0 L/s, the paired fins contribute 100 % of thrust, and the body provides no thrust. As speed increases to over 1.5 L/s, pectoral fins contribute approximately three-quarters of needed thrust, and body undulation occurs to complete the needed thrust to swim at this speed. At the highest steady swimming speeds, the pectoral fins contribute almost nothing to thrust, and the body nearly 100 %. And the bluegill dorsal fin contributes about 12 % of total thrust at speeds of 1.1 L/s [33].

Even in fish such as trout which are known for effective undulatory locomotion, the median fins make a contribution to controlling body position in roll and yaw, and we cannot consider undulatory locomotion without taking this into account. Standen and Lauder [114] estimated that for trout of 16 cm long swimming at 1.0 L/s, the dorsal and anal fins each add about 1 mN to thrust, which is nearly 10 % of total thrust required at this speed.

The use of both the fins and body together is particularly evident during maneuvering, and we recommend abandoning this artificial division of fish swimming modes as BCF versus MPF. Of course, this poses a considerable challenge for designers of whole fish robotic devices who must consider the possibility of adding active fin-like control surfaces to an already complex flexible body, and in the next section we consider a variety of mechanical models for fish fins and bodies.

3 Robotic Models of Fish Locomotion

3.1 Overview of Fish Robotic Systems

Recent years have seen greatly increased interest in robotic models of fish swimming and a wide diversity of approaches has been taken to the mechanical design of fish-like systems. One way of conceptualizing this variety is to consider a fish template and both relatively simple as well as more complex mechanical models derived from these templates (Fig. 6). Simple models have the advantage of more rapid construction and iteration times when changes are desired, and the ability to easily make modifications that represent basic abstractions of fish features. Examples of this approach include modifications of the trailing edge of a flexible flapping foil or panel to represent different fish tail shapes, or changes in body stiffness achieved by changing the panel material (Fig. 6). Examples of using simple mechanical models to understand the dynamics of propulsion include (but are certainly not limited to) the following studies [3, 9, 13, 17, 19, 25, 26, 50, 52, 64, 65, 74, 97, 98, 101].

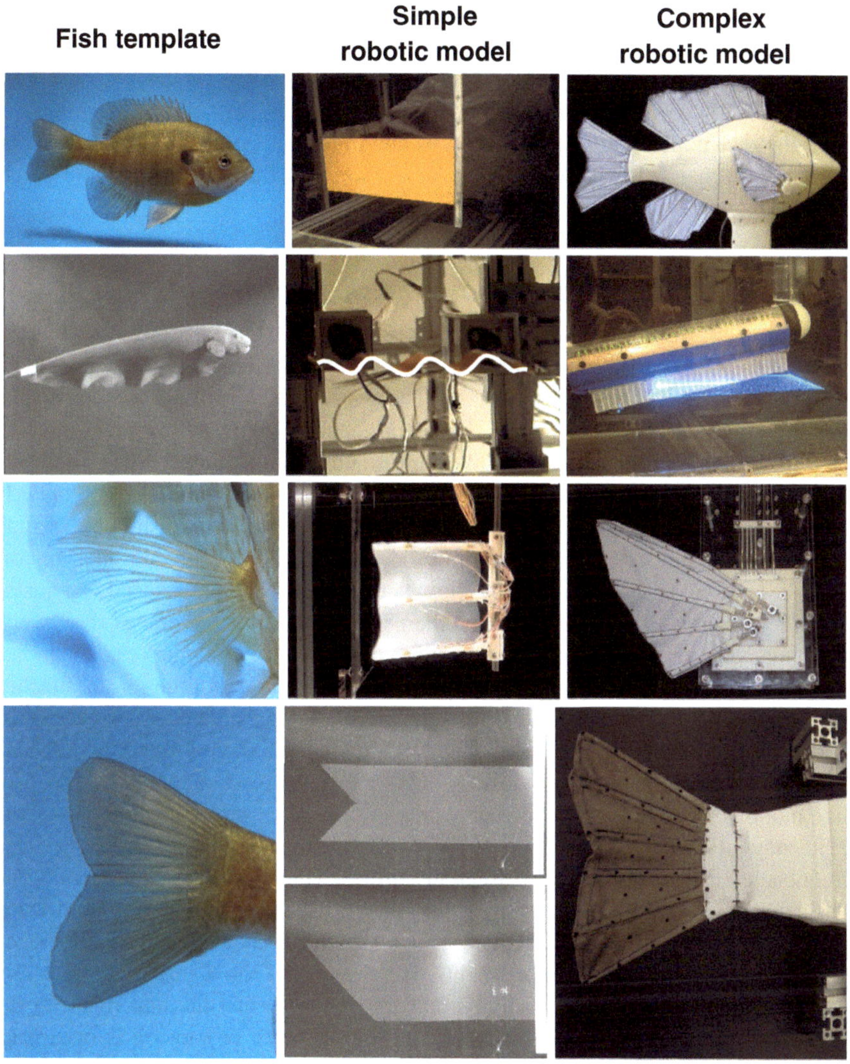

Fig. 6 Robotic approaches to studying fish locomotion can involve relatively simple flexible foil models of propulsion or more complex engineered devices. This diversity of approaches is based on fish templates, and can involve whole fish models (*top two rows*) or mechanical test platforms that represent individual fins such as the pectoral fin (*third row*) or caudal fin (*bottom row*). Individual panels show images from our collaborative work in fish robotics. Other chapters in this volume also present a diversity of fish robotic systems. Images are from research described in [3, 22, 23, 38, 65, 66, 96, 120, 121, 141]

Simple models are convenient but do not represent the complex behaviors inherent in biological systems. Thus, development of more complex mechanical devices that include active control of the body or fins is important for exploring some of the behaviors seen in freely-swimming fish. Recent examples of using complex mechanical devices to understand whole-body or fin-based propulsion include the following studies: [5–8, 20–23], Epps et al. [37, 38, 51, 59, 71, 72, 77–80, 83–85, 90, 91, 93, 96, 104, 118–120].

Below we explore some of the benefits and drawbacks of using both simple and complex mechanical fish-like systems using examples from our previous research.

3.2 Simple Mechanical Models of Fish Locomotion

Simple flexible plastic strips or panels have proven to be useful passive models for flexible body propulsion in the water. These very basic propulsive systems are remarkably fish-like despite their simplicity, and exhibit a traveling wave of bending and self-propulsion at fish-like speeds and Strouhal numbers when actuated at frequencies of from 0.5 to 3 Hz in heave motion at their leading edge [111]. In order to effect propulsion of these simple flexible models of the fish body and tail region, the leading edge is attached to a rod controlled by heave and pitch motors (Fig. 7) and moved in a recirculating flow tank [64, 65, 98]. Flow speed is adjusted to achieve either self-propulsion when the net x-force (thrust) averages zero over a flapping cycle, or propulsive kinematics, forces, and hydrodynamics can be studied under conditions of net thrust or drag, measured with a force/torque sensor attached to the rod. These simple flexible strips (Fig. 7a, b) are convenient subjects for study because stiffness can be altered by changing the material, and different lengths and shapes can easily be attached to focus experiments on these parameters. The fish-like motions (Fig. 7c, d) produce interesting performance surfaces where thrust and efficiency of propulsion can be mapped over a range of heave amplitudes and frequencies for flexible foils of different stiffness (Fig. 7e, f). At low heaves and frequencies, efficiency is negative for the most flexible material as no net thrust is produced, but as stiffness and frequency increase, thrust is developed and efficiency approaches values of 0.2. Experiments like these also show that there can be a complex interaction between stiffness and frequency so that the material with the highest efficiency can depend on the frequency at which propulsion is occurring, and crossover points show where both flexible and stiff materials perform equally well (Fig. 7f).

Mechanical controllers that allow study of propulsion in flexible foils or panels are also convenient for the study of a number of fish structures, including the specialized skin structure in sharks. We have been using this approach to better understand how fish skin functions hydrodynamically, and to study the function of fish skin under conditions of self-propulsion. The vast majority of previous research on the skin of fish has been conducted using relatively simple models in constant velocity flow regimes. But fish bodies bend and flex during locomotion and the skin surface is thus subject to time-varying flows and changing angles of attack. It is our view that in order to understand the diversity of fish skin structures and how these surface structures function during locomotion, testing must be done under dynamic

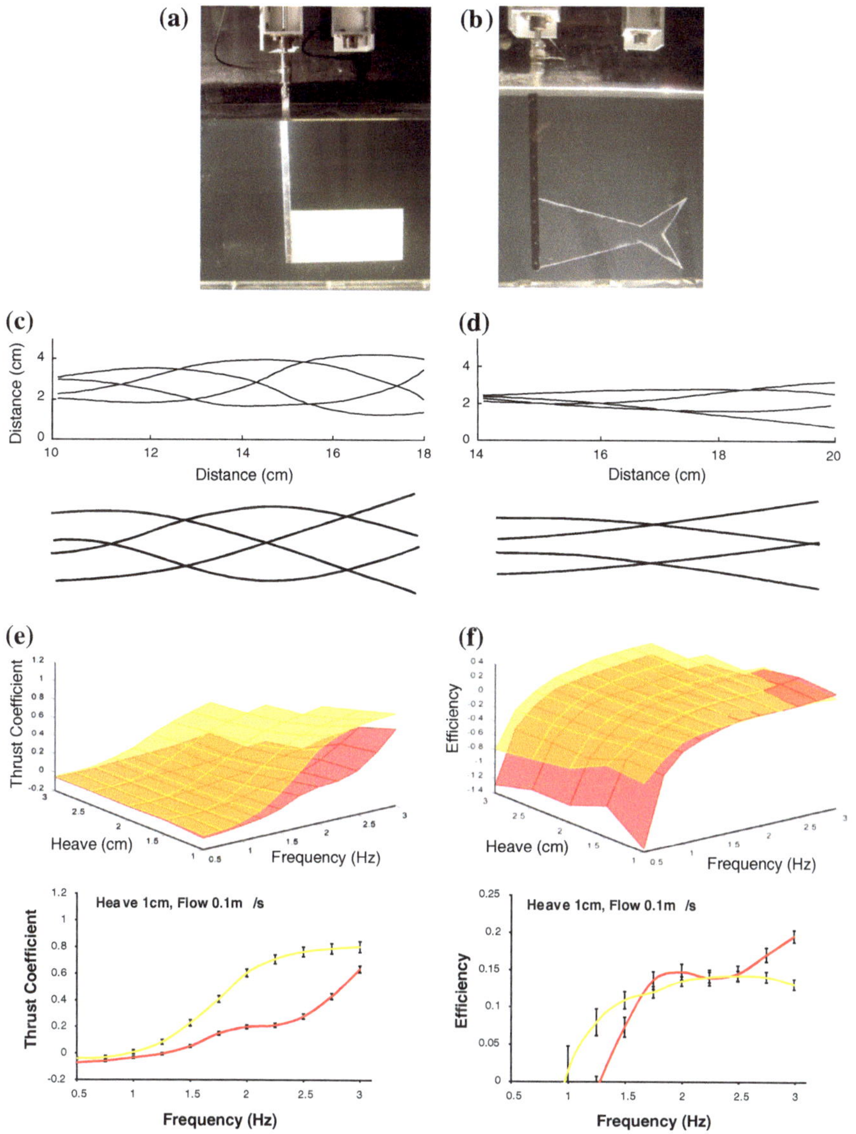

Fig. 7 Use of a flapping foil model system to study the dynamics of undulatory propulsion in fish. **a**, **b** A mechanical controller is used to oscillate flexible panels at the leading edge in heave and pitch motions while forces and torques on the shaft are measured (e.g., [64–66, 97, 98]. Flexible panels can be of any shape, and can be used as a simple model of fish caudal fins. **c**, **d** Motion of these flexible panels is similar to the undulatory body motion of the posterior half of the body of swimming fish and exhibits a range of dynamics that model both flexible (**c**) and stiff (**d**) fish bodies; midline data from clown knifefish and bluegill sunfish are shown on the *left* and *right*, respectively, with midlines from flexible and stiff swimming panels *below*. Use of a mechanical controller allows study of the full parameter space where heave (side to side) amplitude at the leading edge and frequency are used to explore the relationship of these parameters to thrust (**e**) and efficiency (**f**) for both flexible (*reddish* color) and stiff (*yellow*) flexible swimming foils. Plots below each surface show the relationship between thrust and efficiency and heave frequency for a particular heave and flow speed. Panels **c**–**f** modified from Shelton et al. [111]

self-propelled conditions. In addition, the frequencies, Strouhal numbers, and surface curvatures achieved by skin-like test membranes must match those exhibited by swimming fish.

Sharks have a remarkable structure to their skin. The surface of sharks is covered with millions of small tooth-like denticles that can range from 200 μm to 1 mm in size. Each denticle has a complex structure with a base embedded into the skin, and a stalk region that supports a flattened top containing ridges and valleys [58, 89, 92]. There is considerable variation in denticle structure among species of sharks [18], and equally substantial variation in denticle shape on different regions of the body [102, 103]. Adjacent denticles overlap, and changing the angle of the skin changes the angle of each denticle and hence the overlap between adjacent denticles. Previous work has suggested that shark skin functions to reduce drag, and may thus increase locomotor efficiency [10, 11, 24], but these analyses have been conducted under steady flow conditions in which the test surfaces were not moving.

In order to study the function of shark skin structures, we first assembled a flexible membrane composed of two pieces of fresh shark skin into a two-layer "sandwich" so that the skin surface denticles are exposed on each side (Fig. 8a). These membranes were then attached to a rod and moved by our mechanical flapper in a manner that matches the frequencies and surface curvatures of sharks swimming in vivo. By comparing the self-propelled speed of membranes with surface denticles intact and after they had been removed by careful surface sanding, we showed (Fig. 8a) that removal of the surface roughness reduced swimming speeds by an average of about 12 % [94]. Furthermore, the effect of surface roughness on self-propelled speed disappeared when rigid foils with shark skin on the surface are compared, indicating that the surface flexibility and motion of skin denticles is critical to their swimming performance advantage. Finally, particle image velocimetry of flow over the swimming shark skin membrane and the sanded control showed that removal of the surface alters the strength of the leading edge vortex, and suggests that the roughened shark skin surface may enhance leading edge suction and thus increase thrust. Shark skin surface roughness may thus enhancing thrust on oscillating surfaces such as the tail where flow separation occurs [15] as well as reduce drag.

In order to better understand the function of shark skin, [136] designed a biomimetic shark skin membrane using additive manufacturing (Fig. 8b, c). Manufacture of shark-skin-like surfaces permits alteration of biological parameters, and controlled study of individual parameters of shark skin. Manufactured skin-like membranes were assembled into a two-layer sandwich and compared to a smooth control surface (Fig. 8b). These membranes consist of rigid denticles embedded into a flexible membrane and this allows the biomimetic skin membrane to bend and flex during swimming in a manner similar to real shark skin.

Testing under a wide range of parameters revealed improved swimming performance of the membranes with skin-like denticles compared to a smooth control. For example, Fig. 8d shows data from tests of self-propelled speed when the membranes are actuated at 1 Hz, heave values of ±1.5 cm at the leading edge, and

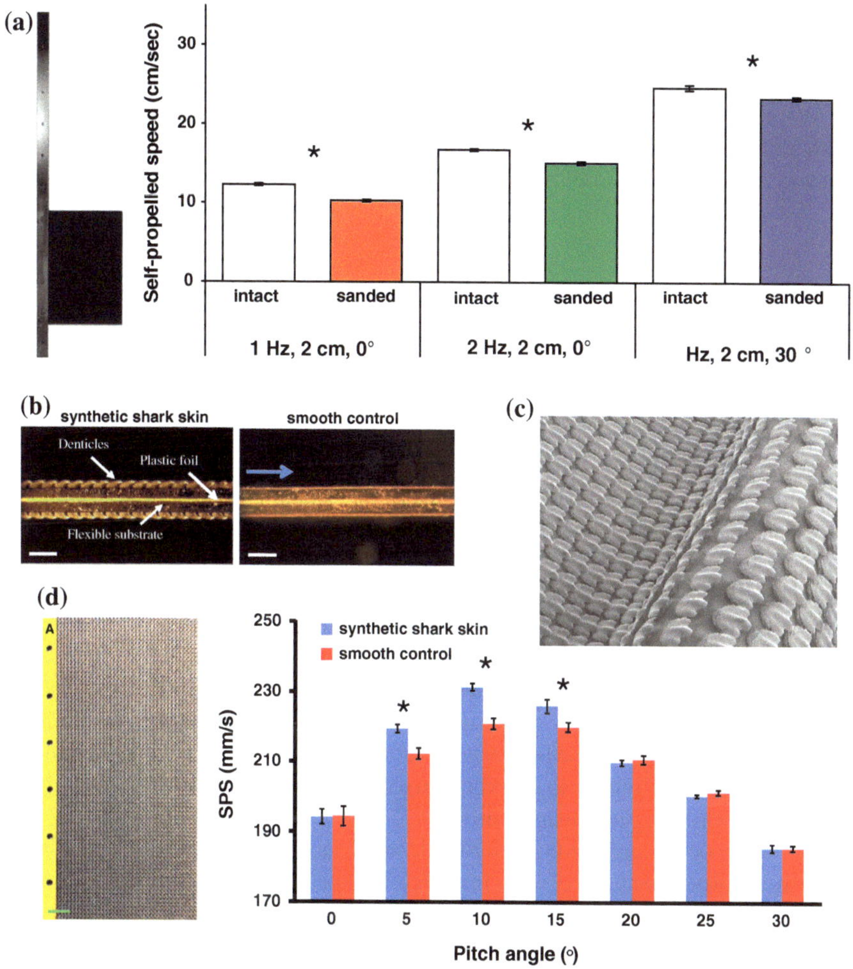

Fig. 8 Use of a mechanical flapping system to study the propulsive effect of shark skin. **a** Dynamic testing of the hydrodynamic function of shark skin denticles using pieces of shark skin that are attached to a flat support (shown on the *left*) which in turn is attached to a mechanical flapping foil device that allows controlled side-to-side and rotational motions of the shark skin membrane. Graph shows the self-propelled swimming speed of the shark skin membrane with intact denticles and after the denticles have been sanded off (to produce a relatively smooth surface) under three different motion programs. Note that in each case the swimming speed of the shark skin with denticles intact is significantly greater (*) than after the denticles have been removed by sanding. **b** Edge-on view of biomimetic shark skin assembled into a two-layer membrane compared to a smooth control of the same mass. *White scale bars* = 3 mm. **c** Scanning electron microscope view of biomimetic 3D-printed rigid shark skin denticles on a flexible membrane substrate. **d** Completed assembly of the tested flexible biomimetic shark skin foil (on *left*). Graph of the results from testing the self-propelled swimming speed of the biomimetic shark skin foil (*blue bars*) compared to the smooth control (*red bars*) at different pitch angles. At pitch angles of 5, 10, and 15° the biomimetic shark skin foils swim significantly faster (*) than the smooth controls. At the other pitch angles, the swimming speeds are similar. Modified from Wen et al. [136], Oeffner and Lauder [94], and Lauder [62]

a variety of pitch angles. Membranes with the roughened denticle-like surface swam faster for pitch angles of 5°, 10°, and 15°, but were not significantly different from controls at both higher and lower pitch values.

We believe that the manufacture of different fish-skin-like surfaces combined with dynamic testing that mimics conditions of fish swimming is a promising avenue for the future study of fish skin surface effects on locomotion.

3.3 Complex Robotic Models of Fish Locomotion

We have focused on bluegill sunfish (*Lepomis macrochirus*) as a model biorobotic system for studies of individual fin function, as well as for the development of more complex biorobotic models of the whole body and multi-fin function. Bluegill sunfish are particularly useful as they represent a generalized fish body shape and anatomy, and a substantial amount of three-dimensional kinematic, behavioral, and experimental hydrodynamic data on locomotor function exist that provide a useful baseline for both the design and analysis of biorobotic performance. The goals of our research using bluegill sunfish as a model system are (1) to better understand the locomotor performance of aquatic propulsive mechanical systems with flexible fish-like fins, (2) to develop manufacturing and assembly methods that allow construction of fins with fin rays having active control of the propulsive surface, (3) to conduct quantitative comparisons of data from mechanical devices to that of live fish, and (4) to use these biorobotic platforms to investigate complex swimming behaviors that are difficult to control in live fish such as multi-fin function during backward swimming and maneuvering.

We have developed several biorobotic models of the flexible fins of bluegill sunfish as experimental tools for investigating the mechanics and control of fin-based swimming. These systems include biorobotic pectoral and caudal fins that can execute the fin motions used during steady swimming, turn maneuvers, and hovering (Fig. 9, see [38, 121]); a freely swimming fish driven by paired pectoral fins [120]; and a model that integrates a moving peduncle with independently driven dorsal, anal, and caudal fins (our PDAC device, shown in Fig. 10). This biorobotic system is being used for investigating multi-fin swimming gaits which are common in fish when they stop, brake, swim backward, and maneuver through obstacles. The most recent of these systems incorporates distributed sensing within the fins (Fig. 9a, [57]), so that closed-loop control and navigation using object contacts can be explored. So that results from experiments conducted with these robotic systems have biological relevance, a priority of our design was to faithfully represent mechanical traits and to reproduce swimming characteristics of the biological fins. Physical features of the engineered fins were thus based on detailed studies of the bluegill sunfish biological system, and each model's dynamics, fin ray kinematics, and wake hydrodynamics were validated against those of the live fish (e.g., [31, 33, 44, 67, 68, 113]).

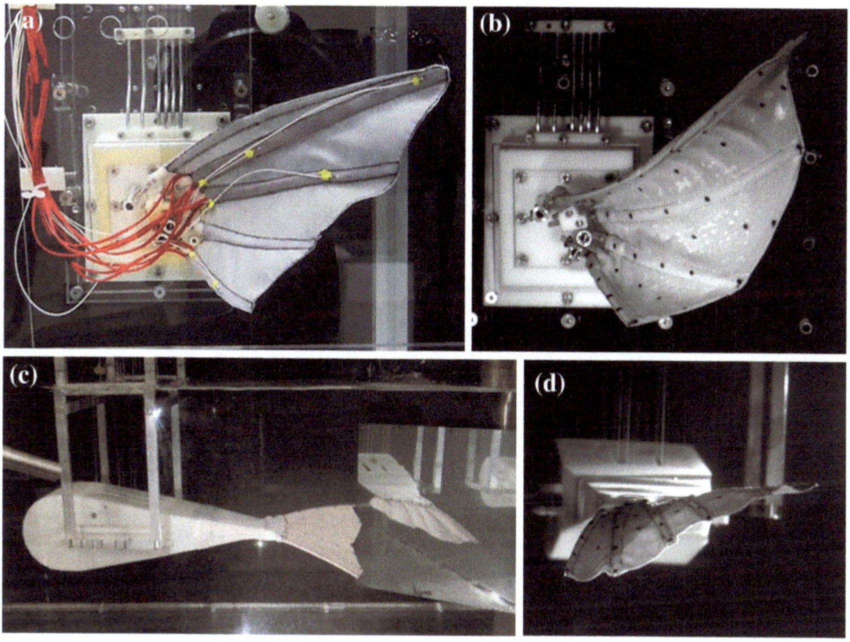

Fig. 9 Biorobotic pectoral (**a**, **b**) and caudal fins (**c**, **d**). Each of the rays of the pectoral fin has two actuated degrees of freedom which make them capable of executing a large repertoire of fin motions. In addition to steady swimming gaits such as that shown in (**b**), the pectoral fin can create the motions used during turns and hovering. The curvature of the fin rays seen in panel **b** is a result of the fluid loading on the compliant fin. Sensors within the pectoral fin (**a**) enable the curvature of the fin rays and the pressure across the fin webbing to be monitored for closed loop control. The rays of the caudal fin (**c**, **d**) are actuated through only one degree of freedom, but complex fin shapes can still be created by controlling phase relations between rays (**d**)

In order to produce good approximations of the complex fin kinematics used by the sunfish during swimming, each biorobotic fin uses multiple, independently actuated, compliant fin rays that are covered by a flexible fin webbing (elastane and polyester, latex, and/or urethane). The base of each fin ray is driven using servomotors (e.g., Maxon Precision Motors, Fall River, MA, USA) that are connected to the fin rays via nylon tendons. The fin rays are mounted to hinged bases which, with the servomotors, are designed to provide each fin ray with up to three degrees of actuated motion. In cases where the fin is designed to execute relatively simple kinematic patterns—such as steady swimming for caudal fins (Fig. 9c, d; [38]) bases with a single degree of freedom are sufficient. The fin rays move back and forth along an arc that is dictated by the orientation of the hinge within the base and complexity can be given to the fin shape by tailoring the compliance, phase relations, and the time course of individual rays. In cases where swimming gaits are more multifaceted such as pectoral fin maneuvering in sunfish, which involves complex motions of the fins and substantial conformational changes in the fin surface [46, 99], or where a single biorobotic model is used to investigate multiple

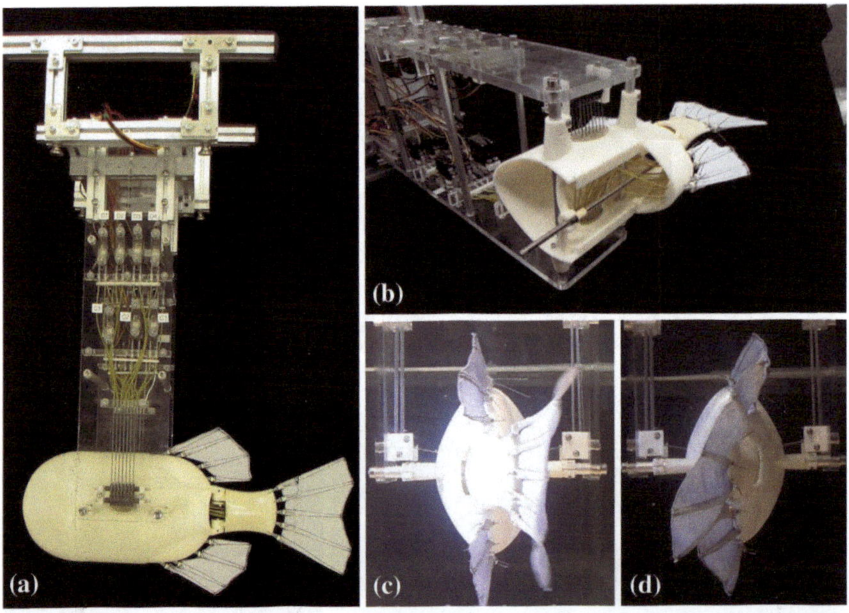

Fig. 10 Biorobotic model of the sunfish caudal peduncle (base of the tail) and the dorsal, anal, and caudal fins. The system provides independent control over the peduncle and 13 rays of the median fins for studies of multi-fin swimming. The location of the anal fin can be changed so that spatial relations among fins can be studied (**b**). Tendons, sheathed in housings, pass from the lateral sides of the fish to the fin ray bases (**b**). Fin motions and interactions are very different during steady swimming (**c**) and maneuvers, such as during backing (**d**) and this model is able to closely replicate the phasing and relative amplitudes of fin motions seen in live fish

kinematic patterns such as during maneuvering and steady swimming [96], the fin rays are also driven to rotate about the axis normal to the base (Fig. 9b). This, for example, enables a fin ray to follow different paths as it moves away from, or toward, the fish body. Active control over a fin ray's curvature, and thus over the fin's shape as it moves through the water, has been enacted in several biorobotic fins as a third degree of freedom using shape memory alloys, or a servomotor, to pull on a tendon that bends the rays shape [95]. In most cases, though, the fin rays are passive and curve only in response to fluidic loading. The cross sections of these passive fin rays are tapered from base to tip so that each ray's flexural rigidity scales to the flexural rigidity of the biological fin. This is key in allowing engineered fins to bend like a biological fin and to create appropriate wake and forces.

The nylon tendons that attach to the base of each fin ray are sheathed in a flexible, but non-collapsible, housing (Fig. 10), as is done for the cables of bicycle brakes. In addition to allowing the actuators to be located away from the fins, which benefits packing, these housings enable the tendons to follow curved paths within the robotic fins. In the PDAC robotic device, this permits the tendons of the robot's caudal fin to pass through the peduncle (Fig. 10), and for the peduncle to be actuated without causing the caudal fin's tendons to bind or to pull accidentally on the fin.

4 Conclusions

These devices based on bluegill sunfish, in addition to the wide variety of interesting complex fish robotic systems developed by other investigators, provide a solid foundation for future work in the area of fish biorobotics. Designing ever more complex fish-like systems will allow us to develop autonomous maneuvering robotic mechanisms that exhibit locomotor performance close to that of live fish. To accomplish this, biorobotic fish systems can make use of a diverse array of fins that allow force vectoring which will permit more complex behaviors than steady forward swimming alone. These devices will also be of great utility in testing biological hypotheses that are challenging to evaluate in living animals.

In the future, the use of simple mechanical systems to test specific hypotheses about the role of body shape, flexibility, and surface structure allows considerable control over experimental parameters, and will be of great value in investigating fundamental principles of aquatic propulsion. Researchers in fish biomechanics will benefit greatly from working with engineers to better understand how to construct a mechanical system that allows biological hypotheses to be tested. Similarly, engineers will continue to benefit from biological input which can provide insights into locomotor behaviors of the body and fins which may be unfamiliar to workers in design and manufacturing. We believe that one of the key challenges in the future is the control system needed to provide fish-like behavior and movement patterns in both steady and unsteady maneuvering propulsion. Developing an open-loop control will be challenging enough, but even more difficult is closed-loop control that integrates information from both the body and multiple fins. Closed-loop control is needed for navigation through complex habitats [43], and yet relatively few mechanical fish-like devices today incorporate such feedback.

Fish possess many interesting structural features of potential significance for locomotion that have yet to be studied, and the use of mechanical devices to explore the function of these features promises to help unlock some of the secrets of fish swimming performance.

Acknowledgments This work was supported by ONR-MURI Grant N000141410533 monitored by Dr. Bob Brizzolara, ONR Grant N00014-09-1-0352 monitored by Dr. Tom McKenna, and by National Science Foundation Grants EFRI-0938043 and CDI 0941674. Many thanks to members of the Lauder and Tangorra Laboratories for numerous helpful discussions.

References

1. Akhtar I, Mittal R, Lauder GV, Drucker E (2007) Hydrodynamics of a biologically inspired tandem flapping foil configuration. Theoret Comput Fluid Dyn 21:155–170
2. Alben S, Madden PG, Lauder GV (2007) The mechanics of active fin-shape control in ray-finned fishes. J R Soc Interface 4:243–256
3. Alben S, Witt C, Baker TV, Anderson EJ, Lauder GV (2012) Dynamics of freely swimming flexible foils. Phys Fluids 24:051901

4. Alexander RM (1967) Functional design in fishes. Hutchinson, London
5. Alvarado PV, Youcef-Toumi K (2005) Performance of machines with flexible bodies designed for biomimetic locomotion in liquid environments. In: Proceedings of the 2005 IEEE international conference on robotics and automation. Barcelona, Spain
6. Bandyopadhyay PR (2002) Maneuvering hydrodynamics of fish and small underwater vehicles. Integr Comp Biol 42:102–117
7. Bandyopadhyay PR, Beal DN, Menozzi A (2008) Biorobotic insights into how animals swim. J Exp Biol 211:206–214
8. Barrett DS, Triantafyllou MS, Yue DKP, Grosenbaugh MA, Wolfgang MJ (1999) Drag reduction in fish-like locomotion. J Fluid Mech 392:183–212
9. Beal DN (2003) Propulsion through wake synchronization using a flapping foil (Thesis). Type, MIT/WHOI
10. Bechert DW, Bartenwerfer M, Hoppe G, Reif WE (1986) Drag reduction mechanisms derived from shark skin. pp 1044–1068
11. Bechert DW, Hoppe G, Reif W-E (1985) On the drag reduction of the shark skin. AIAA J 85–0546:1–18
12. Blake RW (1983) Fish locomotion. Cambridge University Press, Cambridge
13. Blevins EL, Lauder GV (2013) Swimming near the substrate: a simple robotic model of stingray locomotion. Bioinspir Biomim 8:016005
14. Bone Q (1966) On the function of the two types of myotomal muscle fibre in elasmobranch fish. J Mar Biol Assoc UK 46:321–349
15. Borazjani I, Daghooghi M (2013) The fish tail motion forms an attached leading edge vortex. Proc Biol Sci R Soc 280:20122071
16. Bozkurttas M, Mittal R, Dong H, Lauder GV, Madden PGA (2009) Low-dimensional models and performance scaling of a highly deformable fish pectoral fin. J Fluid Mech 631:311–342
17. Buchholz JHJ, Smits AJ (2006) On the evolution of the wake structure produced by a low-aspect-ratio pitching panel. J Fluid Mech 546:433–443
18. Castro JI (2011) The sharks of North America. Oxford University Press, Oxford
19. Clark RP, Smits AJ (2006a) Thrust production and wake structure of a batoid-inspired oscillating fin. J Fluid Mech 562, 415–429
20. Clark RP, Smits AJ (2006b) Visualizations of the unsteady wake of a manta ray model. In: 44 th AIAA aerospace sciences meeting and exhibit, pp 1–8
21. Conte J, Modarres-Sadeghi Y, Watts MN, Hover FS, Triantafyllou MS (2010) A fast-starting mechanical fish that accelerates at 40 m s^{-2}. Bioinspir Biomim 5:035004
22. Curet OM, Patankar NA, Lauder GV, MacIver MA (2011a) Aquatic manoeuvering with counter-propagating waves: a novel locomotive strategy. J R Soc Interface 8:1041–1050
23. Curet OM, Patankar NA, Lauder GV, MacIver MA (2011b) Mechanical properties of a bio-inspired robotic knifefish with an undulatory propulsor. Bioinspir Biomim 6. doi: 10.1088/748-3182/6/2/026004
24. Dean B, Bhushan B (2010) Shark-skin surfaces for fluid-drag reduction in turbulent flow: a review. Phil Trans R Soc A Math Phys Eng Sci 368:4775–4806
25. Dewey PA, Boschitsch BM, Moored KW, Stone HA, Smits AJ (2013) Scaling laws for the thrust production of flexible pitching panels. J Fluid Mech 732:29–46
26. Dewey PA, Quinn DB, Boschitsch BM, Smits AJ (2014) Propulsive performance of unsteady tandem hydrofoils in a side-by-side configuration. Phys Fluids 26:041903
27. Domenici P, Kapoor BG (2010) Fish locomotion: an eco-ethological perspective. Science Publishers, Enfield
28. Dong H, Bozkurttas M, Mittal R, Madden P, Lauder GV (2010) Computational modeling and analysis of the hydrodynamics of a highly deformable fish pectoral fin. J Fluid Mech 645:345–373
29. Donley J, Dickson KA (2000) Swimming kinematics of juvenile Kawakawa tuna (*Euthynnus affinis*) and chub mackerel (*Scomber japonicus*). J Exp Biol 203:3103–3116
30. Donley J, Shadwick R (2003) Steady swimming muscle dynamics in the leopard shark *Triakis semifasciata*. J Exp Biol 206:1117–1126

31. Drucker EG, Lauder GV (1999) Locomotor forces on a swimming fish: three-dimensional vortex wake dynamics quantified using digital particle image velocimetry. J Exp Biol 202:2393–2412
32. Drucker EG, Lauder GV (2000) A hydrodynamic analysis of fish swimming speed: wake structure and locomotor force in slow and fast labriform swimmers. J Exp Biol 203:2379–2393
33. Drucker EG, Lauder GV (2001a) Locomotor function of the dorsal fin in teleost fishes: experimental analysis of wake forces in sunfish. J Exp Biol 204:2943–2958
34. Drucker EG, Lauder GV (2001) Wake dynamics and fluid forces of turning maneuvers in sunfish. J Exp Biol 204:431–442
35. Drucker EG, Lauder GV (2005) Locomotor function of the dorsal fin in rainbow trout: kinematic patterns and hydrodynamic forces. J Exp Biol 208:4479–4494
36. Drucker EG, Walker JA, Westneat M (2006) Mechanics of pectoral fin swimming in fishes. In: Shadwick RE, Lauder GV (ed) Fish biomechanics. Volume 23 in fish physiology, Academic Press, San Diego, pp 369–423
37. Epps B, Valdivia y Alvarado P, Youcef-Toumi K, Techet A (2009) Swimming performance of a biomimetic compliant fish-like robot. Exp Fluids 47:927–939
38. Esposito C, Tangorra J, Flammang BE, Lauder GV (2012) A robotic fish caudal fin: effects of stiffness and motor program on locomotor performance. J Exp Biol 215:56–67
39. Fish F, Lauder GV (2006) Passive and active flow control by swimming fishes and mammals. Annu Rev Fluid Mech 38:193–224
40. Fish F, Lauder GV (2013) Not just going with the flow. Am Sci 101:114–123
41. Fish FE, Howle LE, Murray MM (2008) Hydrodynamic flow control in marine mammals. Integr Comp Biol 48:788–800
42. Flammang BE, Alben S, Madden PGA, Lauder GV (2013) Functional morphology of the fin rays of teleost fishes. J Morphol 274:1044–1059
43. Flammang BE, Lauder GV (2013) Pectoral fins aid in navigation of a complex environment by bluegill sunfish under sensory deprivation conditions. J Exp Biol 216:3084–3089
44. Gibb A, Jayne BC, Lauder GV (1994) Kinematics of pectoral fin locomotion in the bluegill sunfish *Lepomis macrochirus*. J Exp Biol 189:133–161
45. Gillis GB (1998) Environmental effects on undulatory locomotion in the American eel *Anguilla rostrata*: kinematics in water and on land. J Exp Biol 201:949–961
46. Gottlieb J, Tangorra J, Esposito C, Lauder G (2010) A biologically derived pectoral fin for yaw turn maneuvers. Appl Bion Biomech 7:41–55
47. Gray J (1953) How animals move. Cambridge University Press, Cambridge
48. Gray J (1968) Animal locomotion. Weidenfeld and Nicolson, London
49. Hoar WS, Randall DJ (1978) Fish physiology. Locomotion, vol VII. Academic Press, New York
50. Hover FS, Haugsdal O, Triantafyllou MS (2004) Effect of angle of attack profiles in flapping foil propulsion. J Fluids Struct 19:37–47
51. Hu T, Shen L, Lin L, Xu H (2009) Biological inspirations, kinematics modeling, mechanism design and experiments on an undulating robotic fin inspired by *Gymnarchus niloticus*. Mech Mach Theory 44:633–645
52. Hua R-N, Zhu L, Lu X-Y (2013) Locomotion of a flapping flexible plate. Phys Fluids 25:121901
53. Jayne BC, Lauder GV (1993) Red and white muscle activity and kinematics of the escape response of the bluegill sunfish during swimming. J Comp Physiol A 173:495–508
54. Jayne BC, Lauder GV (1994) How swimming fish use slow and fast muscle fibers: implications for models of vertebrate muscle recruitment. J Comp Physiol A 175:123–131
55. Jayne BC, Lauder GV (1995a) Red muscle motor patterns during steady swimming in largemouth bass: effects of speed and correlations with axial kinematics. J Exp Biol 198:1575–1587
56. Jayne BC, Lauder GV (1995b) Speed effects on midline kinematics during steady undulatory swimming of largemouth bass, *Micropterus salmoides*. J Exp Biol 198:585–602

57. Kahn JV, Flammang BE, Tangorra JL (2012) Hover kinematics and distributed pressure sensing for force control of biorobotic fins. In: Intelligent robots and systems (IROS). Vilamoura, Portugal
58. Kemp NE (1999) Integumentary system and teeth. In: Hamlett WC (ed.) Sharks, skates, and rays. The biology of elasmobranch fishes. Johns Hopkins University Press, Baltimore, pp 43–68
59. Kopman V, Laut J, Acquaviva F, Rizzo A, Porfiri M (2014) Dynamic modeling of a robotic fish propelled by a compliant tail. IEEE J Ocean Eng 99:1–13
60. Lauder GV (2006) Locomotion. In: Evans DH, Claiborne JB (eds) The physiology of fishes, 3rd edn. CRC Press, Boca Raton, pp 3–46
61. Lauder GV (2011) Swimming hydrodynamics: ten questions and the technical approaches needed to resolve them. Exp Fluids 51:23–35
62. Lauder GV (2015a) Fish locomotion: recent advances and new directions. In: Annual review of marine science, vol 7, in press
63. Lauder GV (2015b) Flexible fins and fin rays as key transformations in ray-finned fishes. In: Dial KP, Shubin N, Brainerd E (eds) The great transformations: major events in the history of vertebrate life. University of California Press, Berkeley, California
64. Lauder GV, Anderson EJ, Tangorra J, Madden PGA (2007) Fish biorobotics: kinematics and hydrodynamics of self-propulsion. J Exp Biol 210:2767–2780
65. Lauder GV, Flammang BE, Alben S (2012) Passive robotic models of propulsion by the bodies and caudal fins of fish. Integr Comp Biol 52:576–587
66. Lauder GV, Lim J, Shelton R, Witt C, Anderson EJ, Tangorra J (2011a) Robotic models for studying undulatory locomotion in fishes. Mar Technol Soc J 45:41–55
67. Lauder GV, Madden PGA (2007) Fish locomotion: kinematics and hydrodynamics of flexible foil-like fins. Exp Fluids 43:641–653
68. Lauder GV, Madden PGA, Mittal R, Dong H, Bozkurttas M (2006) Locomotion with flexible propulsors I: experimental analysis of pectoral fin swimming in sunfish. Bioinspir Biomim 1: S25–S34
69. Lauder GV, Madden PGA, Tangorra J, Anderson E, Baker TV (2011b) Bioinspiration from fish for smart material design and function. Smart Mater Struct 20. doi:10.1088/0964-1726/20/9/094014
70. Lauder GV, Tytell ED (2006) Hydrodynamics of undulatory propulsion. In: Shadwick RE, Lauder GV (eds) Fish biomechanics, Volume 23 in fish physiology. Academic Press, San Diego, pp 425–468
71. Leftwich MC, Smits AJ (2011) Thrust production by a mechanical swimming lamprey. Exp Fluids 50:1349–1355
72. Leftwich MC, Tytell ED, Cohen AH, Smits AJ (2012) Wake structures behind a swimming robotic lamprey with a passively flexible tail. J Exp Biol 215:416–425
73. Liao JC (2007) A review of fish swimming mechanics and behaviour in altered flows. Philos Trans R Soc B Biol Sci 362:1973–1993
74. Licht SC, Wibawa MS, Hover FS, Triantafyllou MS (2010) In-line motion causes high thrust and efficiency in flapping foils that use power downstroke. J Exp Biol 213:63–71
75. Lighthill MJ (1969) Hydromechanics of aquatic animal propulsion: a survey. Annu Rev Fluid Mech 1:413–446
76. Lindsey CC (1978) Form, function, and locomotory habits in fish. In: Hoar WS, Randall DJ (eds) Fish physiology. Locomotion, vol VII. Academic Press, New York, pp 1–100
77. Liu J, Dukes I, and Hu H (2005) Novel mechatronics design for a robotic fish. In: IEEE/RSJ international conference on intelligent robots and systems, 2005 (IROS 2005), pp 807–812
78. Liu J, Hu H, Gu D (2006) A hybrid control architecture for autonomous robotic fish. In: IEEE/RSJ international conference on intelligent robots and systems, pp 312–317
79. Long J (2012) Darwin's devices: what evolving robots can teach us about the history of life and the future of technology. Basic Books, New York
80. Long J, Koob TJ, Schaefer J, Summers A, Bantilan K, Grotmol S et al (2011) Inspired by sharks: a biomimetic skeleton for the flapping, propulsive tail of an aquatic robot. Mar Technol Soc J 45:119–129

81. Long JH, McHenry MJ, Boetticher NC (1994) Undulatory swimming: how traveling waves are produced and modulated in sunfish (*Lepomis gibbosus*). J Exp Biol 192:129–145
82. Long JH, Nipper KS (1996) The importance of body stiffness in undulatory propulsion. Am Zool 36:678–694
83. Long JH, Schumacher J, Livingston N, Kemp M (2006) Four flippers or two? Tetrapodal swimming with an aquatic robot. Bioinspir Biomim 1:20–29
84. Low KH, Chong CW (2010) Parametric study of the swimming performance of a fish robot propelled by a flexible caudal fin. Bioinspir Biomim 5:046002
85. Low KH, Yu J (2007) Development of modular and reconfigurable biomimetic robotic fish with undulating fin. In: IEEE international conference on robotics, (ROBIO 2007), Sanya, China, pp 274–279
86. Maddock L, Bone Q, Rayner JMV (1994) Mechanics and physiology of animal swimming. Cambridge University Press, Cambridge
87. Maia A, Wilga CD, Lauder GV (2012) Biomechanics of locomotion in sharks, rays and chimeras. In: Carrier JC, Musick JA, Heithaus MR (eds) Biology of sharks and their relatives, 2nd edn. CRC Press, Boca Raton, pp 125–151
88. McHenry MJ, Pell CA, Long JA (1995) Mechanical control of swimming speed: stiffness and axial wave form in undulating fish models. J Exp Biol 198:2293–2305
89. Meyer W, Seegers U (2012) Basics of skin structure and function in elasmobranchs: a review. J Fish Biol 80:1940–1967
90. Moored KW, Bart-Smith H (2007) The analysis of tensegrity structures for the design of a morphing wing. J Appl Mech 74:668–676
91. Moored KW, Dewey PA, Leftwich M, Bart-Smith H, Smits A (2011) Bioinspired propulsion mechanisms based on manta ray locomotion. Mar Technol Soc J 45:110–118
92. Motta P, Habegger ML, Lang A, Hueter R, Davis J (2012) Scale morphology and flexibility in the shortfin mako *Isurus oxyrinchus* and the blacktip shark *Carcharhinus limbatus*. J Morphol 273:1096–1110
93. Neveln ID, Bai Y, Snyder JB, Solberg JR, Curet OM, Lynch KM et al (2013) Biomimetic and bio-inspired robotics in electric fish research. J Exp Biol 216:2501–2514
94. Oeffner J, Lauder GV (2012) The hydrodynamic function of shark skin and two biomimetic applications. J Exp Biol 215:785–795
95. Phelan C, MacDonald RJ, Tangorra JL (2009) Artificial muscle actuators in biorobotic fish fins. In: Annual international conference of the engineering in medicine and biology society. Minneapolis, Minnesota
96. Phelan C, Tangorra J, Lauder GV, Hale M (2010) A biorobotic model of the sunfish pectoral fin for investigations of fin sensorimotor control. Bioinspir Biomim 5:035003
97. Quinn DB, Lauder GV, and Smits AJ (2014a) Flexible propulsors in ground effect. Bioinspir Biomim 9:1–9
98. Quinn DB, Moored KW, Dewey PA, Smits AJ (2014b) Unsteady propulsion near a solid boundary. J Fluid Mech 742:152–170
99. Ramakrishnan S, Bozkurttas M, Mittal R, Lauder GV (2011) Thrust production in highly flexible pectoral fins: a computational dissection. Mar Technol Soc J 45:56–64
100. Ramamurti R, Sandberg WC, Lohner R, Walker JA, Westneat M (2002) Fluid dynamics of flapping aquatic flight in the bird wrasse: three-dimensional unsteady computations with fin deformation. J Exp Biol 205:2997–3008
101. Read DA, Hover FS, Triantafyllou MS (2003) Forces on oscillating foils for propulsion and maneuvering. J Fluids Struct 17:163–183
102. Reif W-E (1985) Squamation and ecology of sharks. Cour Forsch Inst Senckenberg 78:1–255
103. Reif W (1982) Morphogenesis and function of the squamation in sharks. Neues Jb Geol Paläontol Abh 164:172–183
104. Rossi C, Colorado J, Coral W, Barrientos A (2011) Bending continuous structures with SMAs: a novel robotic fish design. Bioinspir Biomim 6:045005
105. Sfakiotakis M, Lane D, Davies JB (1999) Review of fish swimming modes for aquatic locomotion. IEEE J Ocean Eng 24:237–252

106. Shadwick R (2005) How tunas and lamnid sharks swim: an evolutionary convergence. Am Sci 93:524–531
107. Shadwick R, Gemballa S (2006) Structure, kinematics, and muscle dynamics in undulatory swimming. In: Shadwick RE, Lauder GV (eds) Fish biomechanics, Volume 23 in fish physiology. Academic Press, San Diego, pp 241–280
108. Shadwick R, Steffensen J, Katz S, Knower T (1998) Muscle dynamics in fish during steady swimming. Am Zool 38:755–770
109. Shadwick RE, Goldbogen JA (2012) Muscle function and swimming in sharks. J Fish Biol 80:1904–1939
110. Shadwick RE, Lauder GV (2006) Fish biomechanics. In: Hoar WS, Randall DJ, Farrell AP (eds) Fish physiology, vol 23. Academic Press, San Diego
111. Shelton R, Thornycroft P, Lauder G (2014) Undulatory locomotion by flexible foils as biomimetic models for understanding fish propulsion. J Exp Biol 217:2110–2120
112. Standen EM (2008) Pelvic fin locomotor function in fishes: three-dimensional kinematics in rainbow trout (*Oncorhynchus mykiss*). J Exp Biol 211:2931–2942
113. Standen EM, Lauder GV (2005) Dorsal and anal fin function in bluegill sunfish *Lepomis macrochirus*: three-dimensional kinematics during propulsion and maneuvering. J Exp Biol 208:2753–2763
114. Standen EM, Lauder GV (2007) Hydrodynamic function of dorsal and anal fins in brook trout (*Salvelinus fontinalis*). J Exp Biol 210:325–339
115. Syme DA (2006) Functional properties of skeletal muscle. In: Shadwick RE, Lauder GV (eds) Fish biomechanics, Volume 23 in Fish Physiology. Academic Press, San Diego, pp 179–240
116. Taft NK (2011) Functional implications of variation in pectoral fin ray morphology between fishes with different patterns of pectoral fin use. J Morphol 272:1144–1152
117. Taft NK, Taft BN (2012) Functional implications of morphological specializations among the pectoral fin rays of the benthic longhorn sculpin. J Exp Biol 215:2703–2710
118. Tan X, Kim D, Usher N, Laboy D, Jackson J, Kapetanovic A et al (2006) An autonomous robotic fish for mobile sensing. In: Proceedings of the IEEE/RSJ international conference on intelligent robots and systems. Beijing, China, pp 5424–5429
119. Tangorra J, Lauder GV, Madden PG, Mittal R, Bozkurttas M, Hunter I (2008) A biorobotic flapping fin for propulsion and maneuvering. In: International conference on robotics and automation. ICRA, Pasadena, CA
120. Tangorra J, Phelan C, Esposito C, Lauder G (2011) Use of biorobotic models of highly deformable fins for studying the mechanics and control of fin forces in fishes. Integr Comp Biol 51:176–189
121. Tangorra JL, Lauder GV, Hunter I, Mittal R, Madden PG, Bozkurttas M (2010) The effect of fin ray flexural rigidity on the propulsive forces generated by a biorobotic fish pectoral fin. J Exp Biol 213:4043–4054
122. Triantafyllou M, Hover FS, Techet AH, Yue D (2005) Review of hydrodynamic scaling laws in aquatic locomotion and fish swimming. Trans ASME 58:226–237
123. Triantafyllou MS, Triantafyllou GS, Yue DKP (2000) Hydrodynamics of fishlike swimming. A Rev Fluid Mech 32:33–53
124. Tytell ED (2006) Median fin function in bluegill sunfish *Lepomis macrochirus*: streamwise vortex structure during steady swimming. J Exp Biol 209:1516–1534
125. Tytell ED, Lauder GV (2008) Hydrodynamics of the escape response in bluegill sunfish, *Lepomis macrochirus*. J Exp Biol 211:3359–3369
126. Tytell ED, Standen EM, Lauder GV (2008) Escaping flatland: three-dimensional kinematics and hydrodynamics of median fins in fishes. J Exp Biol 211:187–195
127. Videler JJ (1993) Fish swimming. Chapman and Hall, New York
128. Walker JA, Westneat MW (1997) Labriform propulsion in fishes: kinematics of flapping aquatic flight in the bird wrasse *Gomphosus varius* (Labridae). J Exp Biol 200:1549–1569
129. Walker JA, Westneat MW (2000) Mechanical performance of aquatic rowing and flying. Proc R Soc Lond Ser B 267:1875–1881

130. Wardle CS, Videler JJ, Altringham JD (1995) Tuning into fish swimming waves: body form, swimming mode and muscle function. J Exp Biol 198:1629–1636
131. Webb PW (1975) Hydrodynamics and energetics of fish propulsion. Bull Fish Res Board Can 190:1–159
132. Webb PW (1982) Locomotor patterns in the evolution of actinopterygian fishes. Am Zool 22:329–342
133. Webb PW, Blake RW (1985) Swimming. In: Hildebrand M, Bramble DM, Liem KF, Wake DB (eds) Functional vertebrate morphology. Harvard University Press, Cambridge, pp 110–128
134. Webb PW, Weihs D (1983) Fish biomechanics. Praeger Publishers, New York
135. Wen L, Lauder GV (2013) Understanding undulatory locomotion in fishes using an inertia-compensated flapping foil robotic device. Bioinspir Biomim 8:046013
136. Wen L, Weaver JC, Lauder GV (2014) Biomimetic shark skin: design, fabrication, and hydrodynamic function. J Exp Biol 217:1656–1666
137. Westneat MW (1996) Functional morphology of aquatic flight in fishes: kinematics, electromyography, and mechanical modeling of labriform locomotion. Am Zool 36:582–598
138. Westneat MW, Walker JA (1997) Motor patterns of labriform locomotion: kinematic and electromyographic analysis of pectoral fin swimming in the labrid fish *Gomphosus varius*. J Exp Biol 200:1881–1893
139. Wu T, Brokaw CJ, Brennen C (1975) Swimming and flying in nature. Plenum, New York
140. Xiong G, Lauder GV (2014) Center of mass motion in swimming fish: effects of speed and locomotor mode during undulatory propulsion. Zoology 117:269–281
141. Youngerman ED, Flammang BE, Lauder GV (2014) Locomotion of freely-swimming ghost knifefish: anal fin function during four behaviors. Zoology 117:337–348

Wire-Driven Robot Fish

Zheng Li and Ruxu Du

Abstract In this chapter, a class of wire-driven robot fish is presented. It comprises of a streamlined body and a flapping tail, which is designed based on the biomimetic wire-driven mechanism (BWDM). The BWDM mimics not only the fish body skeletal structure but also the muscle arrangement. As a result, the wire-driven robot fish can well replicate various fish flapping motion with minimum number of actuators and thus, has simple structure and excellent propulsion efficiency. Three wire-driven robot fish are presented. The first one is driven by one motor and swims in oscillatory form; the second one is driven by two motors and swims in undulatory form; and the third one can mimic both shark swim and dolphin swim.

1 Introduction

As shown in the other chapters of the book, inspired by fish and other aquatic animals, people have been building robot fish for two decades since MIT's robot tuna [1]. Fish has superior efficiency and agility. It is shown that the propulsion efficiency of the fish can exceed 90 % [1]; the speed of a sailfish can reach 110 km/h [2], and the maximum acceleration of a pike is as high as 249 m/s^2, which is over 25 g [3]. Also, the fish swims silently and can turn around sharply without much deceleration. The outstanding performance of the fish is attributed to multiple factors, such as the streamlined body, mucous skin, etc. Though, the most important one is the flapping tail. It is also what people have been mimicking in making their robot fish.

In fish swimming, there are mainly two types of propulsions, i.e., the Body and/or Caudal Fin (BCF) propulsion and the Medium and/or Paired Fin (MPF) propulsion [4]. BCF is adopted by most of the fast swimmers, such as sailfish, tuna,

Z. Li (✉) · R. Du
The Chinese University of Hong Kong, Hong Kong, China
e-mail: lizheng@cuhk.edu.hk

R. Du
e-mail: rdu@mac.cuhk.edu.hk

© Springer-Verlag Berlin Heidelberg 2015

R. Du et al. (eds.), *Robot Fish*, Springer Tracts in Mechanical Engineering,
DOI 10.1007/978-3-662-46870-8_3

pike, etc. In fact, around 85 % of fish species swim in BCF form. Examples for MPF swimmers are manta ray and box fish. Compared to BCF swimmers, MPF swimmers are more agile. In robot fish development, high speed and high efficiency are the main pursuits. As a result, most of existing robot fish adopt BCF as the way to generate thrust.

Within BCF, the motion modes are further divided into four categories, i.e., Thunniform, Carangiform, Sub-carangiform, and Anguilliform. When the fish swims, its body curve is a wave traveling from the head to the tail. The wave length, measured in body length (BL), increases from Anguilliform to Thunniform. The BCF swimming can also be roughly categorized as oscillatory form swimming and undulatory form swimming. As a comparison, Thunniform is considered as oscillatory form and the other three are viewed as undulatory form.

Fundamental motion of BCF swimming is flapping from side to side. There are a number of ways to generate such a flapping motion. One method is to use direct drive or through traditional mechanisms, e.g., crank, four-bar mechanism, etc. In this method, the unidirectional rotation of the actuator (typically a motor) is transferred to the back-and-forth motion, and then to the flapping motion. A simple example is that the tail is connected to the motor directly. By controlling the motor rotating back and forth, the tail flaps from side to side. In this method, as only one actuator is used, and the tail can only rotate about a joint, they are referred to as the Single Joint (SJ) design. One example is the robot zebra fish in [5]. Robot fish with the SJ design is simple in both mechanical structure and control; also the tail can flap fast and forcefully. Its drawback is that the tail is rigid and the flapping motion is far unlike the fish. Hence, its propulsion efficiency is low. In addition, it can only mimic oscillatory form swimming. To better fit the fish swimming body curve, a passive deflectable fin can be added, such as the robot fish developed in Beihang University [6], and the Tai-robot-kun developed in University of Kitakyushu [7]. By using passive flexible fins, the flapping motion is closer to that of the fish flapping motion, but the improvement is limited.

The second method is to use a chain of serially connected links with revolute joints. Each joint is powered by an actuator. The actuators drive the joints rotating back and forth generating the flapping motion. This method is referred to as the Multi-Joint (MJ) design. In this design, the flapping curve is a polyline, which contains several (usually 3 or 4) straight line segments. Compared to the SJ design, its motion resembles the fish flapping motion better. However, the motor motion synchronization is crucial. Also, it becomes complicated with the increased number of motors and the energy consumption is high. Examples of MJ robot fish include the robot tuna developed in MIT [8], the robot fish developed in Essex University [9], the robot knife fish developed in Northwestern University [10], and the robot fish developed in Chinese Academic of Science Institute of Automation [11].

In addition to the motor-based designs, novel actuators are also used in robot fish designs. Examples include Ionic Polymer-Metal Composite (IPMC) actuated robot fish [12, 13], Shape Memory Alloy (SMA) actuated robot fish [14–16], and Piezoelectric material actuated robot fish [17–19]. For IPMC, when exposed to an electric field, the material deflects. The deflection is close to a circular arc. By controlling the strength and direction of the electric field, the bending amplitude and direction can be controlled. Though, IPMC is soft and the force generated is low. SMA can generate larger force, which is controlled by temperature. It can restore the previous shape when heated above its transit temperature. In other words, by heating and cooling SMA, a flapping motion can be obtained. However, due to the slow heating and cooling process, its flapping frequency is limited. Also, it is difficult to control the flapping amplitude and flapping pattern. Piezoelectric material, such as PZT, is also controlled by electric field. By controlling the voltage applied to the PZT, the mechanical strain can be controlled. However, the strain is usually very small and needs be amplified to generate sufficient flapping. Also, the required voltage is high. Moreover, it should be mentioned that, a common drawback of the smart materials-based flapping methods is that the material itself is highly energy inefficient. This results in the overall efficiency of the propulsion system being rather low.

The drawbacks of the existing flapping systems raise the desire of a new design method, in which the amplitude of the flapping needs to be large, the flapping pattern and flapping frequency need to be easy to control, and most importantly, the efficiency must be high. By reexamining the fish body structure, in this chapter, the Biomimetic Wire Driven Mechanism (BWDM) is first introduced and then used for robot fish design. Three BWDM robot fishes are presented, the first one is capable of oscillatory swim; the second one is capable of undulatory swim and oscillatory swim, and the third one can swim like a dolphin and a shark.

2 Biomimetic Wire-Driven Mechanism

Wires and cables have long been used to transmit motion and power. Examples include lift cranes and marionettes. In these traditional wire-driven systems, the motion and the power are transmitted through the rigid structures by the wires. In this chapter, the BWDM is introduced, in which the body structure is flexible and underactuated. It is shown that this mechanism is well suited for underwater propulsion.

2.1 Nature Inspiration

Most aquatic animals have a flexible body and the body motion is controlled by the attached muscles and tendons. From the structure point of view the flexible bodies are divided into two types. The first type of flexible body is similar to that of the bony fish, such as the carp. As shown in Fig. 1a [20], the carp body has a backbone which contains a chain of short rigid vertebrae. Two successive vertebrae form a joint and each joint can rotate a small angle. Due to the large number of joints the backbone can bend a large angle. A close examination reveals that the joints' rotations are not arbitrary, instead they are constrained by the surrounded muscles.

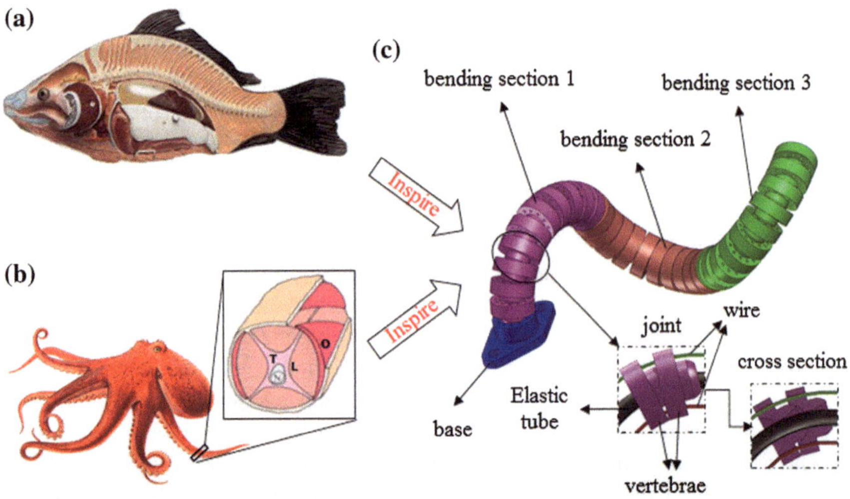

Fig. 1 Examples of flexible body in nature: **a** carp anatomy [20]; **b** octopus arm anatomy [21]; **c** biomimetic wire-driven mechanism (BWDM)

During the flapping motion, the backbone can be viewed as a number of bending sections and the bending follows the fish swimming pattern (i.e., the traveling wave equation). The second type of flexible body is similar to the body of mollusks, such as the octopus. It is soft and boneless. Figure 1b shows the octopus arm structure. It contains three types of muscles: the longitudinal muscle (L), the transversal muscle (T), and the oblique muscle (O). The arm can bend by the longitudinal muscle, elongate by the transversal muscle, and twist by the oblique muscle.

The BWDM design follows these examples. As shown in Fig. 1c, it comprises of a flexible backbone, which mimics the fish backbone, and a number of controlling wires, which mimic the octopus arm muscles. In the following, the design is detailed from three aspects: the backbone, the backbone segmentation, and the wire configuration.

2.2 Backbone Type

Similar to the flexible bodies in nature, the BWDM has two types of backbone. The first one is the serpentine backbone, which is similar to the carp backbone and the second one is named the continuum backbone, which is similar to the octopus arm structure.

Serpentine Backbone

The serpentine backbone comprises of a number of rigid vertebrae and the elastic constraint component, which is usually in the form of an elastic tube or elastic

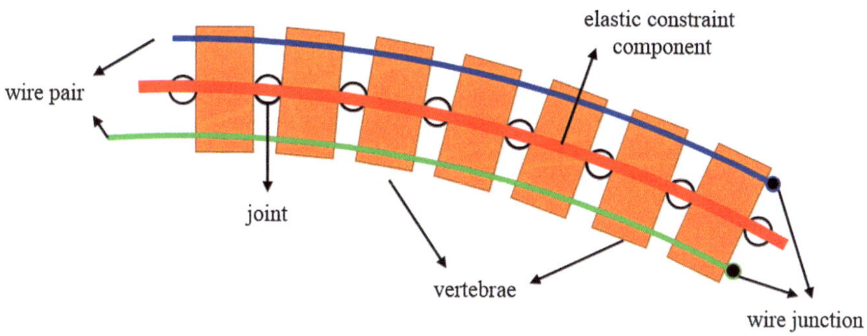

Fig. 2 Illustration of the BWDM with serpentine backbone

beam. Figure 2 shows the serpentine backbone structure. The vertebrae are connected serially, and each pair of adjacent vertebrae forms a joint. The elastic constraint component fits tightly with the vertebrae. It also helps in the vertebrae articulation. The backbone bending is controlled by a pair of wires. By pulling one wire and releasing the other, the joints rotate and the backbone bends. The joints' rotations follow the deflection of the elastic constraint component. Suppose all the vertebrae have identical length and the elastic constraint component has identical bending stiffness along the axis, all the joints' rotations are the same. This is known as the piecewise constant curvature assumption [22].

In the serpentine backbone, the vertebra design is very important as it determines the joint type and therefore the backbone bending. Figure 3 gives two designs. In Fig. 3a, the vertebra has two cylindrical connectors at the two ends. Two successive vertebrae form a revolute joint by the connectors. In the middle of the vertebra there

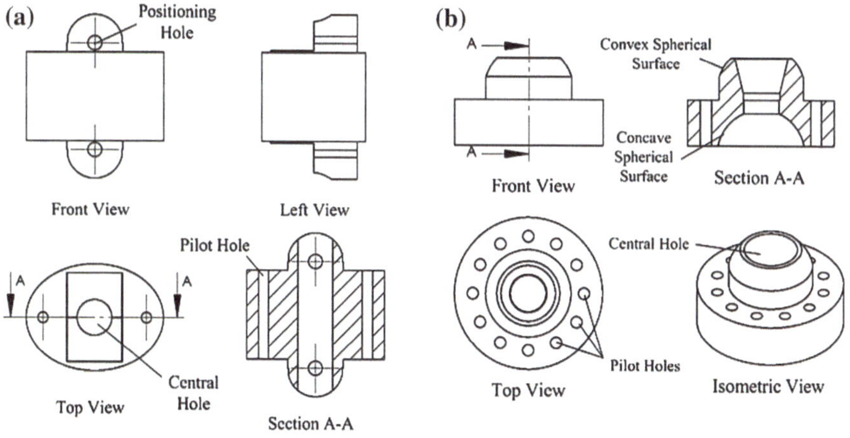

Fig. 3 Two types of vertebra designs: **a** vertebra with revolute joint (for 1D bending); **b** vertebra with spherical joint (for 2D bending)

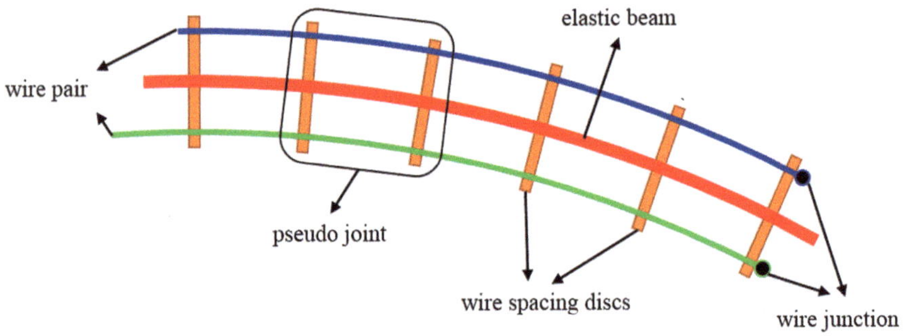

Fig. 4 Illustration of the BWDM with continuum backbone

is a through hole, which houses the elastic constraint component. Another two holes are placed at the two sides of the vertebra. They allow the controlling wires to pass through. This design can only bend in a plane. Figure 3b shows another vertebra design. In this design, one end of the vertebra has a concave spherical surface and the other end has a matching convex spherical surface. In the articulation, the convex spherical surface of the vertebra fits the concave spherical surface of the successive vertebra. With this design, each joint can rotate in 3D. However, the rolling DOF is confined by the constraint component. Hence, it can bend in all directions with limited twisting.

Continuum Backbone
As in nature, some flexible bodies are entirely made of soft tissues. Inspired by this, a continuum backbone is designed which replaces the aforementioned rigid vertebrae by a continuous elastic beam. The wires are guided by the wire spacing disks attached to the elastic beam. There is no joint in the continuum backbone, however, one can treat the beam segment between two wire spacing disks as a pseudo-joint as shown in Fig. 4. Similarly, the backbone bending is controlled by pulling one wire and releasing the other in the wire pair. Ideally, without considering the gravity and other external loads, the deformation of the continuum backbone is a circular arc, i.e., the piecewise constant curvature assumption applies as well.

2.3 Backbone Segmentation

The bending of the BWDM backbone is controlled by the wire pairs. When there is only one pair of wires as shown in Fig. 5a, all the vertebrae are in the same bending section. The backbone bends to a C-shape as shown in Fig. 7a. The backbone can also be segmented into multiple sections. The vertebrae within a segment form a bending section. Figure 5b shows a two-section serpentine backbone. In the figure, the first bending section is between the base and the junctions of the first wire pair; the second bending section contains the vertebrae between the junctions of wire pair 1 and wire pair 2. Each bending section can bend independently. When the two

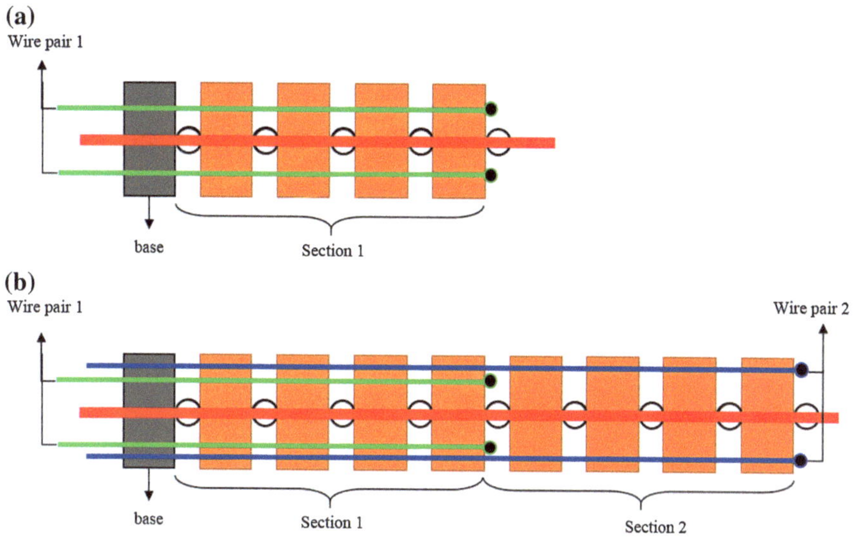

Fig. 5 Backbone segmentation: **a** serpentine backbone with one bending section; **b** serpentine backbone with two bending sections

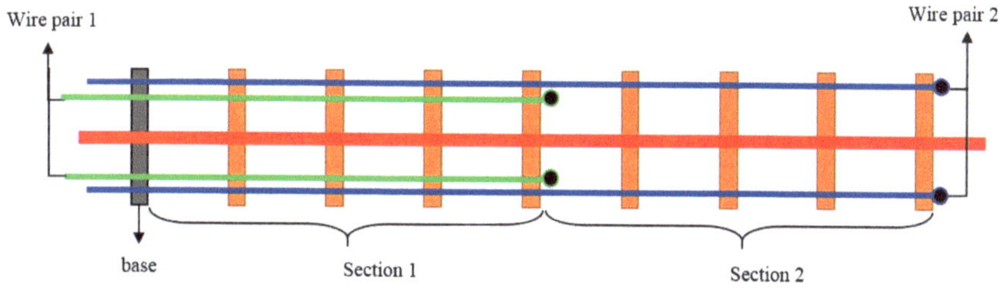

Fig. 6 Continuum backbone with two bending sections

bending sections bend to the same direction the shape of the backbone is C-shape; when they bend in the opposite direction the backbone is in S-shape, which is shown in Fig. 7b. There could be more bending sections, and the backbone segmentation is similar. For the continuum backbone, the segmentation is the same. Figure 6 shows the continuum backbone with two bending sections.

2.4 *Wire Configuration*

The backbone bending motion is controlled by the wire pairs. Therefore, the bending is affected by the wire configuration. In this section, three wire configurations are considered, i.e., the parallel wire configuration, the tapered wire configuration, and the twisted wire configuration.

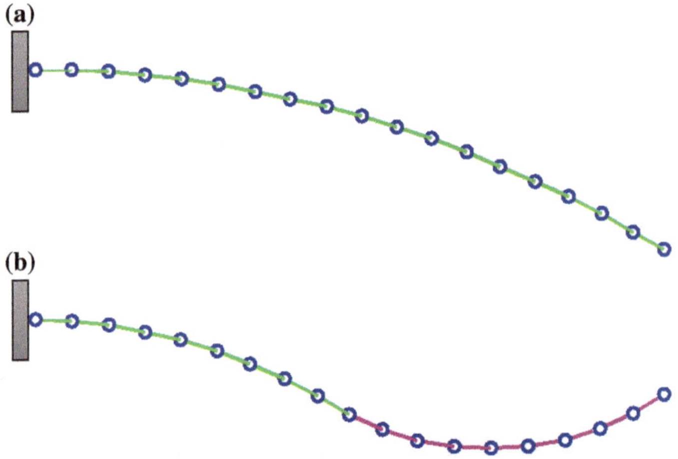

Fig. 7 Bending motion of the BWDM backbone: **a** C-shape bending; **b** S-shape bending

Parallel wire configuration is straightforward. In this configuration, the wires with the same length as the elastic constraint component are placed along the neutral axis. Figure 5a, b shows the parallel wire configuration of the serpentine BWDM with one bending section and two bending sections, respectively. Figure 6 shows the parallel wire configuration of the continuum BWDM with two bending sections. It is noted that when there are multiple bending sections, the motions of the two wire pairs are coupled, i.e., the motion of the wire pair 2 is affected by the motion of the wire pair 1. Figure 8 shows another parallel wire configuration, in which the motions of the wire pairs are decoupled. In this configuration, the wire pair 2 converges at the end of the first bending section, goes through the neutral axis, and diverges at the beginning of the first bending section. When the backbone is deflected, the length of the neutral axis does not change. Therefore, the motion of the second wire pair is not affected by the bending of the first section. In the parallel wire configuration, the wires in the pair are coplanar. Hence, the bending of the

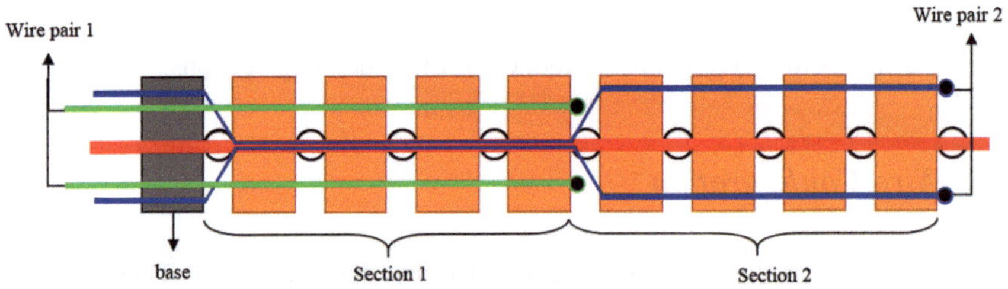

Fig. 8 Parallel wire configuration with decoupled wire motions

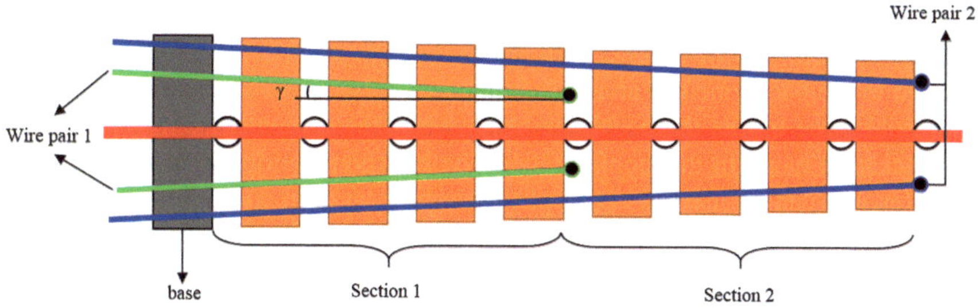

Fig. 9 Tapered wire configuration

backbone is in the same plane. Three-dimensional bending can be achieved by the control of multiple wire pairs, such as two orthogonally placed wire pairs.

Figure 9 shows the tapered wire configuration, which is similar to the parallel wire configuration. In this configuration, the wires in a pair are coplanar as well. The only difference is that there is a taper angle between the wires and the backbone neutral axis. With this wire configuration, the vertebrae are wider when close to the base and are narrower when close to the distal end. In this configuration, the moment exerted on the backbone is related to the wire spacing and the wire tension, but is not affected by the taper angle [23, 24]. Therefore, to generate the same amount of bending angle with the tapered wire configuration the mass at the distal end is smaller, which is helpful in reducing the energy consumption in flapping. It should be noted that, the backbone bending with the tapered wire configuration is still planar.

In the previous two configurations the wires and the backbone are coplanar. The wires control the backbone bending in the plane. In the twisted wire configuration as shown in Fig. 10, the wires spiral around the backbone, which is similar to the

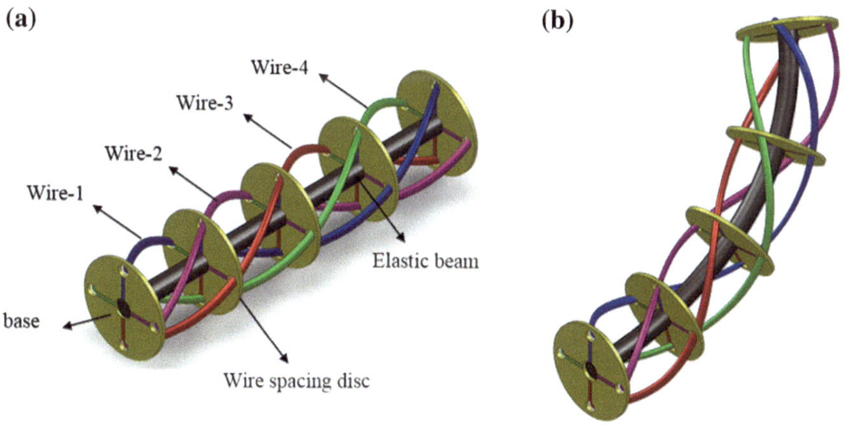

Fig. 10 Twisted wire configuration: **a** the wire configuration in resting position; **b** bending and twisting

octopus arm muscle arrangement. The backbone bending with the twisted wire configuration is shown in Fig. 10b. When one of the wires contracts, the force exerted to the backbone has an axial component as well as a tangential component. The axial force makes the backbone bend, while the tangential component makes the backbone twist. In this configuration, the backbone deformation is much more complex and can be used to accomplish complex tasks in addition to bending.

By properly choosing the backbone type, the backbone segmentation, and the wire configuration, one can design the BWDM for a variety of applications, such as underwater propulsion.

3 Wire-Driven Robot Fish

The BWDM is well suited for designing flapping-based underwater propulsion [25]. Figure 11 shows a family of wire-driven robot fish [24–28]. The BWDM is also applicable for other propulsions, such as the double hull boat [29] and the LTAV [23], or work in confined space [30–34]. In this chapter, three BWDM robot fish and experimental results are presented. The first BWDM robot fish has an oscillatory tail and a pair of pectoral fins, the second one has an undulatory tail, and the third one can mimic both shark swim and dolphin swim.

3.1 Design Example 1: Serpentine Oscillatory Wire-Driven Robot Fish with Pectoral Fins

This section introduces the serpentine oscillatory wire-driven robot fish. It uses one-section, planar joint, and tapered wire configuration. It is capable of making oscillatory swims. To improve the stability, it also has a pair of pectoral fins.

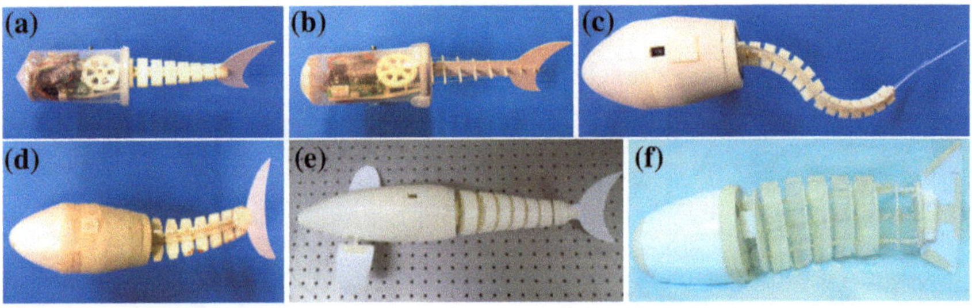

Fig. 11 A family of BWDM robot fish: **a** serpentine oscillatory wire-driven robot fish; **b** continuum oscillatory wire-driven robot fish; **c** serpentine undulatory wire-driven robot fish; **d** wire-driven robot fish with vector propulsion; **e** serpentine oscillatory wire-driven robot fish with pectoral fins; **f** twisting wire-driven robot fish

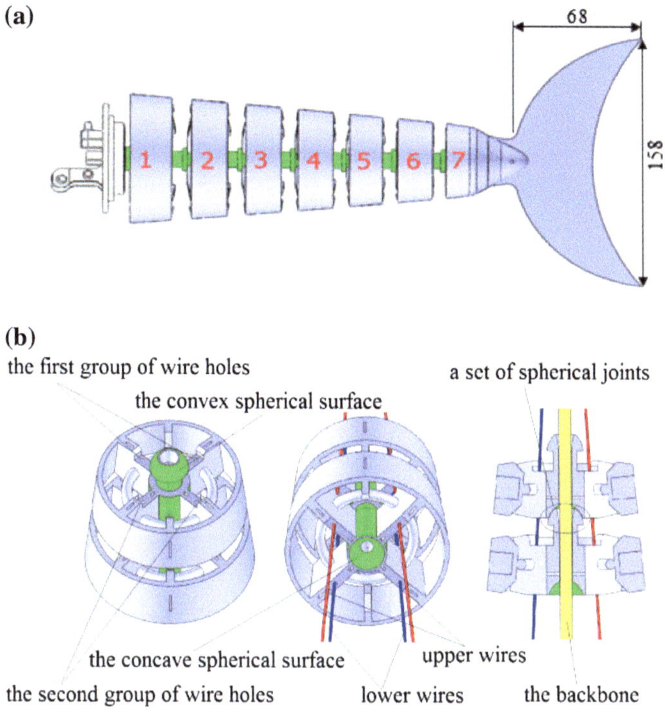

Fig. 12 The design of flapping tail: **a** the overall structure; **b** the joint

The Design

The oscillatory wire-driven robot fish comprises of three parts [35]: the wire-driven oscillatory flapping tail, the pectoral fins, and the fish body.

(A) The Wire-Driven Oscillatory Flapping Tail

In this design, the single segment serpentine BWDM is used. The design is shown in Fig. 12a. It is composed of seven vertebrae, two pairs of wires, and an elastic tube. The tube and the vertebrae make up the backbone of the tail. All the vertebrae, except the last one, have a similar structure. Two successive vertebrae form a joint, as shown in Fig. 12b. The vertebrae parameters are shown in Table 1. Each vertebra has four through holes for the two pairs of wires. The first pair uses the lower holes, and the second pair uses the upper holes. The elastic tube goes through all the vertebrae to confine the joints' rotations. A lunate fin plate is mounted onto the last vertebra. Its chord length is 68 mm and the width is 158 mm. The oscillatory flapping tail is driven by two servomotors through the two pairs of wires. Each motor controls a reel, which in turn drives the wire pairs as shown in Fig. 13.

When the motor rotates, one wire is shortened while the other wire is released; as a result, a torque is applied to the backbone. The backbone will bend to a circular arc. As an approximation, the rotation angles of the joints are considered the same [25, 27, 28, 30, 31]. Note that the two parallel wire pairs are used to control the

Table 1 Vertebrae parameters

Vertebra no.	H_i (mm)	h_0 (mm)	L_i (mm)	θ_{max} (°)
1	25	6	68.6	10.00
2	24	6	42.7	16.00
3	23	6	42.7	16.00
4	22	6	42.7	16.00
5	21	6	42.7	16.00
6	20	6	42.7	16.00
7	19	6	42.7	16.00

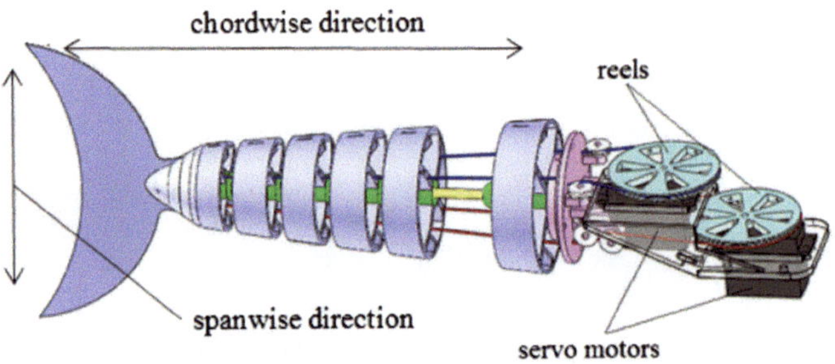

Fig. 13 The wire connection of the oscillatory wire-driven flapping tail

bending motion of the tail. When the rotations of the two reels are the same the tail bends to a circular arc in a plane as shown in Fig. 14a. When the rotations of the two reels are different, the tail will not only bend but also twist as shown in Fig. 14b. This makes the wire-driven tail flaps in a 3D manner, which resembles the fish flapping in some special occasions, such as slow cruising.

Fig. 14 Illustration of the tail movement: **a** pure bending in plane; **b** bending as well as twisting

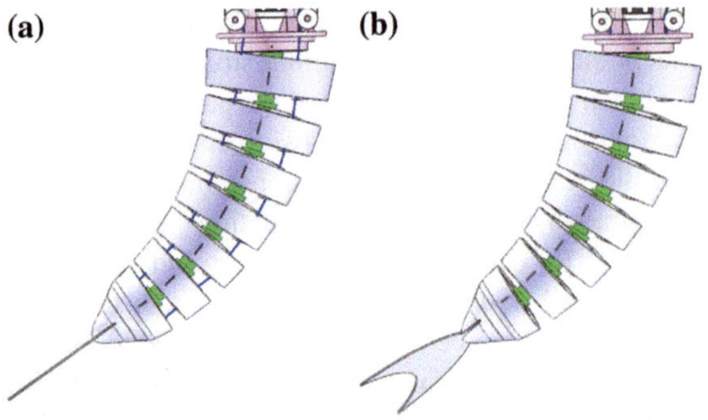

Fig. 15 The design of the pectoral fin

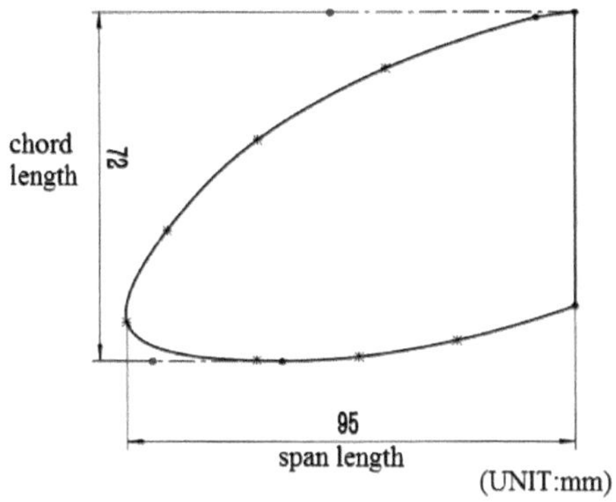

(B) Pectoral Fins
Pectoral fins are used mainly for maneuvering. Some fish also use pectoral fins to cruise. In this robot fish, the pectoral fins can be used for both propulsion and maneuvering. It can also be used in coordination with the caudal fin.

The shape of the pectoral fin is shown in Fig. 15. Based on Blake's theory, such a wedge-shaped blunt fin is hydrodynamically more efficient for drag-based propulsion than that of a rectangular shape due to the reduced interference drag near the body [36].

The movement of the pectoral fin is similar to that of the bird's wing. From Azuma's theory, the movement of the bird's wing can be decomposed to four basic motions: rowing, feathering, flapping, and spanning as shown in Fig. 16 [37]. From the figure, it is seen that the feathering motion is a 3D motion. The range of the feathering motion is from 0° (parallel to the horizontal plane) to 90° (perpendicular to the horizontal plane). The former is the flapping motion while the latter is the rowing motion. The main difference among the rowing, flapping, and feathering is the feathering angle.

To realize the feathering motions, the designed pectoral fin is shown in Fig. 17. It is driven by two motors. One motor controls the flapping of the pectoral fin through a gear rack, while the other controls the feathering angle through a circle gear drum. By controlling the two motors, it can achieve rowing, feathering, and

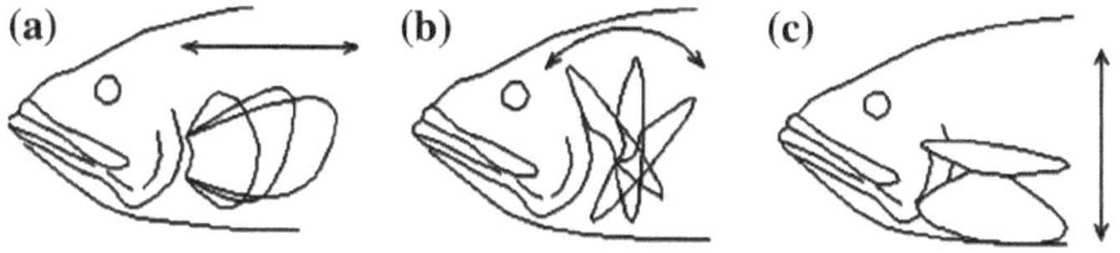

Fig. 16 Illustration of the pectoral fin motions: **a** rowing; **b** feathering; **c** flapping

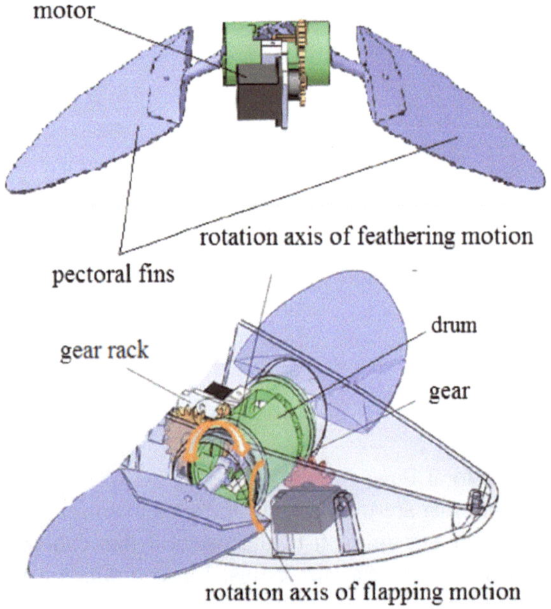

Fig. 17 The design of the pectoral fin

flapping motions. Specifically, the feathering angle ranges from −15° to 105° relative to the horizontal plane and hence, it can do rowing, flapping, and feathering.

(C) The Body
The body is the base of the wire-driven flapping tail and pectoral fins. It also holds the servomotors, the control system, the power supply (batteries), etc. It is known the fish body is the main source of water drag. Mimicking the shape of the fish, the cross section of the designed robot fish body is ellipsoid. The size changes gradually along the body length direction. This streamlined design helps to reduce the water resistance. Figure 18 shows the design and the photo of the robot fish. The overall body length is 555 mm and the pectoral fins span 290 mm.

Modeling and Analysis

(A) The Flapping Tail Propulsion Model
The wire-driven flapping tail has seven joints, whose rotations are assumed to be identical as indicated previously. The rotation of a joint is illustrated in Fig. 19. In the figure, H_i is the height of vertebra i, h_0 is the initial joint gap, h_β is the initial wire length through the joint gap, L_i is the diameter of collision edge, d_i is the top wire eyelet distance, D_i is the bottom eyelet distance, β is the taper angle, and θ is the joint rotation angle. Assuming the joint rotations are small (i.e., $\sin(\theta) \approx \theta$), the relationship between the wire length variation and joint rotation angle can be found by simple geometric manipulations:

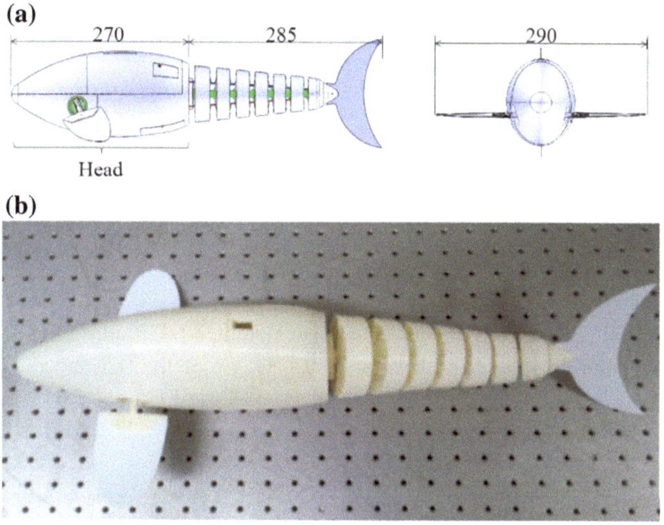

Fig. 18 The oscillatory wire-driven robot fish: **a** the simplified CAD model; **b** the photo of the prototype

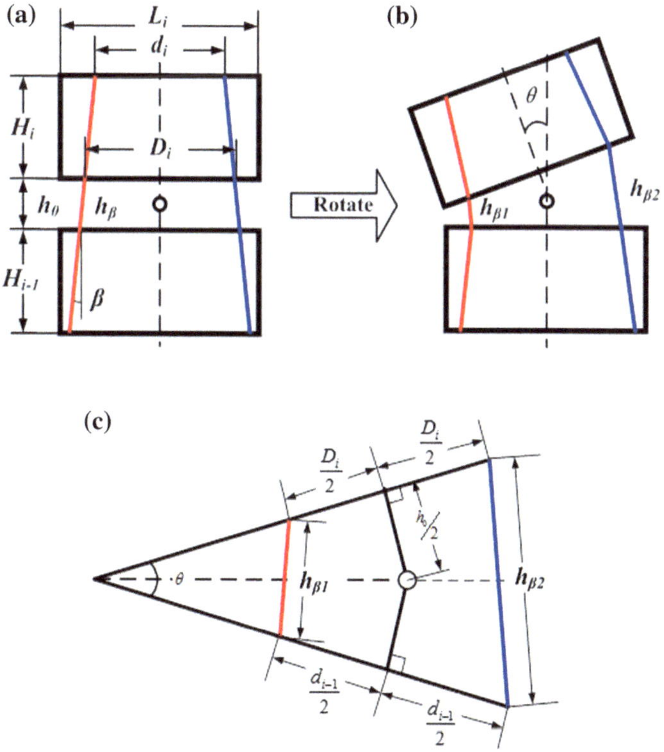

Fig. 19 Illustration of the wire displacement with respect to joint rotation. **a** Joint at rest, **b** joint rotated, **c** wire displacement in joint rotation

$$\begin{cases} h_{\beta 1} = h_0 - \Delta h_{\beta 1} = \sqrt{(h_0 \cdot \tan \beta)^2 + \left(h_0 - \frac{D_i\theta}{2}\right)\left(h_0 - \frac{d_i\theta}{2}\right)} \\ h_{\beta 2} = h_0 - \Delta h_{\beta 2} = \sqrt{(h_0 \cdot \tan \beta)^2 + \left(h_0 + \frac{D_i\theta}{2}\right)\left(h_0 + \frac{d_i\theta}{2}\right)} \end{cases} \tag{1}$$

where $d_i = D_i - 2H_i \cdot \tan \beta$. When the taper angle is small, it can be further simplified as follows:

$$h_{\beta 1} \approx -h_{\beta 2} \approx \frac{D_i}{2} \cdot \theta \cdot \cos \beta \tag{2}$$

The rotation angle of the driving motor is calculated by dividing the wire motion by the wheel radius. The overall bending angle, Θ, of the tail and the maximum joint rotation, $\theta_{\max}$, can be computed as in Eqs. (3) and (4) below:

$$\Theta = N\theta \tag{3}$$

$$\theta_{\max} = 2 \arctan\left(\frac{h_0}{L_i}\right) \tag{4}$$

where N is the number of joints. Note that the vertebra parameters and $\theta_{\max}$ are shown in Table 1. Moreover, the positions of the joints can be found by Eq. (5), where x is the position along the fish body, and y is the lateral displacement of the joint.

$$\begin{cases} y_i(t) = \sum_{j=1}^{i} \left(\left(H_j + h_0\right) \cdot \sin(j \cdot \theta)\right) \\ x_i(t) = \sum_{j=1}^{i} \left(\left(H_j + h_0\right) \cdot \cos(j \cdot \theta)\right) \end{cases} \tag{5}$$

Based on the elongated body theory (EBT) [38, 39], the cruising speed is dependent on the tail tip lateral velocity and slope. From Eq. (5), the lateral displacement of tail tip with respect to the tail curve length q is:

$$y(q,t)|_{x=L} = \sum_{i=1}^{N} ((H_i + h_0) \cdot \sin(i \cdot \theta)) \tag{6}$$

Accordingly, the tail tip slope and lateral velocity are as follows:

$$\left.\frac{\partial y(q,t)}{\partial q}\right|_{x=L} = \sin\left(\sum_{i=1}^{N} \theta\right) \tag{7}$$

$$\left.\frac{\partial y(q,t)}{\partial t}\right|_{x=L} = -\sum_{i=1}^{N} (H_i + h_0) \cos(i \cdot \theta) \cdot \left(i \cdot \dot{\theta}\right) \tag{8}$$

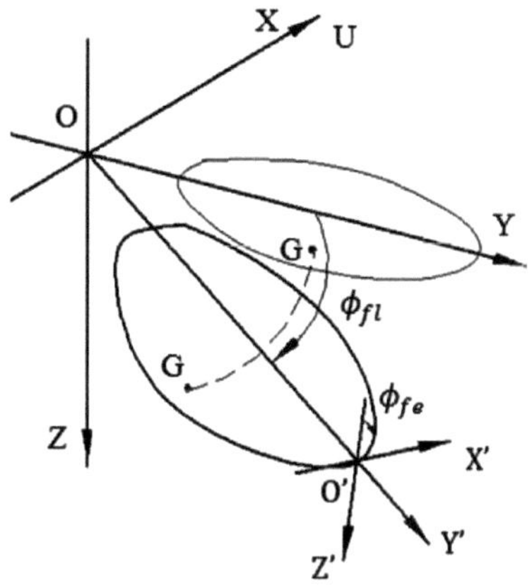

Fig. 20 Coordinate system of the pectoral fin

We can also compute the cruising speed, the thrust, as well as the Froude efficiency and the results are shown in the experiment section.

Note that in this case, it is assumed that the pectoral fins are kept still and are parallel to the horizontal plane so that its influence is ignored.

(B) Pectoral Fins Propulsion Model

The two pectoral fins of the robot fish have the same synchronous motions. Thus, only one pectoral fin is analyzed herein. The coordinate system is shown in Fig. 20. The distance between the distal end of the fin and the center of flapping motion is $L = 122.5$ mm, the area of each fin is $S_p = 4683.4\ \text{mm}^2$, the flapping angle is ϕ_{fl}, and the feathering angle is ϕ_{fe}. The two angles are independently controlled by two servomotors.

Assuming that $P_G = (x', y', z')$ is an arbitrary point on the fin plane ignoring its thickness, then the position of the point is:

$$^{o}P_G = {}^{o}_{o'}T\, {}^{o'}P_G \tag{9}$$

where $\{O\}$ represents the coordinate system attached to the fish body and $\{O'\}$ is the coordinate system attached to the distal end of the pectoral fin, and $OO' = L$. The homogeneous transformation matrix ${}^{o}_{o'}T$ is defined in Eq. (10).

$${}^{o}_{o'}T = \begin{bmatrix} {}^{o}_{o'}R & {}^{o}P_{o'} \\ 0_{1\times 3} & 1 \end{bmatrix} \tag{10}$$

where ${}^{o}_{o'}R$ is the rotation matrix and ${}^{o}P_{o'}$ is the position vector.

$${}^{o}_{o'}R = R(y, \phi_{\text{fe}}) \cdot R(x, \phi_{\text{fl}}) = \begin{bmatrix} \cos\phi_{\text{fe}} & 0 & \sin\phi_{\text{fe}} \\ 0 & 1 & 0 \\ -\sin\phi_{\text{fe}} & 0 & \cos\phi_{\text{fe}} \end{bmatrix} \begin{bmatrix} 1 & 0 & 0 \\ 0 & \cos\phi_{\text{fl}} & -\sin\phi_{\text{fl}} \\ 0 & \sin\phi_{\text{fl}} & \cos\phi_{\text{fl}} \end{bmatrix} \tag{11}$$

$$^{o}P_{o'} = [-L\sin\phi_{\rm fl}\sin\phi_{\rm fe} \quad L\cos\phi_{\rm fl} \quad L\sin\phi_{\rm fl}\cos\phi_{\rm fe}]^{T} \tag{12}$$

Combining Eqs. (10)–(12), it follows that:

$$^{o}_{o'}T = \begin{bmatrix} \cos\phi_{\rm fe} & \sin\phi_{\rm fe}\sin\phi_{\rm fl} & \sin\phi_{\rm fe}\cos\phi_{\rm fl} & -L\sin\phi_{\rm fl}\sin\phi_{\rm fe} \\ 0 & \cos\phi_{\rm fl} & -\sin\phi_{\rm fl} & L\cos\phi_{\rm fl} \\ -\sin\phi_{\rm fe} & \cos\phi_{\rm fe}\sin\phi_{\rm fl} & \cos\phi_{\rm fe}\cos\phi_{\rm fl} & L\sin\phi_{\rm fl}\cos\phi_{\rm fe} \\ 0 & 0 & 0 & 1 \end{bmatrix} \tag{13}$$

When the pectoral fins move, the displacement of point P_G is determined by Eq. (9). Furthermore, its velocity, acceleration, and force or torque can also be calculated. Specifically, from Eq. (9), taking the first time derivative, the velocity is:

$$^{o}V_G = {^{o}\dot{P}_G} \tag{14}$$

The thrust force of the fin is:

$$P_t = \int_0^{t_0}\int \mathrm{d}F_t\mathrm{d}t = \int_0^{t_0}\iint_S \frac{1}{2}\rho((^{o}V_{\rm G})_x - U)^2 C_{D2}\mathrm{d}S'_p\mathrm{d}t \tag{15}$$

where $(^{o}V_G)_x$ is the component along the forward direction, C_{D2} is the drag coefficient of the fin. For a plate, C_{D2} is 1.18. $S'p$ is the projection of S_p on the YOZ plane.

$$S'_p = S_p\cos\phi_{fl}\cos\phi_{fe} \tag{16}$$

When the fish is cruising, the thrust force is equal to the overall drag force, namely, $2F_t = F_D + 2F_{Dp}$, where F_D is the drag from the body, and F_{Dp} is the drag force of the pectoral fins. The equilibrium equation is as Eq. (17):

$$\int_0^T\iint_S \rho((^{o}V_G)_x - U)^2 C_{D2}\mathrm{d}S'_p\mathrm{d}t = \frac{1}{2}\rho U^2 C_{D1}ST + \rho U^2\xi C_{D2}S_P T \tag{17}$$

where ξD_2 is the average drag coefficient of the pectoral fins during one flapping cycle. The cruising speed U can then be resolved from Eq. (17).

Motion Control

The robot fish is controlled by the operator through Bluetooth. The control scheme is shown in Fig. 21. The command is sent to the micro control unit (MCU) by the operator via a remote controller. Upon receiving the command, the MCU generates

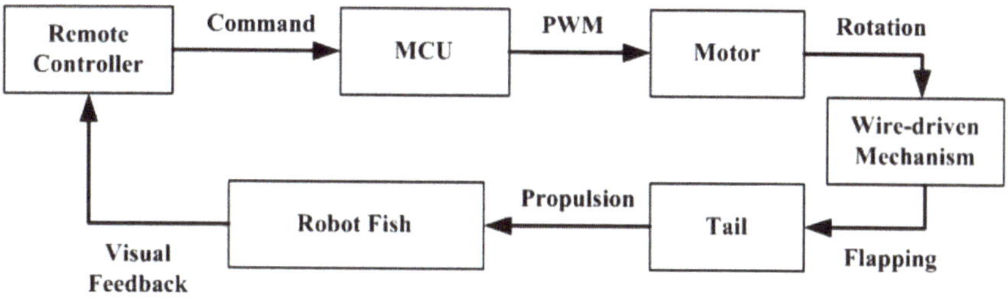

Fig. 21 The swimming control scheme of the robot fish

a pulse width modulation (PWM) sequence which controls the velocity and position of the servomotor. The rotation of the motor is transmitted to the oscillatory flapping motion of the tail through the BWDM. The flapping motion controls the direction and magnitude of the robot fish. This is a human in-loop control system in which visual-based feedback is needed.

Each flapping cycle is divided into four stages as shown in Fig. 22. Stage I: Flap from resting position to the right limit; Stage II: Flap from right limit back to the resting position; Stage III: Flap from resting position to the left limit; Stage IV: Flap from left limit back to the resting position. The forward speed and turning performance (i.e., turning speed and turning radius) are controlled by the flapping velocity in the four stages and the amplitudes of the two limits.

For the pectoral fins, the control is carried out in a similar manner.

For the oscillatory flapping robot fish, there are three basic motion modes, i.e., swimming by the tail only, swimming by the pectoral fins only, and swimming by the tail and the pectoral fins. In each mode, the robot fish can swim forward, turn leftward, and turn rightward. In the first motion mode, the robot swims forward by flapping the tail symmetrically, and turns around by flapping the tail asymmetrically. In the second motion mode, the robot swims forward by rowing and

Fig. 22 Illustration of an oscillatory flapping cycle

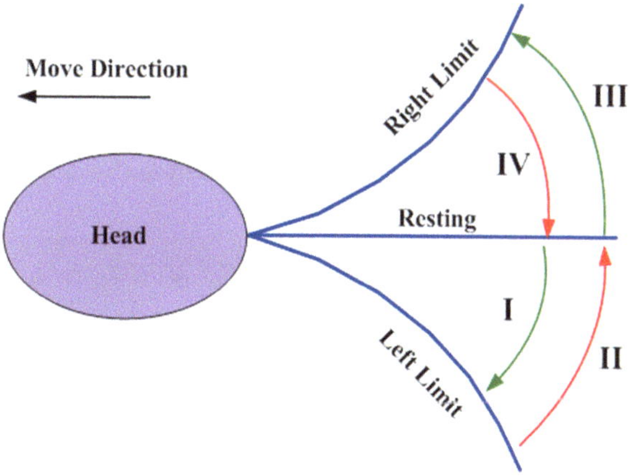

feathering the pectoral fins, and turns around with the tail serving as the rudder. In the third motion mode, the two fins work together to enhance the performances. The experiment results are shown in the next section.

Experiment Results

The prototype of our robot fish is shown in Fig. 11. Its tail and shell are fabricated by 3D printing. The material used is ABS plastic. The caudal fin and pectoral fins are made of 1 mm thick ABS plastic plate. The wires are steel wires with 0.5 mm diameter. Four AAA batteries are used to drive the motors and the controller (AVR ATmega 128 [20]). Motion commands are sent to the controller via Bluetooth. Waterproof of the fish is done by covering the body with a latex skin. The total weight is about 1.5 kg. Tests are carried out in a water tank of 1.9 × 0.9 × 0.8 m. As mentioned in the previous section, three sets of tests are conducted: (A) swimming using the wire-driven caudal fin alone; (B) swimming using the pectoral fins alone; and (C) swimming using a combination of all the fins.

(A) Swim by the Wire-Driven Caudal Fin

In this experiment, the pectoral fins are kept still in horizontal plane and the robot fish is propelled by the tail only. During the test, the robot fish is controlled to swim with different flapping frequencies and amplitudes. The cruising speed and turning performance were estimated from the videos as shown in Fig. 23.

Figure 24 shows the cruising speed of the robot fish. The red dashed lines are predicted results and the blue solid lines are the experimental results. Figure 24a shows the cruising speed as the function of the flapping amplitude when the flapping frequency is constant of 1 Hz. Note that the amplitude is the total bending angle of the tail. From the figure, it is seen that the cruising speed increases as the flapping amplitude increases. Though, the rate of the increase is decreasing as the flapping amplitude increases. Moreover, the experiment results match the theoretical results well. Figure 24b shows the cruising speed as the function of the flapping frequency when the flapping amplitude is 60°. From the figure, it is seen that the

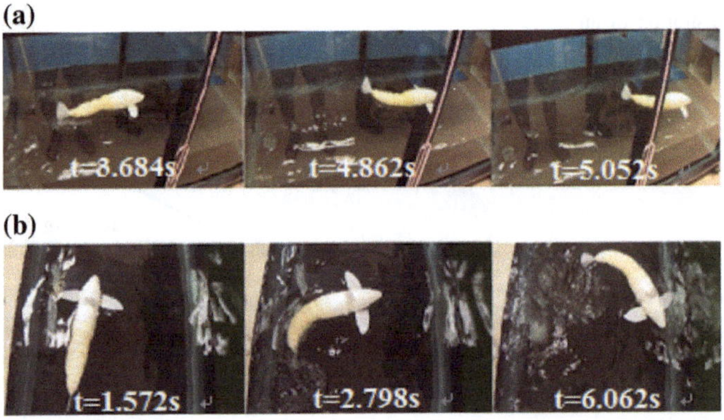

Fig. 23 Swimming with the wire-driven tail alone: **a** swimming forward; **b** turning

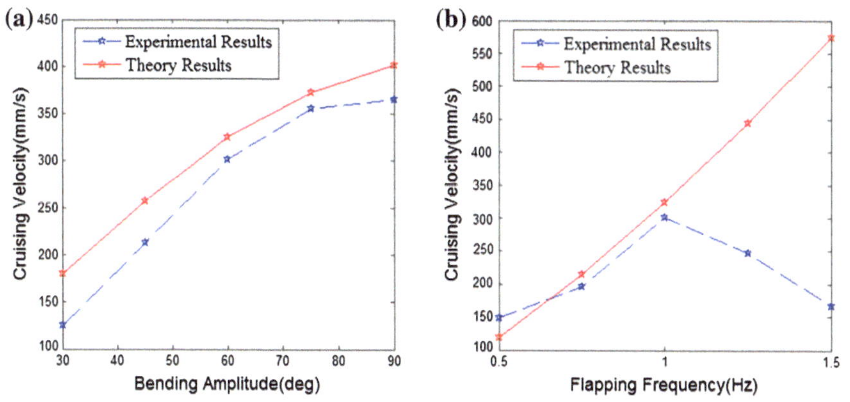

Fig. 24 Swimming speeds of the robot fish against **a** the flapping amplitude; **b** the flapping frequency

cruising speed increases as the flapping frequency increases. Though, unlike the theoretical prediction, the cruising speed actually decreases when the flapping frequency is higher than 1 Hz. This may be attributed to several reasons. First, the tank is small. When the flapping is fast, the water becomes turbulent, which influences the cruising performance. Second, the dynamic response of the motor and the tail is slow. As a result, the actual flapping amplitude does not reach 60°. In fact, it is found that when the flapping frequency is 1.25 Hz, the actual amplitude is about 40°; when the lapping frequency is 1.5 Hz, the actual amplitude is only 30°. This impairs the performance of the robot fish. In general, the theoretical model (EBT) predicts the cruising speed reasonably well except in the high flapping frequency. The average error is 13.7 %.

In addition, among all the tests, the maximum cruising speed was 365 mm/s (0.66 BL/s) and the maximum turning speed was about 65°/s; the mean turning speed was about 38°/s and the turning radius was about 0.60 BL. This performance can be further improved when using a larger motor and more stiff tail.

(B) Swim by the Pectoral Fins

In this test, the caudal fin is kept still. The pectoral fins are controlled by the two angles, the flapping angle, ϕ_{fl}, and the feathering angle, ϕ_{fe}. In each flapping circle, the pectoral fins start in the rest positions, that is $\phi_{\mathrm{fl}} = 0^{\circ}$ and $\phi_{\mathrm{fe}} = 0^{\circ}$. First, the feathering angle moves to $\phi_{\mathrm{fe}} = 90^{\circ}$, then the flapping angle ϕ_{fl} increases to $\phi_{\mathrm{fl}} = 60^{\circ}$. In this way, the water is pushed back generating thrust and driving the fish forward. The recovery stroke starts with changing the feathering angle from 90° to 0° to reduce the water resistance. Finally, the flapping angle is controlled from 60° to 0° completing the flapping circle. Note that using the pectoral fins, the robot fish can swim backward by reversing the aforementioned flapping circle. Also, this flapping pattern is fixed and the swimming speed is controlled by the flapping frequency.

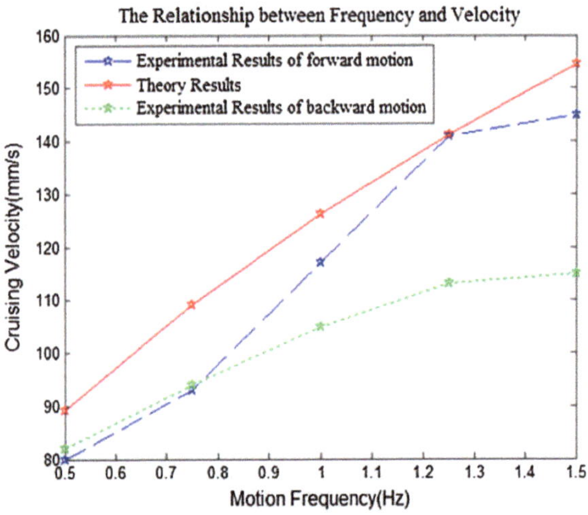

Fig. 25 Experiment results when swimming with pectoral fin alone

The test results are shown in Fig. 25. The blue dashed line shows the speed of forward swimming, and the green dashed line shows the speed of backward swimming. The drag model prediction is shown by the red solid line. From the figure, it is seen that the cruising speed increases with the increase of the flapping frequency. The forward swimming speed is faster than the backward swimming speed. This is because of the difference between the head and tail of the robot fish. In general, the drag model predicts the performance reasonably well.

Among the tests, the maximum forward speed of the robot fish is 145 mm/s and the maximum backward speed is 115 mm/s. As expected, this is significantly slower than that of using the caudal fin (365 mm/s).

(C) Swim by Both Wire-Driven Flapping Tail and Pectoral Fins

In nature, when a fish is swimming, especially maneuvering, all the fins cooperate. In this test, the fin cooperation is studied, especially for turning. Three motion modes are tested. The first two are using the tail as a rudder, and the pectoral fin driving the fish body, one forward and the other backward; the third is the tail flapping in turning mode and the pectoral fins in backward mode.

Figure 26 shows the video frames in the three turning cases. In Fig. 26a, the pectoral fins are used to propel the body swimming forward while the tail is used as a rudder; in Fig. 26b, the pectoral fins are used to propel the body swimming backward while the tail is used as a rudder; in Fig. 26c, the tail is used to conduct turning motion while the pectoral fins are moving backward. The robot fish turns effectively in all three modes. In the first turning mode, the turning radius is about 0.25 BL, and the mean turning speed is about 31°/s. In the second turning mode, the turning radius is about 0.70 BL, and the mean turning speed is about 13°/s. In the third turning mode, the turning radius is about 0.45 BL, and the mean turning speed is about 23°/s. For comparison, as shown in the previous test, when turning by the tail only, the turning radius is about 0.6 BL, and the turning speed is about 38°/s.

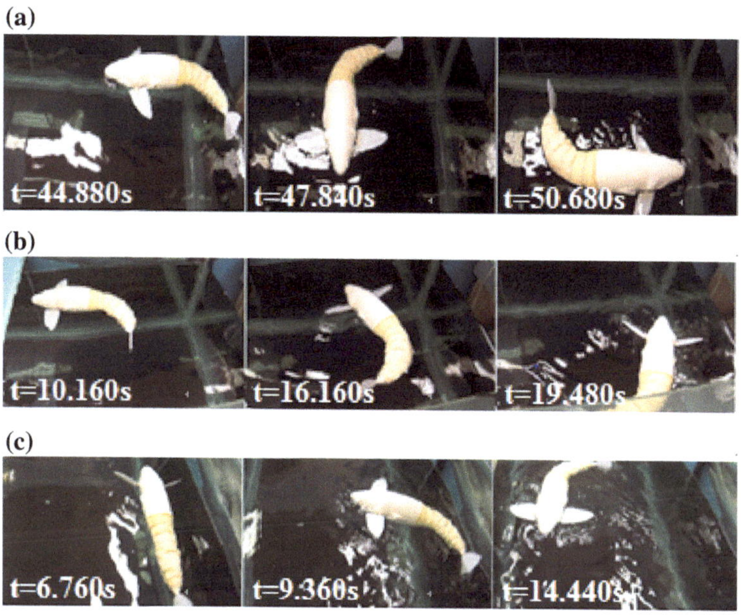

Fig. 26 Performance of fin coordination: **a** turning by the pectoral fin and the tail serves as a rudder; **b** backward turning by the pectoral fins and the tail serves as a rudder; **c** turning by the two types of fins together

Therefore, it can be concluded that with the help of the pectoral fins, the robot fish can turn around with a smaller radius, but makes no improvement in turning speed. This is because while the pectoral fins help the turning, it also incites water resistance.

3.2 Design Example 2: Undulatory Wire-Driven Robot Fish

In this section, an undulatory wire-driven robot fish is presented. It is made of a two-segment serpentine BWDM and hence, is capable of undulatory swimming.

The Design
The undulatory wire-driven robot fish is composed of the wire-driven flapping tail and the fish body [28, 40].

(A) The Flapping Tail
The undulatory wire-driven flapping tail is composed of 13 vertebrae divided into two segments. All the vertebrae are serially linked as shown in Fig. 27a. The first segment has six vertebrae (numbered from 1 to 6); the second segment has seven vertebrae (numbered from 7 to 13). The profile of the tail is shaped by the fin plate.

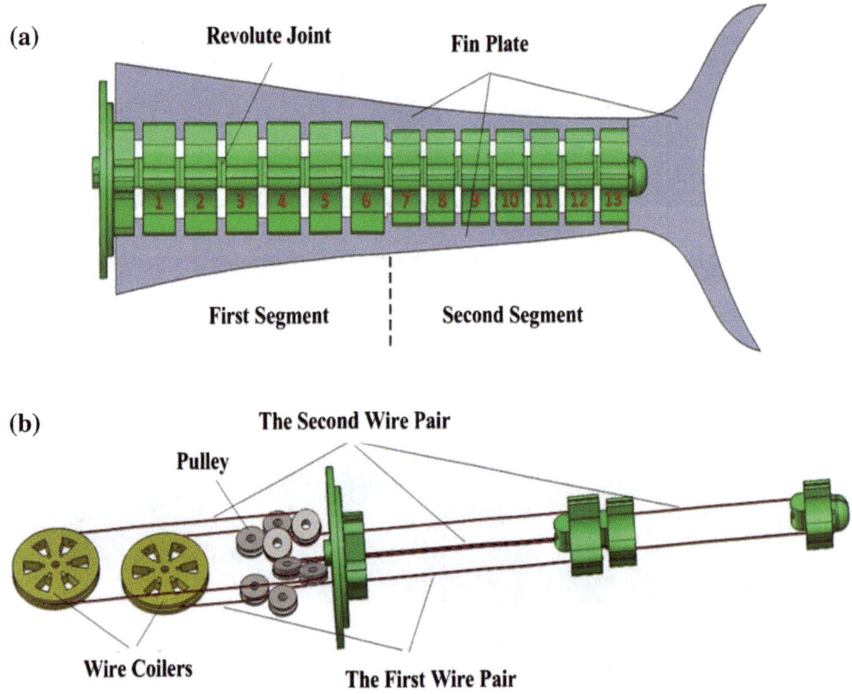

Fig. 27 The design of the two segment wire-driven tail

There are three plates in total. The upper plate and the lower plate define the tail shape as well as articulating the joints and constrain the joints rotations. The end plate defines the caudal fin shape. The overall profile of the tail is similar to that of a slender fish tail. The wire route in the tail is shown in Fig. 27b. The two pairs of wires are coplanar. The first wire pair goes through the first six vertebrae via the pilot holes and is connected to the end of the 6th vertebra. The second wire pair passes by the first six vertebrae via the central cavity and then goes through the left seven vertebrae via the pilot holes. Its end is connected to the end of the 13th vertebra. Each pair of wire is driven by a wire coiler, which is driven by a servomotor (not shown in the figure).

The design of the vertebra and joint connection is shown in Fig. 28. Specifically, the first three figures in Fig. 28 show the details of the 6th vertebra. The other vertebrae are similar. For vertebrae 1–6, there are two pilot holes used to guide the wires; two cylindrical surfaces at the end are used to form a revolute joint; one guide hole is used to pass the wires of the second group; two slots (upper slot and lower slot) connect the plastic plate; and one central cavity to hold the elastic tube, which is optional. For vertebrae 7–13, the structure is similar except that there is no guide hole. As shown in the figure, two successive vertebrae form a revolute joint. The wire routing in the 6th and the 7th vertebrae is also shown in the figure. Detailed vertebra parameters are shown in Table 2. In the table H is the height of

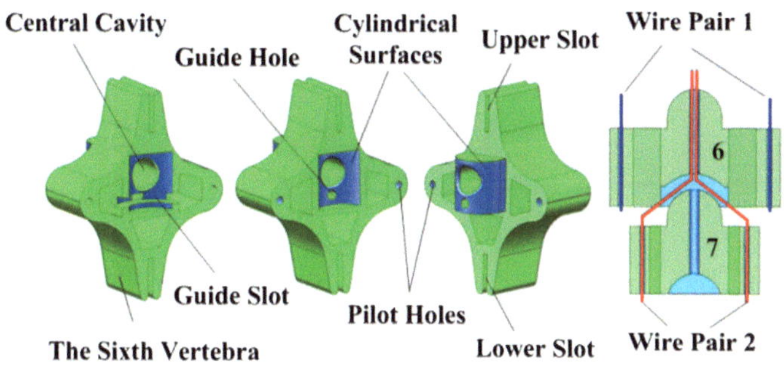

Fig. 28 The design of the vertebra

Table 2 Vertebra structure parameters

Vertebra no	D (mm)	d (mm)	h_0 (mm)	θ_{max} (°)
1–6	31	27	4	14.70
7–13	23.5	19	3	14.55

vertebra, d is the wire spacing distance within the pair, h_0 is the initial joint gap distance between adjacent vertebrae, and D is the diameter of the collision circle. The maximum joint rotation angle and the flapping amplitude are determined in similar as that of the oscillatory wire-driven tail.

When the motor rotates, the wheel winds one wire and unwinds the other wire applying a torque onto the backbone. It is known that the deformation of a uniform elastic beam under pure moment is a circular arc. In this design, the plastic plates act as the elastic beam. Since the friction between the joints is small and evenly distributed, the rotation of each joint within a segment is the same. The vertebrae are rigid and the axial deformation can be neglected. The length of the neutral axis remains unchanged. Therefore, for each segment, the bending is close to a circular arc. For the two segments, their motions are independently controlled. Thus, the resultant motion is a combination of the two motions. In particular, two motions are most common: One is that the two segments bend in the same direction; and hence, the tail forms a C-Shape. The other is that the two segments bend in the opposite direction and hence, the tail forms an S-shape. Figure 29 shows these two basic motions. Consequently, the robot fish can swim in both oscillatory form and undulatory form. Moreover, there are other motions, such as keeping the first segment relaxed and flapping the second segment only. This is called the Small-C-Motion, resulting in another form of oscillatory swimming.

The rotation of each joint is as similar to Fig. 19. At the resting position, the vertebrae are parallel and the wire lengths on both sides are the same, as shown in Fig. 19a. When the joint rotates, for instance, the joint rotates leftward, the lengths of the two wires will then change, as shown in Fig. 19b. Maximum rotation happens when the adjacent vertebrae collide with each other. Assume that there are N joints in the segment. The wire length change ΔL with respect to the rotation angle θ can

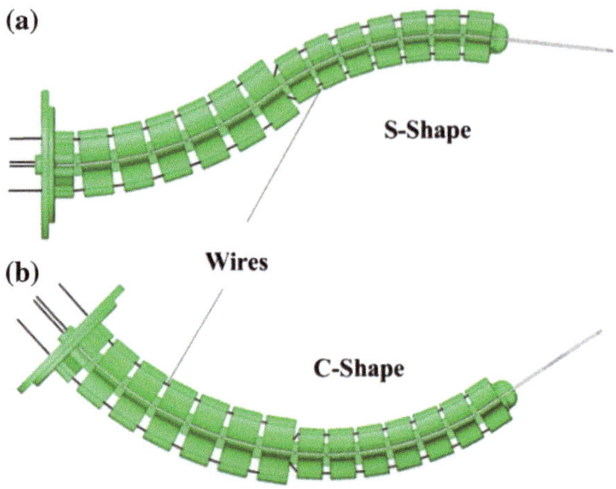

Fig. 29 Illustration of the bending in **a** S-shape; **b** C-shape

be calculated by Eq. (18). The maximum rotation angle of each joint θ_{max} and the segment $\Theta_{\max}$ is shown in Eq. (19).

$$\Delta L = 2N \cdot d \cdot \sin\left(\frac{\theta}{2}\right) \tag{18}$$

$$\Theta_{\max} = N \cdot \theta_{\max} = 2N \cdot \arctan\left(\frac{h_0}{D}\right) \tag{19}$$

In this design, the vertebra parameters are given in Table 2. For the first segment, the collision circle diameter is 31 mm, and the joint gap distance is 4 mm. By Eq. (19), for the first segment, the maximum joint rotation is 14.7°. The maximum bending angle is 88.20°. For the second segment, the collision circle diameter is 23.5 mm and the joint gap distance is 3 mm. The maximum joint rotation is 14.55°. The maximum bending angle is 101.85°. The overall maximum bending angle is 190.05°.

The motion of the undulatory wire-driven tail is compared to the fish undulatory swimming body curve, as shown in Fig. 30. The red dashed curves are the fish body curves at four time instances in a flapping cycle according to the traveling wave model. In the model, parameters are chosen as $c_1 = 0.2$, $c_2 = 0.4175$, $k = 2.4$, and $\omega = 2\pi$. The blue solid curves are the curves of the undulatory wire-driven tail at the same time instances. The rotation of each angle is controlled as per Eq. (20). The phase lag of the second segment is -0.3π. From the figure, the two set of curves match reasonably well, especially at the tail tip, which, according to Lighthill's EBT, is the major factor that affects the fish's swimming performance.

$$\theta_i = \begin{cases} 8\sin(2\pi t) = \theta_a & i = 1, 2, \ldots 6 \\ -12\sin(2\pi t - 0.3\pi) = \theta_b & i = 7, 8, \ldots 13 \end{cases} \tag{20}$$

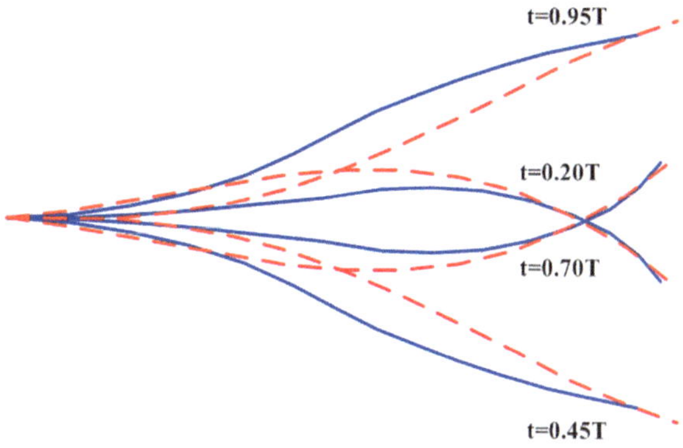

Fig. 30 A comparison between the travel wave equation (*red dash line*) and the flapping of the undulatory wire-driven tail (*blue solid line*)

(B) Fish Body Design

Figure 31 shows the design of the fish body with the tail assembled. It comprises an airtight hull, a central supporting board, a controller, two motors, and the DC power supply (battery). The motors, the controller, and the battery are mounted on the central board. The hull is axisymmetric. The shape of the hull front is parabolic, which helps to reduce water resistance. At the end of the hull, pinholes are opened to let the wires pass through. Each pair of wires is connected to a drum wheel, which rotates with the motor shaft. The undulatory wire-driven tail is connected to the central board by screws. Waterproofing is achieved by a silicone covering. The tail is purely mechanical and hence, no waterproof is needed. Though, waterproof at the connecting holes is crucial. The pinhole is coated with Teflon and has the same diameter as the wire. The wire is passed through the pinhole.

The vertebrae were fabricated using 3D printing. The material used was ABS plastic. Steel wires with ϕ 0.475 mm diameter were used as the controlling wires. Two MG995 servomotors are mounted in the fish body, and their maximum torque

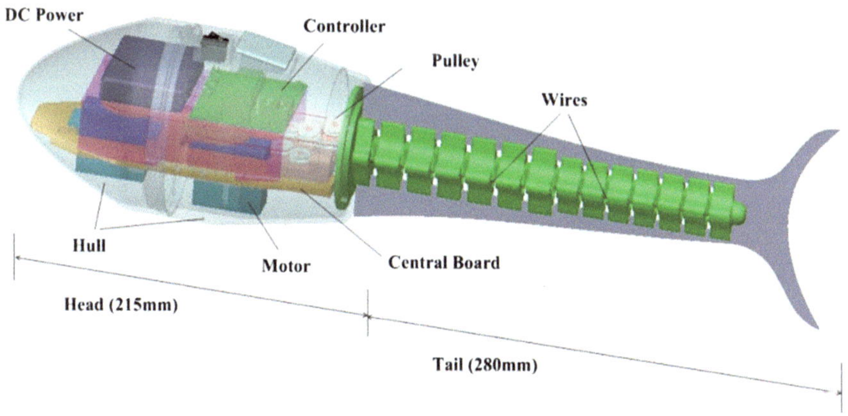

Fig. 31 The design of the undulatory wire-driven robot fish

is 13 kg cm. The motors are driven by 4 AAA batteries and controlled by a MCU. Control commands are sent to the MCU via Bluetooth. The overall length of the robot fish is 495 mm. The tail length is 280 mm. The weight of the robot fish is 1256 g including the balance weight. The overall density of the robot is smaller than water and hence, the robot fish floats on water surface with about 85 % submerged.

Figure 32 shows the robot fish with the inner structures. In Fig. 32a the robot fish tail is at the resting position. In Fig. 32b, the two segments of the tail bend to the same direction and the tail bends to a C-Shape. The maximum bending angle is 190.05°. In Fig. 32c, the two segments bend in opposite direction, and the tail is in S-shape. The form of S-shape can be controlled by the bending amplitude of the

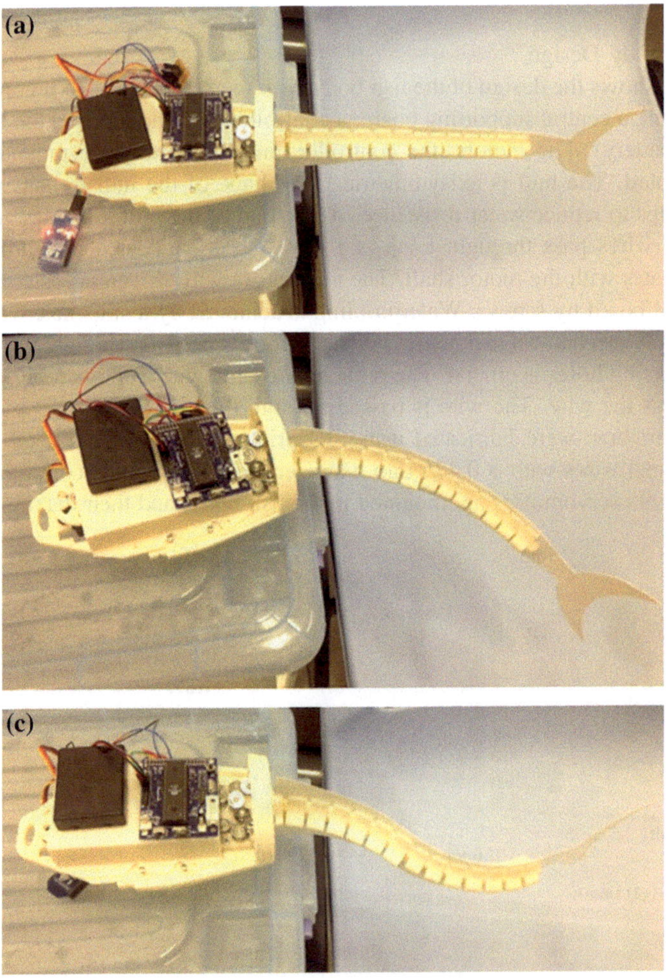

Fig. 32 The prototype of the undulatory wire-driven robot fish: **a** tail at resting position; **b** tail bends to C-shape; **c** tail bends to S-shape

two segments and the phase angle as in Eq. (20). It shall be pointed out that when the first segment is at rest, a C-Shape can be generated by controlling the bending of the second segment. To distinguish the two C-Shapes, we call the former as the Big-C and the later as the Small-C.

Modeling and Analysis

The propulsion model of the undulatory wire-driven tail is developed in a similar way: substituting the tail tip lateral displacement, lateral velocity, and tail tip slope into EBT. The undulatory wire-driven tail is a two-segment WDM. Here, the piecewise constant curvature assumption is still adopted. Let θ_a be the joint rotation in the first segment, θ_b be the joint rotation in the second segment, and l_i be the length of each vertebra central line. From the kinematics model, the lateral displacement of the tail tip is:

$$y(q,t)|_{q=L} = \sum_{i=1}^{6} l_i \cdot \sin(i \cdot \theta_a) + \sum_{i=7}^{13} l_i \cdot \sin[6\theta_a + (i-6) \cdot \theta_b] \tag{21}$$

It is a function of both time and x displacement. By partial differentiation, the slope and lateral velocity of the tail tip are as follows:

$$\left.\frac{\partial y(q,t)}{\partial q}\right|_{q=L} = \sin(6\theta_a + 7\theta_b) \tag{22}$$

$$\begin{aligned} \left.\frac{\partial y(q,t)}{\partial t}\right|_{q=L} &= \sum_{i=1}^{6} l_i \cdot \left[\cos(i \cdot \theta_a) \cdot \left(i \cdot \dot{\theta}_a\right)\right] \\ &\quad + \sum_{i=7}^{13} l_i \cdot \left(\cos(6\theta_a + (i-6) \cdot \theta_b) \cdot \left(6\dot{\theta}_a + (i-6) \cdot \dot{\theta}_b\right)\right) \end{aligned} \tag{23}$$

Substituting Eqs. (22) and (23) into the EBT, the cruise speed, thrust, and Froude efficiency of the robot fish can be predicted.

Experiment Results

The experiments were performed in a small inflated water tank. The length of the tank is 1.4 m and the width is 0.9 m. The robot fish swims in the water tank under the control of the operator. Figure 33 shows the two basic swimming motions of the robot fish: oscillatory swimming (C-Motion) and undulatory swimming (S-Motion).

Four experiments were performed, i.e., “Big-C-Motion” forward, “Small-C-Motion” forward, “S-Motion” forward, and turning. The first three groups of experiments were designed to test robot’s cruising speeds and efficiencies in various motions. The last experiment was designed to test the maneuverability of robot fish.

Experiment 1—“Big-C-Motion” forward

In this swimming mode, the two segments flap in synchronization and there is no phase lag between the two segments. In the experiment, the cruising velocity of the

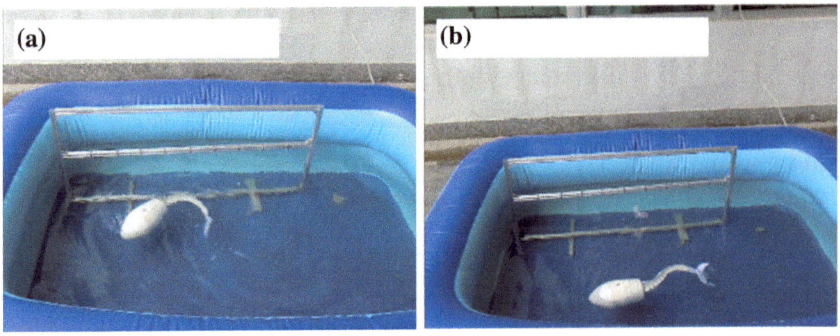

Fig. 33 Undulatory robot fish swimming in **a** oscillatory form; **b** undulatory form

Table 3 Big-C-motion results

A_1 (°)	A_2 (°)	f (Hz)	V_m (mm/s)	V_m (BL/s)	V_P (mm/s)	η (%)
20	30	1	201.73	0.408	252.50	61.99
30	45	1	260.51	0.526	325.90	62.14
40	60	1	300.75	0.608	353.60	63.04
40	60	0.5	198.87	0.402	176.80	65.62
40	60	1.5	154.00	0.311	522.30	55.56

robot fish was recorded under various bending amplitudes and flapping frequencies. The average speed is estimated by the traveling distance and time from the video frames. The traveling distance is estimated using a ruler set in the tank, as shown in Fig. 33. The results are shown in Table 3, where A_1 is the amplitude of the first segment, A_2 is the amplitude of the second segment, f is the flapping frequency, V_m is the average cruising velocity, and V_p is the predicted velocity. From the table, it is seen that the cruising speed increases with the increase of the bending amplitude. When the flapping frequency is low, the cruising speed also increases in proportion to the frequency. However, it is not applicable when the frequency is increased further. In this test, the maximum velocity was 300.75 mm/s, which is 0.608 BL/s, when the bending amplitudes of the first segment and the second segment are 40° and 60°, respectively, and the flapping frequency is 1 Hz. The velocities predicted by the propulsion model are generally larger than the measured one. This may be due to the size limitation of the pool, the measured velocity is the average speed, including the acceleration stage. When the flapping frequency is 1.5 Hz, the error is significant. This may again attribute to the small size of the water tank, by which the water resistance reduces the flapping frequency. In the test, the measured Froude efficiency of the robot fish swimming in oscillatory form is between 55.56 and 65.62 %.

Experiment 2—"Small-C-Motion" forward

In the experiment, only the second segment flaps in oscillatory form while the first segment remains still. Table 4 shows the experiment results. The relationship

Table 4 Small-C-motion results

A_1 (°)	A_2 (°)	f (Hz)	V_m (mm/s)	V_m (BL/s)	V_P (mm/s)	η (%)
0	30	1	91.74	0.185	134.60	61.99
0	45	1	140.85	0.285	190.10	62.75
0	60	1	162.76	0.329	233.30	62.11
0	60	0.5	143.54	0.290	116.70	68.07
0	60	1.5	136.75	0.276	346.80	57.52

between the cruising speed and the bending amplitude is similar to that in the "Big-C-Motion", and so is the speed–frequency relationship. The maximum speed is 162.76 mm/s (0.329 BL/s), when the bending angle of the second segment is 60° and the flapping frequency is 1 Hz. This is lower than that of the "Big-C-Motion" as expected. The propulsion model predicts the swimming velocity reasonably well. The prediction is generally larger than the measurement. The measured Froude efficiency is between 57.52 and 68.07 %. This is a little bit higher than that of "Big-C-Motion". In other words, it seems that the efficiency is lower when the flapping tail is more flexible.

Experiment 3—"S-Motion" forward
In this case, because the motion is the superposition of the two segments motions, the modes of "S-Motion" is dependent on the phase difference between the two segments. Table 5 shows the experiment results, where t_p is the number of periods that the second segment precedes the first segment. According to the experiment results, when t_p is 5/8 the performance of "S-Motion" is optimal. For this reason, in the test, t_p is chosen as 5/8. The maximum speed in this experiment was 333.33 mm/s (0.673 BL/s), when the bending angles of the two segments are 40° and 60°, respectively, the flapping frequency is 1 Hz. This is higher than the "Big-C-Motion." The velocity prediction is reasonably good. The measured Froude efficiency is much higher than the "Big-C-Motion" and "Small-C-Motion" as well. The maximum efficiency is 92.85 % and the averaged efficiency is 75.56 %. In the tests, it is noted that the head of the robot fish is more stable. This reduces the drag force and hence, improves the efficiency.

Table 5 S Motion results

A_1 (°)	A_2 (°)	f (Hz)	t_p (T)	V_m (mm/s)	V_m (BL/s)	V_P (mm/s)	η (%)
20	30	1	5/8	207.29	0.419	194.00	81.50
30	45	1	5/8	278.88	0.563	280.60	79.83
40	60	1	5/8	333.33	0.673	356.60	78.27
40	60	0.5	5/8	308.37	0.623	178.30	88.51
40	60	1.5	5/8	200.57	0.405	537.00	63.54
40	60	1	1/8	289.26	0.584	578.50	66.14
40	60	1	3/8	121.95	0.246	356.60	62.47
40	60	1	1/2	32.26	0.065	8.40	92.85
40	60	1	7/8	303.33	0.613	578.50	66.84

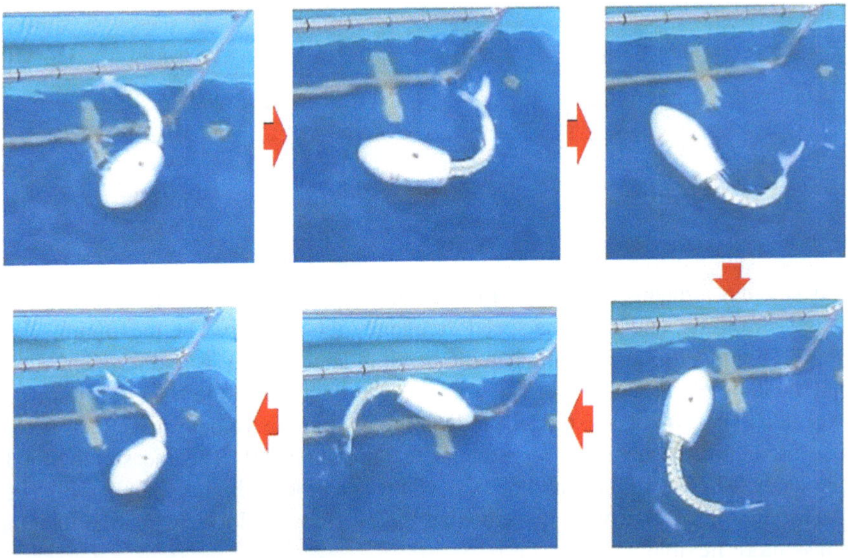

Fig. 34 Undulatory wire-driven robot fish turning

Experiment 4—"Turning"
In the fourth experiment, the turning of the robot fish is tested. In the test, the robot fish turned using the "Big-C-Motion" mode, but flapping only in one side. Figure 34 shows the rightward turning cycle of the robot. The turning radius of the robot fish is around 0.7 BL and the turning speed is 51.4°/s.

In summary, the undulatory wire-driven robot fish swims efficiently in both oscillatory form and undulatory form. In the undulatory form of swimming, the swaying of the fish body is reduced. As a result, the drag force is reduced and the efficiency is improved. This helps to increase the swimming speed. However, compared to the oscillatory swim, the undulatory swim is inferior in turning.

3.3 Design Example 3: Wire-Driven Robot Fish with Vector Propulsion

A close examination reveals that many fish can flap in a 3D manner so that it can swim forward, downward and upward. This is because the water world is 3D. The wire-driven robot fish can mimic the 3D flapping as well. In the section, such a design is presented.

The Design
This robot fish comprises of the vector propulsor and the fish body. The propulsor provides the controllable thrust (in both amplitude and direction), and the fish body houses the actuator, the controller, and the power unit (battery).

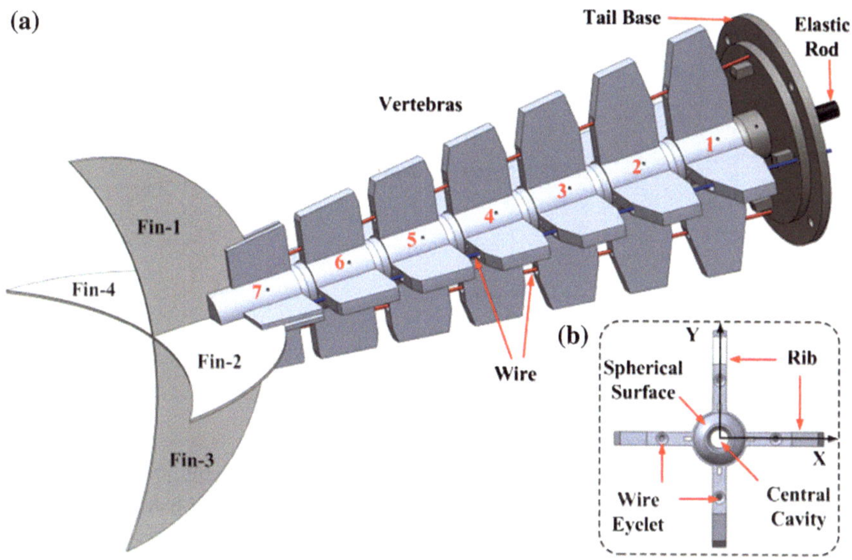

Fig. 35 The design of the vector propulsor in **a** isometric view; **b** a section view

(A) The Vector Propulsor

Figure 35a shows the design of the propulsor. It comprises of the tail base, seven vertebrae, a tail fin, an elastic rod, and two pairs of controlling wires. Four fin pieces are inserted to the last vertebra making the tail fin as shown in the figure. Two opposite fin pieces form a crescent caudal fin. The vertebra structure is shown in Fig. 35b. It has four orthogonally distributed ribs, where the wire eyelets are located. Two successive vertebrae form a spherical joint by the convex and concave spherical surfaces at the ends of the vertebrae. A center through hole is used to hold the elastic rod, whose function is twofold: help articulating the joints and confining the joints' rotations. Each joint can rotate freely about *X*- and *Y*-axes as shown in Fig. 35b, but the rotation about *Z*-axis is constrained. Therefore, the propulsor can bend horizontally and vertically, with limited twisting. The bending of the propulsor is controlled by two pairs of wires, which are guided by the eyelets on the rib of the vertebra as shown in Fig. 35b. Since the two wire pairs are orthogonally arranged the bending in horizontal and vertical planes are independent. The elastic rod has uniform cross section, thus when the wire is pulled, the propulsor will bend to a circular arc.

Figure 36 shows the details of the joint. The wire eyelet is tilted with respect to the propulsor axis, which is helpful in reducing the wire tension [23, 24]. In the figure, H is the rib height; h_0 is the joint gap distance; r_1 is the top wire eyelet central distance; R_1 is the vertebra top width; r_2 is the bottom eyelet central distance; R_2 is the vertebra bottom width; R is the radius of the stopper. The joint can independently rotate about the *X*-axis and the *Y*-axis, and the two rotations are

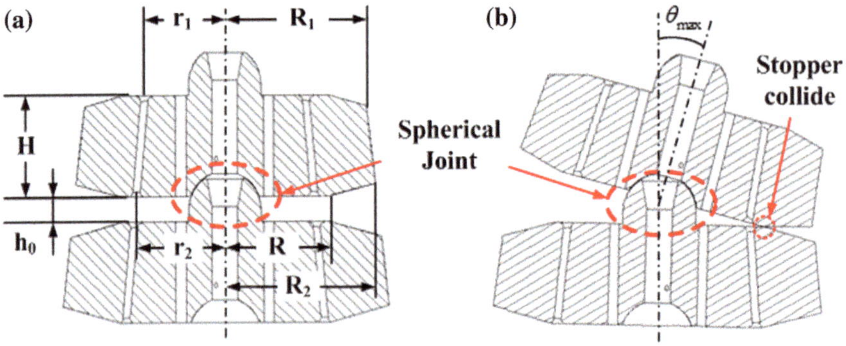

Fig. 36 Illustration of the joint **a** at rest; **b** rotating rightward

identical. The maximum joint rotation angle is determined by the joint gap distance h_0 and the stopper radius R as shown in Fig. 36b. The relationship is shown in Eq. (24).

$$\theta_{max} = 2\arctan\left(\frac{h_0}{2R}\right) \tag{24}$$

The design parameters of the vertebrae are shown in Table 6. In this design, each joint can rotate up to 13.5° in both X and Y directions. The maximum bending angle of the propulsor is 94.5°.

(B) Fish Body

The fish body design is shown in Fig. 37. It comprises of a shell, a main board, an auxiliary board, two servomotors, two wheels, a controller, a battery, two pulleys, etc. The shell has three pieces: hull-1, hull-2, and hull-3. Hull-1 serves as the base of the robot fish. The main board and tail base are fastened to Hull-1. Hull-2 is positioned to Hull-1 by four pegs. It is used to facilitate the robot fish assembly. Hull-3 has a parabolic front surface, which is helpful in reducing the water resistance. The assembly procedure is also shown in the figure: Step I, connect Hull-2 to

Table 6 The key parameters of the vertebrae (in mm)

Number	H	h_0	R	R_1	R_2	r_1	r_2
1	20	5	21.12	20.00	22.29	10.00	11.71
2	20	5	21.12	22.86	25.14	12.14	13.86
3	20	5	21.12	25.71	28.00	14.29	16.00
4	20	5	21.12	28.57	30.86	16.43	18.14
5	20	5	21.12	31.43	33.71	18.57	20.29
6	20	5	21.12	34.29	36.57	20.71	22.43
7	20	5	21.12	37.14	39.43	22.86	24.57

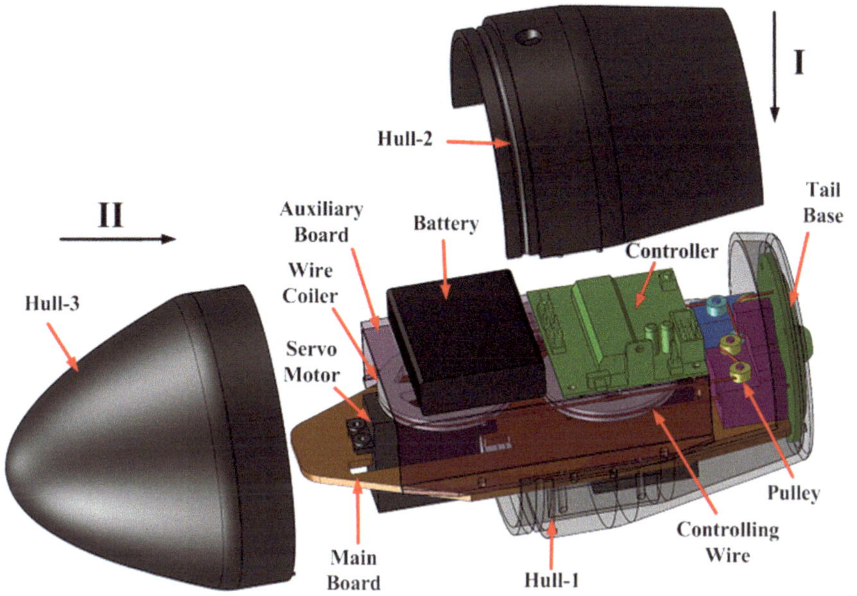

Fig. 37 The design of the fish body

Hull-1; step II, cover the fish body by Hull-3. The three pieces form the shell of the robot fish, which has a similar shape as that of the design example 2.

Two servomotors are used to drive the two wire pairs. The one in front controls the vertical wire pair, while the other one controls the horizontal wire pair. The wires are guided by the pulleys. One end of the wire is fixed at the last vertebra, and the other end is connected to the wire coiler, which rotates with the servomotor. The rotation of the servomotor is controlled by the MCU controller, same as in the design example 2. The control scheme is also the same. The command is sent out using a remote controller or using comport via Bluetooth. Upon receiving the signal, the MCU generates a 50 Hz PWM sequence. The rotation of the servomotor is controlled by the duty cycle of the PWM, while the velocity is controlled by setting the time delays between rotary positions.

(C) Robot Fish Prototype

The robot fish prototype is built as shown in Fig. 38. The key design parameters of the vertebrae are listed in Table 6. The vertebrae are connected by a silicon rubber rod with a diameter of ϕ5 mm. To improve the elasticity, four carbon sticks with ϕ0.5 mm diameter are connected to the tail. To reduce the friction, lubricating grease is added to all the joints. Four plastic fins are orthogonally mounted onto the last vertebra. Two opposite fins make a lunate shape, which is similar to the caudal fin of the dolphin. Two Tower Pro servomotors are used to drive the propulsor bending. The length of the robot fish is 425 mm, and the overall mass is 1.65 kg. Figure 38b shows the fish tail bending in horizontal plane. In this mode, the

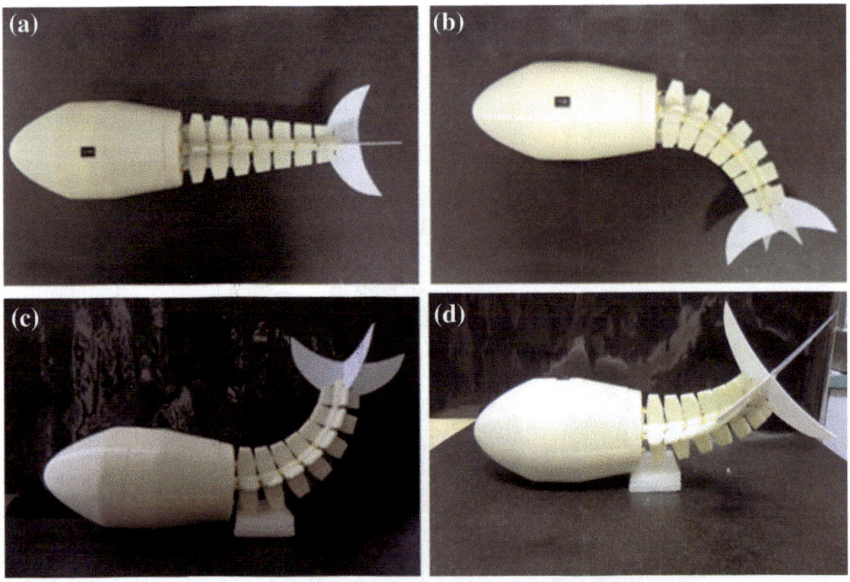

Fig. 38 The robot fish capable of vector propulsion: **a** in the rest position; **b** bending *horizontally*; **c** bending *vertically*; **d** bending to an arbitrary direction

bending is controlled by the horizontal wire pair, and the vertical wire pair remains still. Figure 38c shows the fish tail bending in the vertical plane. In this mode, the tail motion is controlled by the vertical wire pair only. Figure 38d shows the robot fish tail bending to an arbitrary direction. The bending magnitude and direction are controlled by the components in horizontal and vertical plane.

Modeling and Analysis

As shown in Fig. 39, the displacement of the propulsor is determined by the flapping angle Θ and flapping direction Φ. The former is defined as the angle between the bended tail and the Z-axis (it is zero when the tail is at resting position). The latter is defined as the angle between X-axis (horizontal direction) and the flapping plane. The wire configuration is shown in Fig. 39b, where P_1, P_2, P_3, P_4 denote the wire location. P_1 and P_3 are the horizontal wire pairs, which control the bending about Y-axis (flap in horizontal plane). P_2 and P_4 are the vertical wire pairs, which control the bending about X-axis (flap in vertical plane). When the flapping direction is Φ, it is conceived that the bending is about a virtual axis Y' as shown in the figure. The distance from the wires in vertical group to the virtual axis is a, while the distance from the wires in horizontal group to the virtual axis is b.

The joint rotations are confined by the elastic rod. As the vertebrae are rigid, the backbone length is constant. Also, the load acted on the propulsor is decoupled during wire pulling. Theoretically, the deformed shape of the rod under pure moment is a circular arc. Hence, we assume that the rotations of all the joints are the

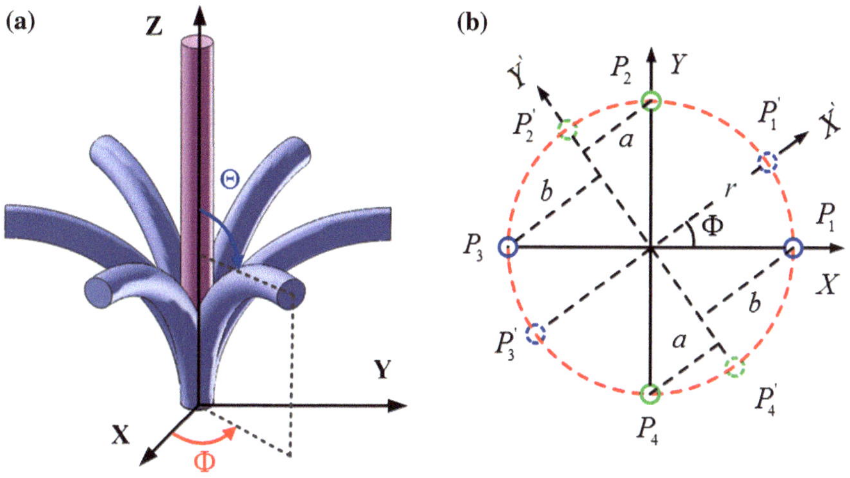

Fig. 39 Kinematics illustration. **a** Tail bending illustration, **b** wire configuration

same during the flapping process. The relationships between the wire lengths and the propulsor displacement are as follows [31]:

$$P_1: \quad L_1 = L_0 + 2N\left[b \cdot \sin\left(\frac{\theta}{2}\right) - h_0 \cdot \sin^2\left(\frac{\theta}{4}\right)\right] \tag{25}$$

$$P_2: \quad L_2 = L_0 + 2N\left[a \cdot \sin\left(\frac{\theta}{2}\right) - h_0 \cdot \sin^2\left(\frac{\theta}{4}\right)\right] \tag{26}$$

$$P_3: \quad L_3 = L_0 - 2N\left[b \cdot \sin\left(\frac{\theta}{2}\right) + h_0 \cdot \sin^2\left(\frac{\theta}{4}\right)\right] \tag{27}$$

$$P_4: \quad L_4 = L_0 - 2N\left[a \cdot \sin\left(\frac{\theta}{2}\right) + h_0 \cdot \sin^2\left(\frac{\theta}{4}\right)\right] \tag{28}$$

where, $a = r \cdot \sin(\Phi)$ and $b = r \cdot \cos(\Phi)$; r is the average distance between the wire and Z-axis; L_0 is the wire initial length; N is the number of vertebrae; and $\theta = \Theta/N$ is the rotation of the joint. From Eqs. (25)–(28), the flapping angle and direction can be represented as:

$$\Phi = \arctan\left(\frac{L_2 - L_4}{L_1 - L_3}\right) \tag{29}$$

$$\Theta = N \cdot \theta = 2N \cdot \arcsin\left[\frac{\sqrt{(L_1 - L_3)^2 + (L_2 - L_4)^2}}{4N \cdot r}\right] \tag{30}$$

The propulsor tip position with respect to the first joint rotate center is as in Eq. (31). In the equation, H is the rib height, h_0 is the joint initial gap distance, and H_T is the fin length. The positions along the propulsor can be obtained accordingly. It is noted that the excursion of the propulsor tip from the Z-axis is $\sqrt{x^2+y^2}$. When all the joints rotate from one limit θ_{min} to the other limit θ_{max}, the propulsor flaps from side to side. The flapping motion is in the plane defined by the flapping angle Φ, which is independent on the flapping angle Θ. The displacement of the propulsor tip is determined as follows:

$$\begin{cases} x = \left\{ (H_T - h_0/2) \cdot \sin(N \cdot \theta) + (H + h_0) \sum_{i=1}^{N} \sin[i \cdot \theta] \right\} \cos(\Phi) \\ y = \left\{ (H_T - h_0/2) \cdot \sin(N \cdot \theta) + (H + h_0) \sum_{i=1}^{N} \sin[i \cdot \theta] \right\} \sin(\Phi) \\ z = (H_T - h_0/2) \cdot \cos(N \cdot \theta) + (H + h_0) \sum_{i=1}^{N} \cos[i \cdot \theta] \end{cases} \tag{31}$$

Equation (31) is the kinematics model, from which the excursion, the slope and the traversing speed of the propulsor can be derived as per Eqs. (32–33), where q is the curve length.

$$y(q,t)\big|_{q=L} = (H_T - h_0/2) \cdot \sin(N\,\theta) + (H + h_0) \sum_{i=1}^{N} \sin(i \cdot \theta) \tag{32}$$

$$\left.\frac{\partial y(q,t)}{\partial q}\right|_{q=L} = \sin(\Theta) \tag{33}$$

$$\left.\frac{\partial y(q,t)}{\partial t}\right|_{q=L} = \left\{ (H_T - h_0/2) \cdot N \cdot \cos[N \cdot \theta] + (H + h_0) \sum_{i=1}^{N} i \cdot \cos[i \cdot \theta] \right\} \cdot \dot{\theta} \tag{34}$$

By substituting these into Lighthill's EBT, the propulsion model is therefore obtained.

Experiment Results

Two basic swimming modes are tested: one mimics the shark swimming and the other mimics the dolphin swimming. The experiment results are presented as follows:

(A) Shark Swimming

In this experiment, the vector propulsor flaps horizontally like a shark. In this mode, the back motor controls the horizontal wire pair and the front motor keeps still. Fin 1 and fin 3 provide thrust, while fin 2 and fin 4 are held still. In the test, the flapping frequency is $f = 1$ Hz, and the flapping amplitude is 45°. At first, the robot fish is placed in the swimming pool and the water is still. The robot fish is controlled via Bluetooth to flap horizontally. One flapping cycle is shown in Fig. 40a–e. Figure 40a shows the robot fish at rest position. Then, it flaps to the left as shown in

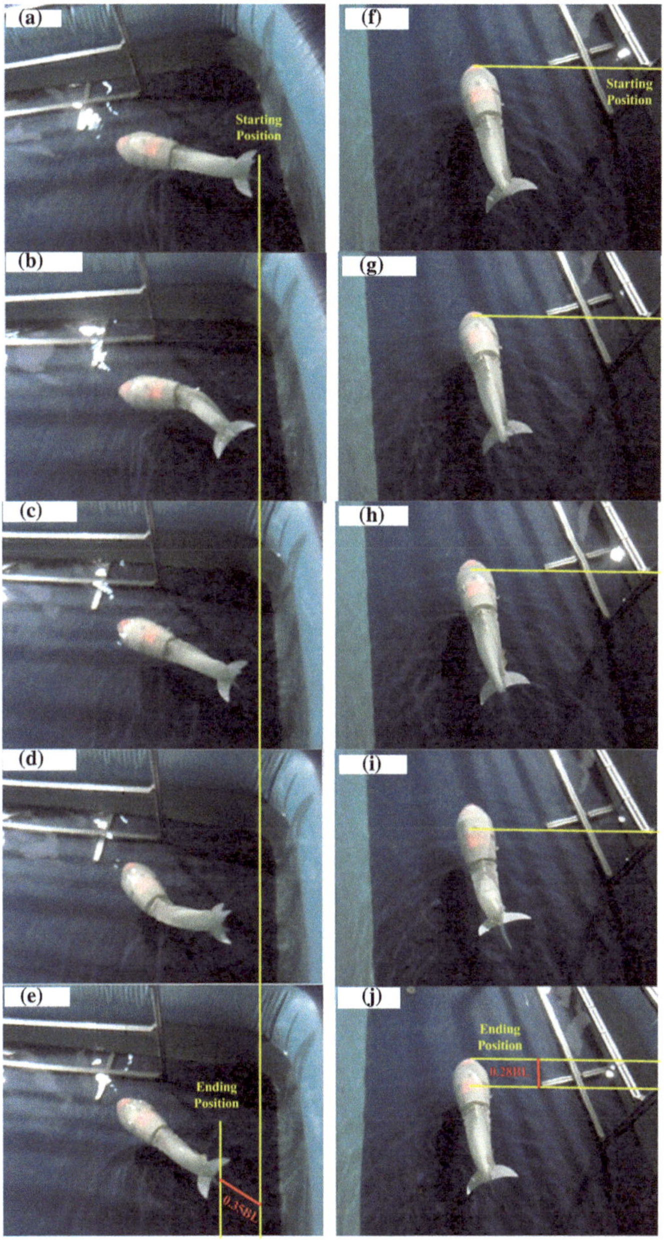

Fig. 40 Experiment results: **a–e** is the flapping cycle of shark form swimming; **f–j** is the flapping cycle of dolphin form swimming

Fig. 40b. When it reaches to the left most position, it flaps back to the rest position as shown in Fig. 40c. The other half cycle is done in a similar way. The propulsor flaps rightward at first, as shown in Fig. 40d, and then flaps back to the resting position. In this experiment, the left flapping amplitude and right flapping amplitude are both 45°. Also, the flapping frequencies in the four stages are the same. From the results, it is shown that after one flapping cycle, the robot fish moves forward by 148 mm, i.e., 0.35 BL. From the previous propulsion model, when the propulsor flapping frequency is 1 Hz, and the amplitude is 45°, i.e., the tail end excursion is 65 mm or 0.153 BL, the cruise speed of the robot fish is 170.4 mm/s. The prediction error is about 13 %. In the simulation, the drag coefficient is selected as 0.5.

(B) Dolphin Swimming
In this experiment, the vector propulsor flaps vertically like a dolphin. Fin 2 and fin 4 provide thrust in this mode, while fin 1 and fin 3 are kept still. Similar to the previous case, the flapping frequency of the propulsor is $f = 1$ Hz and the flapping amplitude is 45°. Same as in the previous test, the robot fish is placed in still water. When the fish received the command, it starts flapping vertically. In this mode, the front motor controls the vertical wire pair and the back motor keeps still. Figure 40f–j shows one flapping cycle. As shown in Fig. 40f, the robot fish is at rest position. Once receiving the command, the tail flaps downward at first, as shown in Fig. 40g. After reaching the down-most position, the tail flaps back as shown in Fig. 40h. The tail continues to flap passing the rest position and reaching the upmost position as shown in Fig. 40i. Finally, the tail flaps back the rest position finishing a cycle. From the experiment, the distance the robot fish travels in one flapping cycle is around 0.28 BL, slightly slower than that of the shark swimming. The prediction error for the dolphin form swimming is 30 %. The large error is mainly attributed to the fact that the robot fish is not fully under water, i.e., part of the flapping cycle is out of water.

It should be pointed out that many existing robot fish flap either in horizontal plane or in vertical plane. Therefore, they are confined in a single flapping direction. In comparison, the presented robot fish can turn (using horizontal flapping) and dive (using vertical flapping) by the tail only.

4 Conclusions

In this chapter, the BWDM is introduced and used to build three different robot fishes. The BWDM follows both the backbone structure and the muscle arrangement of the aquatics. Moreover, it is underactuated and thus, requires only one or two motors to generate various flapping motions. Therefore, the BWDM robot fish is simple in structure, easy to control, and most importantly highly energy efficient. Three robot fish examples are presented. The first robot fish has an oscillatory wire-driven propulsor and a pair of pectoral fins. The second robot fish is controlled by two motors capable of undulatory swimming and ocillatory swimming. The third robot fish can generate thrust in all directions mimicking both shark swimming and dolphin swimming. The experiments demonstrate that the BWDM robot fishes are well suited for underwater propulsion. They can reach a speed of 330 mm/sec with an efficiency of 80 %. It is expected that they will find many applications in the near future.

References

1. Triantafyllou MS, Triantafyllou GS (1995) An efficient swimming machine. Sci Am 272:64–71
2. Nature Culture Discover. http://australianmuseum.net.au/Sailfish-Istiophorus-platypterus
3. Domenici P, Blake R (1997) The kinematics and performance of fish fast-start swimming. J Exp Biol 200:1165–1178
4. Colgate JE, Lynch KM (2004) Mechanics and control of swimming: a review. IEEE J Ocean Eng 29:660–673
5. Abaid N, Bernhardt J, Frank JA, Kapila V, Kimani D, Porfiri M (2013) Controlling a robotic fish with a smart phone. Mechatronics 23:491–496
6. Liang J, Wei H, Wang T, Wen L, Wang S, Liu M (2009) Experimental research on biorobotic autonomous undersea vehicle. In: Inzartsev AV (ed) Underwater vehicles. InTech, Osaka
7. Gu Z, Yamamoto I, Hiratsuka T (2013) Research on bio-mechanism robotics by robotic fish fin technology. In: Kacprzyk J (ed) Intelligent autonomous systems 12, vol 193. Springer, Berlin, pp 577–584
8. Kumpt JM (1996) The design of a free swimming robot pike. Bachelor of Science, Department of Mechanical Engineering, MIT
9. Hu H (2006) Biologically inspired design of autonomous robotic fish at essex. In: Presented at the proceedings of the IEEE SMC UK-RI chapter conference 2006 on advances in cybernetic systems, Sheffield
10. Malcolm EF, MacIver A, Burdick JW (2004) Designing future underwater vehicles principles and mechanisms of the weakly electric fish. IEEE J Ocean Eng 29:651–659
11. Yu JZ, Tan M, Wang S, Chen E (2004) Development of a biomimetic robotic fish and its control algorithm. IEEE Trans Syst Man Cybern Part B Cybern 34:1798–1810
12. Kim B, Kim D-H, Jung J, Park J-O (2005) A biomimetic undulatory tadpole robot using ionic polymer–metal composite actuators. Smart Mater Struct 14:1579
13. Mbemmo E, Chen Z, Shatara S, Tan XB (2008) Modeling of biomimetic robotic fish propelled by an ionic polymer-metal composite actuator. In: IEEE international conference on robotics and automation, Pasadena, pp 689–694
14. Wang ZL, Hang GR, Wang YW, Li J, Du W (2008) Embedded SMA wire actuated biomimetic fin: a module for biomimetic underwater propulsion. Smart Mater Struct 17:025039
15. Rossi C, Colorado J, Coral W, Barrientos A (2011) Bending continuous structures with SMAs: a novel robotic fish design. Bioinspir Biomim 6:045005
16. Le CH, Nguyen QS, Park HC (2012) A SMA-based actuation system for a fish robot. Smart Struct Syst 10:501–515
17. Chu WS, Lee KT, Song SH, Han MW, Lee JY, Kim HS et al (2012) Review of biomimetic underwater robots using smart actuators. Int J Precis Eng Manuf 13:1281–1292
18. Wiguna T, Heo S, Park HC, Goo NS (2009) Design and experimental parametric study of a fish robot actuated by piezoelectric actuators. J Intell Mater Syst Struct 20:751–758
19. Nguyen QS, Heo S, Park HC, Goo NS, Kang T, Voon KJ et al (2009) A fish robot driven by piezoceramic actuators and a miniaturized power supply. Int J Control Autom Syst 7:267–272
20. Carp. http://www.somso.de/english/zoologie/zos105.htm

21. Octopus. http://www.scandfish.com/ig/gallery.asp?action=viewimage&categoryid=33&text=&imageid=1245&box=&shownew=
22. Jones BA, Walker ID (2006) Kinematics for multisection continuum robots. IEEE Trans Robot 22:455
23. Li Z, Du R, Yao Y (2012) Flying octopus: a LTAV with wire-driven flapping wings. In: International mechanical engineering congress and exposition (IMECE 2012), Houston
24. Li Z, Du R, Zhang Y, Li H (2013) Robot fish with novel wire-driven continuum flapping propulsor. Appl Mech Mater 300:510–514
25. Li Z, Du R (2012) Design and analysis of a biomimetic wire-driven flapping propeller. In: 4th IEEE RAS and EMBS international conference on biomedical robotics and biomechatronics (BioRob), Roma, pp 276–281
26. Li Z, Du R, Zhang Y, Li H (2012) Robot fish with novel wire-driven continuum flapping propulsor. In: Presented at the 2nd international conference on mechatronics and applied mechanics (ICMAM), Hong Kong
27. Li Z, Gao W, Du R, Liao B (2012) Design and analysis of a wrie-driven robot tadpole. In: International mechanical engineering congress and exposition (IMECE 2012), Houston
28. Liao B, Li Z, Du R (2012) Robot tadpole with a novel biomimetic wire-driven propulsor. In: Presented at the IEEE international conference on robotics and biomimetics (ROBIO 2012), Guang Zhou
29. Li Z, Du R (2014) A novel double-hull boat with biomimetic wire-driven flapping propulsors. In: IEEE international conference on robotics and automation (ICRA 2014), Hong Kong
30. Li Z, Du R (2013) Design and analysis of a bio-inspired wire-driven multi-section flexible robot. Int J Adv Robot Syst 10
31. Li Z, Du R, Lei MC, Yuan SM (2011) Design and analysis of a biomimetic wire-driven robot arm. In: Proceedings of the ASME 2011 international mechanical engineering congress and exposition, Denver, pp 11–17
32. Li Z, Du R (2014) Expanding workspace of underactuated flexible manipulator by actively deploying constrains. In: IEEE international conference on robotics and automation (ICRA 2014), Hong Kong
33. Li Z, Du R, Ren H, Yu H (2014) Statics modeling of an underactuated wire-driven flexible robotic arm. In: Presented at the IEEE international conference on biomedical robotics and biomechatronics, Sao Paulo
34. Li Z, Du R (2012) Design and implementation of a biomimetic wire-driven underactuated serpentine manipulator. Trans Control Mech Syst 1:250–258
35. Sugihara T (2011) Solvability-unconcerned inverse kinematics by the Levenberg–Marquardt method. IEEE Trans Robot 27:984–991
36. Kim J-S, Lee D-Y, Kim K, Kang S, Cho K-J (2014) Toward a solution to the snapping problem in a concentric-tube continuum robot grooved tubes with anisotropy. In: Presented at the IEEE international conference on robotics and automation (ICRA), Hong Kong
37. Orin DE, Schrader WW (1984) Efficient computation of the Jacobian for robot manipulators. Int J Robot Res 3:66–75
38. Lighthill MJ (1970) Aquatic animal propulsion of high hydromechanical efficiency. J Fluid Mech 44:265–301
39. Lighthill MJ (1971) Large-amplitude elongated-body theory of fish locomotion. Proc R Soc B Biol Sci 179:125–138
40. Buss SR (2004) Introduction to inverse kinematics with jacobian transpose, pseudoinverse and damped least squares methods. IEEE J Robot Autom 17:1–19

Design and Control of a Multi-joint Robotic Fish

Junzhi Yu and Min Tan

Abstract This chapter is devoted to the development and control issues of a multi-joint robotic fish, with emphasis on creating a controlled kinematic and dynamic centered environment to further shed light on designing control methods. By virtue of the hybrid propulsion capability in the body plus the caudal fin and the complementary maneuverability in accessory fins, a synthesized propulsion scheme involving a caudal fin, a pair of pectoral fins, as well as a pelvic fin is proposed. To aid the systematic analysis of the multi-joint tail, a multilink Digital Fish Simulator (DFS) is developed, enabling simulations of various fictive swimming patterns. To achieve flexible yet stable motions in aquatic environments, both body wave-based control and central pattern generator—(CPG)-based control are proposed and compared in terms of oscillatory signals and swimming stability. Furthermore, a series of multi-joint robotic prototypes with diversified functions have been built to validate the well-formed ideas and to attain a new level of swimming performance close to real fish.

1 Introduction

There has been increasing interest in the development and application of the bio-inspired robots in academia and industry over the last 20 years, particularly due to the growing vitality in biomimetics [1–3]. As a small branch, extensive work on designing and controlling a variety of swimming robots has been done since the first fish robot (also known as robotic fish, or robot fish), RoboTuna, was developed at MIT [4]. In particular, astonishing swimming ability including high efficiency and maneuverability inspired us to improve the propulsive performance of man-made aquatic robotic systems [5, 6]. Rising missions like military defense and

J. Yu (✉) · M. Tan
The State Key Laboratory of Management and Control for Complex Systems, Institute of Automation, Chinese Academy of Sciences, Beijing 100190, China
e-mail: junzhi.yu@ia.ac.cn

© Springer-Verlag Berlin Heidelberg 2015

R. Du et al. (eds.), *Robot Fish*, Springer Tracts in Mechanical Engineering,
DOI 10.1007/978-3-662-46870-8_4

marine protection require high-performance autonomous underwater vehicles (AUVs), especially in the use of the fast propulsion and multi-directional maneuverability. As a paradigm of bio-inspired AUV, robotic fish will be more competent than current AUV propelled by rotary propellers. Besides advancing the understanding of fish swimming mechanisms, as an aquatic mobile platform, robotic fish will play an important role in fulfilling real-world tasks such as underwater exploration, mobile sensing, patrol, wreck surveying, search and rescue, and environmental monitoring [7, 8].

As the earliest vertebrate in the world, fish are undoubtedly the king from the perspective of underwater propulsion. In particular, fins are widely used by aquatic animals. The propulsion modes of swimming fish, from a functionality standpoint, are classically divided into body and/or caudal fin propulsion (BCF) and median and/or paired fin (MPF) propulsion [9]. Specifically, fish classes that use varying degrees of body undulation and/or caudal fin oscillations for thrust generation are examples of the BCF mode, while fish relying on paired fins (pectoral fins and pelvic fins) and/or median fins (dorsal fin and anal fin) for thrust production are categorized under the MPF mode. It is generally thought that no absolutely superior model exists among these modes since body shape and motor function level tightly depend on the fish habitats.

More recent evidence has suggested that fish actually rely on multiple control surfaces including the caudal fin, the pectoral fins, the pelvic fins, the dorsal fin, the anal fin as well as the body to achieve fast and maneuverable propulsion [10]. Particularly, the position, mobility, and hydrodynamic characteristics of the control surfaces are linked with the propulsive performance. These well-integrated, configurable multiple control surfaces provide an excellent paradigm to create and control high-performance underwater vehicles [11]. However, it is unrealistic to totally replicate a real fish due to the vast difference between the biological system and the engineering counterpart. Tradeoffs in engineering practice will have to be struck among biological mechanism, engineered method, feasibility, cost/gain, and so forth. The existing robotic fish, however, have predominantly used BCF, or pectoral fins (PF), or undulatory fins (UF) for propulsion and maneuvering. There have been few or limited studies related to simulating and constructing a robotic fish with multiple different fins, which are desirable for enhanced controllability and maneuverability. For example, inspired by the boxfish with multiple fins, Lachat et al. designed a miniature boxfish-like swimming and crawling robot with three actuated fins [12]; Kodati et al. [13] developed a robotic ostraciiform with a self-correcting mechanism to emulate the maneuverability of the boxfish. In these two boxfish-like robots, only two types of fins including pectoral fin and caudal fin were incorporated and examined, making them unavailable for the extensive exploration of various fin–fin and body–fin interactions.

This chapter, as a summary of our previous research on fishlike swimming [14–17], aims at creating self-propelled, multi-joint robotic fish with multiple complemented gaits and improved swimming performance. Specifically, a design framework considering both mechatronic constraints in physical realization and feasibility of control methods is presented. Although there have been numerous studies of numerical optimal control methods to find a time- or energy-efficient state-to-state trajectory in motion planning, there have been few optimum procedures used to make an artificial fish system as effective as possible in propulsion, especially in thrust generation. Contrary to identical links with the same length, mass, and inertia in the physical model of anguilliform locomotion [18], our models focus more on an attribute of spindle-shaped fish body; i.e., the length of segments along spinal column is shorter and shorter in the direction from nose to tail. A locally optimal link-length-ratio is numerically calculated. Meanwhile, a central pattern generator—(CPG) centered control framework is applied to enhance the swimming performance. Particularly, the body wave-based control and the CPG-based control are comparatively investigated in terms of oscillatory signals and swimming stability. Finally, the proposed models and methods are verified in the developed robotic fish prototypes.

In the following sections, we will present our effort to the design and control of multi-joint robotic fish. We start by surveying fish-inspired biomimetic research in Sect. 2. We then proceed to investigate design framework and simulator for mimicry of fish swimming with a multi-joint configuration in Sect. 3. The CPG-centered swimming control is tackled in Sect. 4. Experiments and results are provided in Sect. 5. Finally, Sect. 6 concludes the chapter with the outline of future work.

2 Review of Bio-inspired Fish Swimming

In this section, we will restrict our attention to the main problem of mechanical design, swimming kinematics as well as motion generation and modulation for a class of self-propelled, multi-joint robotic fish. Note that the developed robotic fish in this work is only suited for shallow waters.

2.1 Ichthyology Basis

Generally, as illustrated in Fig. 1, there exist two distinct propulsion modes for technical inspiration in developing robotic fish: BCF mode and MPF mode. The former is favorable for the cases requiring greater thrust and accelerations, while the latter for the cases demanding higher maneuverability. Meanwhile, in terms of movement's temporal features, swimming locomotion can be categorized into periodic swimming characterized by a cyclic repetition of the propulsive movements and transient movement involving rapid starts, escape maneuvers, and turns.

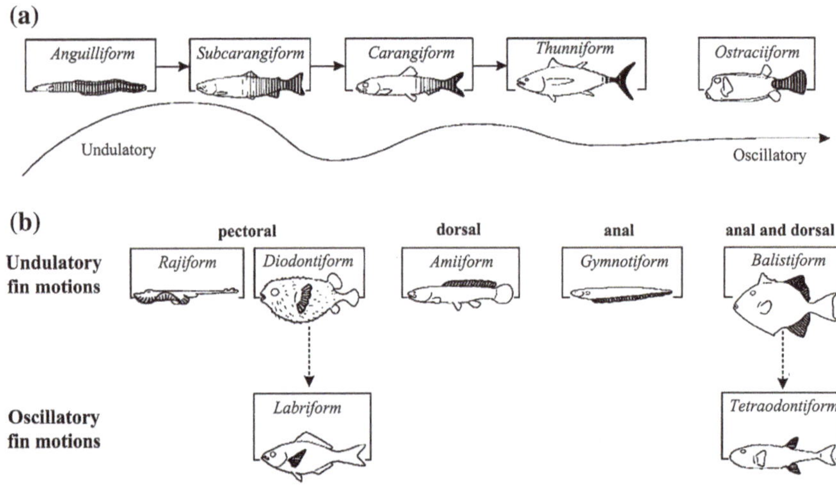

Fig. 1 Propulsion modes in fish swimming (Adapted from [9])

Fig. 2 Illustration of the skeleton of a common bony fish

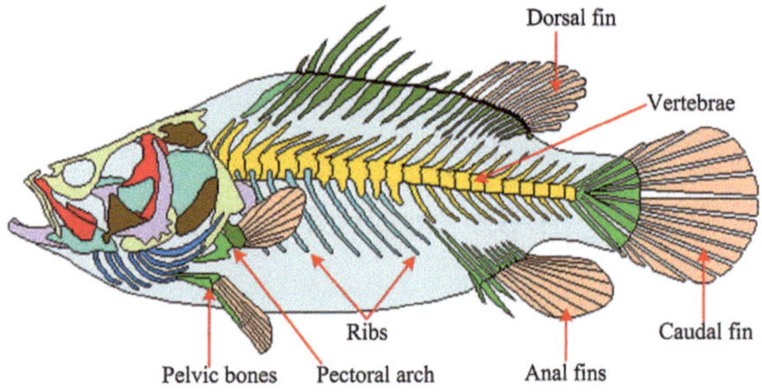

Meanwhile, studies of the dynamics of fish locomotion show that most fish synthetically use multiple control surfaces (e.g., tail plus caudal fin, pectoral fins, pelvic fin, dorsal fin, anal fin) to accomplish efficient and effective propulsion. Figure 2 shows the skeleton of a bony fish, which involves functionally complementary control surfaces. From the structural design standpoint, the vertebrae, cranium, jaw, ribs, and intramuscular bones make up the bony fish skeleton. Basically, the skeleton provides a foundation for the body and the fins, encases and protects the brain and the spinal cord, and serves as an attachment for muscles. Meanwhile, the tail is laterally compressed and corresponding tail vertebrae become smaller distally. That is, the lengths of skeleton elements, from the skull to the last caudal vertebra, tend to be smaller and smaller, offering some clues to the structural optimization.

2.2 Biomimetic Principles

As an effective and efficient underwater propulsive system, fish are of some technological interest in developing innovative AUVs. Typically, this involves the following four aspects.

- **Hydrodynamics aspect** Fish in nature vary greatly in body shape with significant hydrodynamic consequence. An important and intriguing mechanism associated with high-performance swimming is the shedding of vortex rings and recycling of vortex energy exploited by fish. For instance, a pair of abducted pectoral fins cause the formation of a drag wake, and the fish tail will recycle the energy of the pectoral fin vortices. Vortex interaction among different control surfaces (e.g., pectoral fins and tail) facilitates the generation of thrust. The fish tail shape also exerts certain effects on vortex formation patterns [19].
- **Propulsive mechanism** Fish are propelled through the water by fins, body movements, or both. A fish can swim even if its fins are removed, though it usually has difficulty in governing direction and balance [20]. While swimming, the fins are driven by muscles attached to the base of the fin spines and the rays. In particular, fish with fairly rigid bodies depend mostly on active fin action for propulsion. Notice also that fish fins are flexible and move in a complex three-dimensional (3-D) manner.
- **Motion control** So far, the control mechanisms of fish bodies and fins are not fully understood. In general, fish swim using multiple body segments and organizing left–right alternations in each segment so as to produce the body wave that propels them through water. These rhythmic motor patterns are internally produced by CPGs, i.e., central neuronal circuits whose activation can produce rhythmic patterns in the absence of sensory or descending inputs that carry specific timing information [21, 22]. Thus, neural system can generate and control a variety of motor behaviors via coordination among segmental CPGs. Typically, CPGs has the advantage of rhythmicity, stability, adaptability, and variety [23]. These fascinating properties make CPGs suitable for locomotion control of robots with multiple joints or DOFs and even of hyper redundant robots. Remarkably, CPGs eliminate the need for trajectory planning and precise knowledge of mechanical system properties.
- **Stability** Another point to be emphasized for fish swimming is stability, a significant issue in real-world applications. With the center of buoyancy both below and anterior or posterior to the center of mass, most fish with swim bladders are hydrostatically unstable. This instability necessitates the use of hydrodynamic lift to control posture in concert with hydrostatic forces. Therefore, a fundamental tradeoff in locomotion design exists between stability and maneuverability. Although rhythmic patterns of body undulations are very similar in steady swimming, fish realistically carry out more individual maneuvers rather than swimming steadily.

Note that the locomotor methods employed by fish cannot necessarily be considered optimal, even if the biologically evolved designs are highly effective for fish adapting to their habitats. So we should not blindly copy biological structures and control mechanics in bio-inspired engineering, but selectively absorb the biological advantages in a hybrid fashion.

3 Design Framework for Multi-joint Robotic Fish

In this section, we focus our attention on the radical problem of fishlike swimming generation and modulation, as well as their robotic implementation.

3.1 Functional Design

Consider a multilink robotic fish, the key to this high-quality bio-mimicry is how to simplify the mechanism and generate the reasonable control data. That is, to quantify the lateral body motions of swimming fish, kinematic and anatomical data of vertebral column and tail should be paid much more attention. Typical of steady swimming is the contraction from head along the midline of the fish body. A widely used body wave is described by

$$y_{\text{body}}(x,t) = (c_1 x + c_2 x^2)\sin(kx + \omega t) \tag{1}$$

where y_{body} represents the transverse displacement of moving tail unit, x denotes the displacement along the head-tail axis, k indicates the body wave number ($k = 2\pi/\lambda$), λ is the body wave length, c_1 is the linear wave amplitude envelope, c_2 is the quadratic wave amplitude envelope, and ω is the body wave frequency ($\omega = 2\pi f = 2\pi/T$).

As shown in Fig. 3, the oscillatory part of the robotic fish is commonly discretized as a multilink (or N-link) mechanism composed of several oscillating hinge joints actuated by motors in bio-inspired engineering. It can be modeled as a planar, serial chain of links along the axial body displacement, and the endpoints of the links in the chain can be achieved by numerical fitting to a discretized, spatial- and time-varying body wave. In the interest of simplicity, we consider the following discrete form of (1):

$$y_{\text{body}}(x,i) = (c_1 x + c_2 x^2)\sin\left(kx \pm \frac{2\pi}{M} i\right) \tag{2}$$

where i denotes the ith variable of the sequences $y_{\text{body}}(x,i)$ ($i = 0, 1, \ldots, M-1$) in one oscillation period, M indicates the discrete degree of the traveling wave, and the signs "+" and "–" represent different initial movement directions, which are dependent on different initial values. For more details on link-based body wave fitting, please refer to [14].

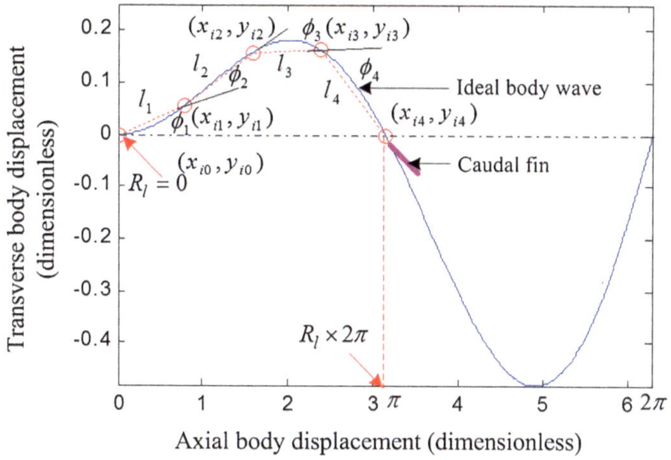

Fig. 3 Illustration of the multilink-based fishlike swimming design

Taking more diverse sinusoidal motions exhibited in fishlike or snake-like locomotion into consideration, a generalized body wave that facilitates engineering realization is proposed below:

$$y_{\text{body}}(x,t) = (c_1x + c_2x^2)\sin(k_1x + k_2x^2 + t) \tag{3}$$

where k_1 denotes the linear body wave number and k_2 indicates the quadratic body wave number. The determination of k_1 and k_2 depends on the desired oscillation type and function.

When altering the swimming direction, the real fish usually uses the tail in collaboration with pectoral fins. We remark that the pectoral fins of the carangiform swimmer have minimal effects on propulsion and steering, which are neglected in the current robotic model. For the multilink robotic fish, various turning modes can be implemented by commanding specific deflected angle in each oscillation cycle to the part or all of moving links. So a corrected turning body wave is hypothesized to be yielded as follows:

$$y_{\text{body}}(x,i) = (c_1x + c_2x^2)\sin(kx - \frac{2\pi}{M}i) + \sqrt{\frac{1}{4}D'^2 - x^2} - \frac{1}{2}D' \tag{4}$$

where D' is the diameter of the curved tail axis, having a bearing on the turning diameter D. An example of corrected turning body wave is demonstrated in Fig. 4, where $c_1 = 0.05$, $c_2 = 0.09$, $k = 2\pi/3$, $M = 9$, $D' = 8$.

In this sense, fishlike movements, either forward propulsion or turning maneuvers, can be produced within a well-integrated body wave framework. The morphological parameters relevant to swimming motion mainly consist of the following:

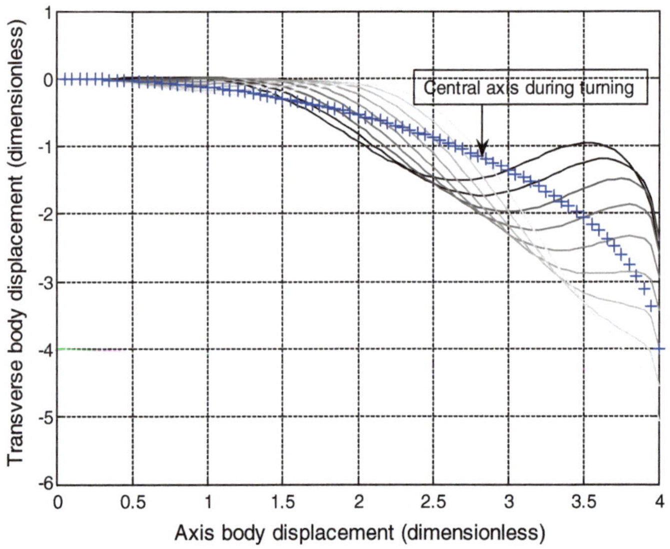

Fig. 4 Demonstration of a corrected body wave in turning maneuvers

(1) The length ratio of the fish's oscillatory part to that of the fish body (R_l). With the decrease of R_l, in general, efficiency and speed of fish swimming remarkably increase, but maneuverability reduces to a certain extent.
(2) The number of simplified joints in oscillatory part (N). In principle, larger value of N will lead to better maneuverability and redundancy, but harder construction and control of the robot.
(3) The length ratio of each link (segment) in oscillatory part (l_1:l_2:…:l_N). The length of each link in the direction from nose to tail, as a rule, is getting smaller and smaller. Oscillatory amplitude, however, increases gradually and reaches its maximum at tail peduncle of the fish.
(4) Characteristics of caudal fin [24].

On the other hand, the kinematic parameters are those fundamental physical quantities describing swimming performance and propulsive efficiency, which primarily include the following:

(1) The form of the body wave $y_{\text{body}}(x, t)$, determining the type of fish swimming.
(2) Transverse oscillatory amplitude of each joint in the body.

3.2 Dynamic Modeling of Robotic Fish

To simulate the dynamic behavior of the robotic fish, dynamic analysis is generally carried out in advance. We consider a three-dimensional model of robotic fish,

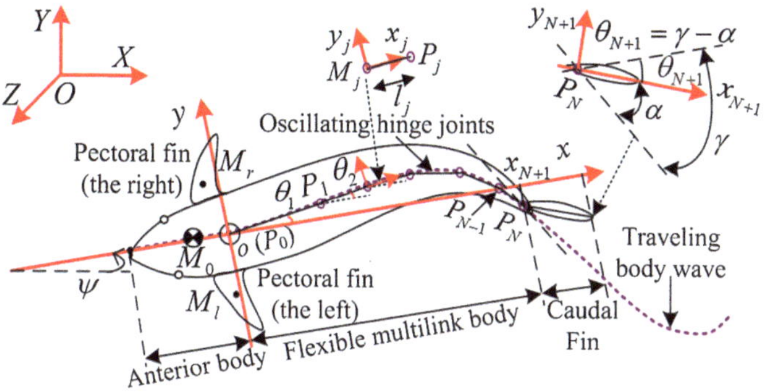

Fig. 5 Demonstration of a robotic fish capable of 3-D swimming

taking into account both the carangiform and anguilliform natural swimming modes. Since the carangiform swimmer bears the virtue of keeping high-speed swimming in calm waters, whereas the anguilliform swimmer exhibits remarkable maneuverability especially in cluttered environment, we assume that the constructing robotic fish has a flexible rear body for maneuverability and an oscillating caudal fin for propulsion. As illustrated in Fig. 5, a robotic fish structurally consists of three parts: a stiff anterior body with a pair of pectoral fins for up-and-down motion, a flexible rear body, and an oscillating lunate caudal fin. Especially, the flexible rear body is designed as a multilink mechanism capable of anguilliform undulation, while the caudal fin moves in a fashion of oscillating fin mimicking carangiform oscillation. As stated previously, the length ratio of the fish's oscillatory part to that of the fish body R_l is a vital morphological parameter, and not all the body or the oscillatory portion takes part in thrust production at all time, it may be a feasible way to employ different length of the oscillatory part at various speeds. For a multilink robotic fish, this method can be easily operated by locking or unlocking some links. Therefore, the robotic fish can switch between anguilliform mode and carangiform one in order to achieve various motions and better swimming performance.

To formularize the dynamics of the fish system, we consider that 3-D motion of the robotic fish can be divided into 2-D (two-dimensional), planar swimming plus diving/climbing motion in the vertical plane. Mechanically, the robotic fish can essentially be viewed as an open, tree-like multi-body system. The kinematic analysis is then integrated with hydrodynamic analysis on multiple moving elements to derive complete dynamic equations in a form suited for computer implementation as well as controller design. The following matrix representation of the equations of motion is derived as follows [25]:

$$\left(M(\underline{q},t) + M_I\left(\underline{q},t\right)\right)\ddot{\underline{q}} + K(\underline{q},\dot{\underline{q}},t) = \hat{Q}\left(\underline{q},\dot{\underline{q}},t\right) \tag{5}$$

where $M(\underline{q}, t) + M_I\left(\underline{q}, t\right)$ indicates the mass matrix incorporating all masses and inertias of the multilink robot, which also contains the virtual terms associated with the accelerated surrounding fluid, and $K(\underline{q}, \dot{\underline{q}}, t)$ includes the matrix of Coriolis/Centripetal term. Considering (5) is a nonlinear ordinary differential equation (ODE) in essence, we can apply a standard integration technique to solve the initial value problems for ODE in the Matlab/Simulink environment. The velocity and the trajectory can further be obtained via numerical tools; thus propulsive characteristics of the robot can therefore be predicted and assessed.

3.3 Design Framework for Robotic Fish

After conducting the dynamic analysis, it is necessary to verify the theoretical model by engineered robots. An integrated design scheme demonstrated in Fig. 6 is presented, which primarily involves morphological modeling, kinematic modeling, dynamic simulation, and parameter optimization. The procedure is detailed as the following steps:

Step 1. Abstract morphological and swimming characteristics from referenced biological model. Both morphological and kinematic models, in this step, are derived, which can provide a configuration frame to building prototypes.

Step 2. Incorporate the derived morphological and kinematic parameters into dynamic analysis of the employed propulsive model. Some biological control principles, at the same time, can be borrowed to accomplish fishlike motion. These

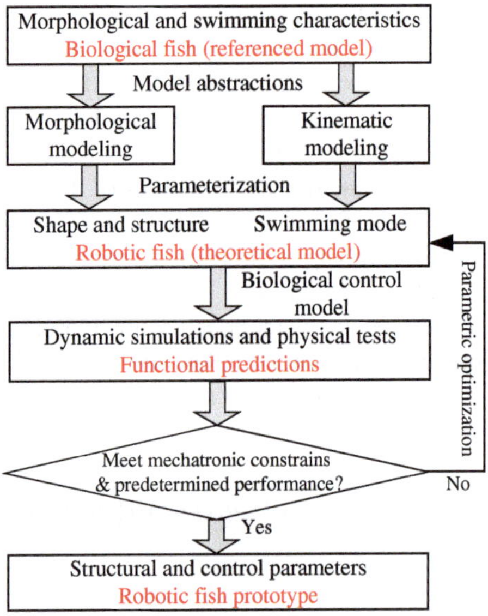

Fig. 6 Design framework of robotic fish in flowchart

simulations or empirical tests will provide a basis for functional predictions and evaluations.

Step 3. Adjust the hardware-based morphological parameters (e.g., R_l, N, and l_1: l_2:…:l_N) conditioned by both constraints of the mechatronic system and predetermined performance. That is, optimize characteristic parameters according to the simulated results from step 2, and return to step 2 to reevaluate until satisfactory performances (e.g., good forward speed and small turning diameter) are achieved. Note that the adjustment of kinematic parameters mainly depends on prototype testing, which can be performed in an online or offline fashion.

Step 4. Conduct analyses of geometric space configuration and mass distribution of the robot, to empirically ensure a balance of maneuverability and stability.

Step 5. Develop the robot prototype, as well as motion control. Since hydrodynamic coefficients are generally determined experimentally, dynamic simulations can only provide a rough estimate, and rigorous hydrodynamic experiments on the prototype are needed to achieve desired performances.

3.4 Development of Digital Fish Simulator

To facilitate the design and control of robotic fish, a fish-inspired simulator platform called Digital Fish Simulator (DFS) is developed. The desired functions of the DFS mainly include the following three aspects:

- Comparison between the multilink oscillations and the referenced body wave: The graphics of the moving multilink and the theoretical body wave can be comparatively displayed in one oscillation period, which provides an instructive guide to observe approximation degree.
- Dynamic status display: Through sequentially display the motion states of moving links in one oscillation period, one can visually observe oscillatory amplitude and swimming trajectory.
- Motion simulation: Motion animation embodies the most direct manifestation of swimming effect. A rendered fish body and a virtual swimming pool with obstacles, static or dynamic, will be devised. Motion control methods such as turning, obstacle avoidance, and other maneuvering controls can also be loaded and tested on site.

As a final step, the proposed fish-inspired steady and maneuvering swimming mechanisms, together with conceived control methods, are blended into the DFS via an Object-Oriented software engineering methodology (see Fig. 7). That is, we developed a custom-built executive routine to account for both theoretical and experimental factors. Specifically, basic input parameters involve fish body wave part and motor control part in the DFS. The former mainly includes link number (ranging from 2 to 10), discrete degree in one oscillation (ranging from 8 to 72), relative wave length (ranging from 0.3 to 1.0), phase difference (ranging from 75° to 90°), and link-length ratio. We remark that a strategy that is based on the

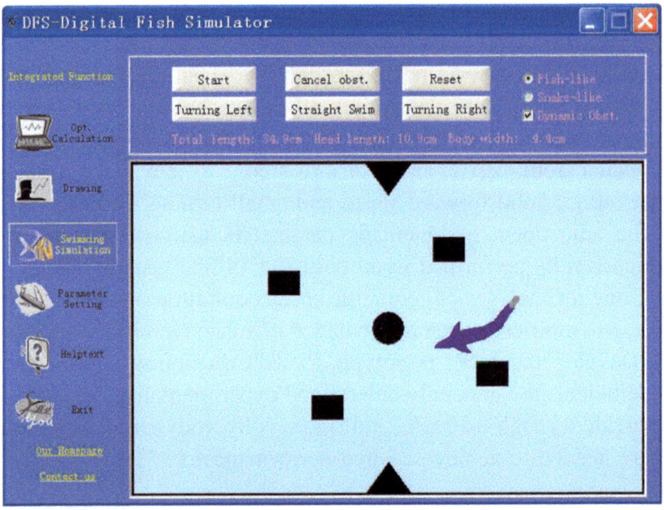

Fig. 7 Graphical user interface of the DFS

geometric optimization of relative link lengths to approximate a given smooth, spatial- and time-varying body wave curve for enhanced swimming performance has been added to the DFS. Please refer to [15] for more geometric optimization details. The latter comprises maximum rotary angle of used motors, left rotary limit (LA), right rotary limit (RA), minimum link length, etc. Through body wave fitting-based optimal calculation, the swimming data is automatically generated from the simulator, which can directly be fed into the fish robots for control purpose. The supposed data in a specified form, in turn can directly fed into the simulator for visual verification. Therefore, a two-way swimming data exchange interface is achieved, facilitating subsequent robotic development.

Besides steady swimming, fish in nature applies more maneuvering swimming. Typical maneuvering mechanisms include body-tail deflection, pectoral-fin stroke, stabilization control in pitch, fast-turn, backward swimming, and so on. Our current emphasis is restricted to the body-tail deflection-based maneuvers. By adding different deflections (i.e., dynamic offsets) to the straight, symmetric swimming gaits, various turns can be easily achieved. As investigated previously [16], the characteristic parameters associated with turning performance involve magnitude, position, and time of the deflections applied to the links. This turning control method now is employed to accomplish flexible obstacle negotiation with the aid of sensory perception. As demonstrated in Fig. 7, a controlled simulation environment with static and dynamic obstacles is created. Different obstacle avoidance approaches can then be loaded and tested in the DFS flexibly.

Furthermore, with the well-integrated DFS, many kinematics studies can be emulated and evaluated. For example, specific parameter combination $P = \{c_1, c_2, k_1, k_2\}$ for diversified swimming motions can be defined as a predominant kinematic feature. According to the obtained simulation results, P_1 = {0.2, 0, 2.0, 0}, P_2 = {0.05, 0.09, 0.5, 0.1} and P_3 = {0.35, 0, 3.0, 0} are representative of anguilliform, carangiform, and snake-like swimmers, as depicted by Fig. 8. It implies that carangiform, anguilliform, and snake-like swimmers share multi-

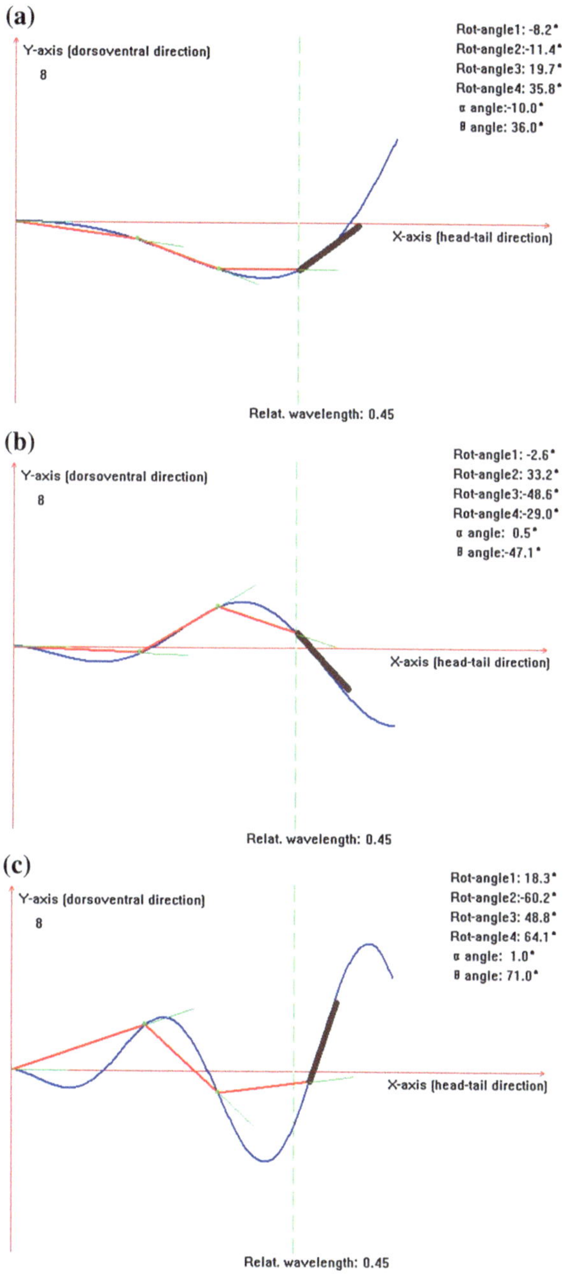

Fig. 8 Comparison of different swimmers in the DFS. **a** Anguilliform swimming, **b** carangiform swimming, **c** snake-like swimming

segment mechanical attribute despite their great difference in morphology. Further parameter optimization in conjunction with hydrodynamic analysis can be achieved and applied to the design of novel robotic fish.

4 CPG-Based Swimming Control

To mimic the BCF-type swimming, the link-based body wave fitting is intuitively employed to produce fishlike motions. For simplicity, the movements of joints (corresponding to control surfaces in function) are considered as oscillatory in a harmonic (sinusoid) fashion that can be described below.

$$\theta_i(t) = \overline{\theta_i} + A_i \sin(2\pi f_i t + \phi_i)\ (i = 1, 2, \ldots, N) \tag{6}$$

where $\theta_i(t)$ represents the angular position of the ith control surface at time t, $\bar{\theta}_i$ indicates the angular bias for asymmetric oscillations deviating from the median position, A_i stands for the oscillatory amplitude, f_i is the oscillatory frequency, and ϕ_i is the phase difference. As a response to the input data, the swimming speed can be mediated by the magnitude of amplitude (A_i) and frequency (f_i), while the direction is influenced by the bias (ϕ_i). The control commands are then wirelessly transmitted to the robotic fish in a remote control manner.

However, this body wave-based swimming control should be properly discretized and parameterized for a specific swimming gait. Moreover, the swimming stability of a desired gait and the smooth transition between two different gaits are hardly guaranteed. Inspired by the lamprey, an eel-like fish whose propulsion is governed by activity in its spinal neural network, some CPG-based models have been built to produce fishlike swimming. As summarized by Ijspeert, CPG-based control presents several attractive features including distributed control, the ability to deal with redundancies, fast control loops, and permitting the modulation of locomotion by simple control signals [22]. For the multi-joint robotic fish, as shown in Fig. 9, a complete CPG-centered control architecture is proposed. By coupling a set of nonlinear oscillators, a CPG network is built, involving one caudal CPG (O_1), three body CPGs (corresponding to O_2–O_4), two pectoral CPGs (O_5 and O_6), and one pelvic CPG (O_7). Each swimming gait is then encoded with a standardized "template" corresponding to a CPG parameter set. With the proper tuning of feature parameters, diverse swimming gaits in three dimensions result. Notice that, in the interests of simplicity, a weak coupling scheme is adopted, where all self-couplings are eliminated. Considering that the pectoral fins and the pelvic fin can be individually controlled, the couplings to O_5, O_6 and O_7 are not intrinsic but optional. The dynamics of the ith oscillator is described as follows [26]:

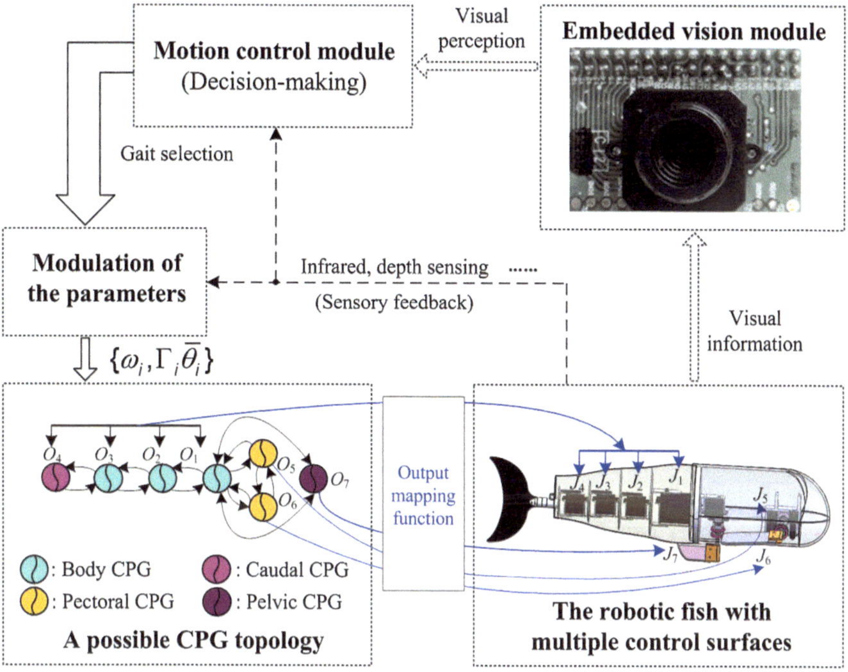

Fig. 9 Diagram of a complete CPG-centered control architecture

$$\begin{cases} \dot{u}_i = -\omega_i v_i + u_i\left(\Gamma_i^2 - u_i^2 - v_i^2\right) + \sum\limits_{j=1, j\neq i}^{n} a_{ij} v_j \\ \dot{v}_i = \omega_i u_i + v_i\left(\Gamma_i^2 - u_i^2 - v_i^2\right) + \sum\limits_{k=1, k\neq i}^{n} b_{ik} u_k \end{cases} \tag{7}$$

where n indicates the total number of nonlinear oscillators in the CPG network. The state variables u_i and v_i denote the membrane and adjustment potential, respectively. ω_i and Γ_i stand for the intrinsic oscillatory frequency and amplitude. $\sum_{j=1, j\neq i}^{n} a_{ij} v_j$ and $\sum_{k=1, k\neq i}^{n} b_{ik} u_k$ are the coupling relationships of the ith oscillator with other oscillators in the CPG network. a_{ij} and b_{ik} are the corresponding coupling weights. It should be remarked that this coupling is just an assumption of the CPG model, and that whether the coupling scheme relates to biological existence or optimality is currently unproved.

To translate the rhythmic output signal of the ith oscillator to the motor actuating signal, a mapping function $f_i(u_i)$ is defined as follows:

$$\theta_m^i = f_i(u_i) = \begin{cases} \gamma_i u_{\max}^i + u_b^i, & u_i \geq u_{\max}^i \\ \gamma_i u_i + u_b^i, & 0 < u_i < u_{\max}^i \\ 0, & u_i = 0 \end{cases} \tag{8}$$

Table 1 The relations between swimming gaits and control surfaces

Swimming gaits	Control surfaces	Control parameters
BCF forward	Body + caudal fin	ω_i, Γ_i (i = 1, 2, …, 4)
Hybrid forward	Body + caudal fin + PF	ω_i, Γ_i (i = 1, 2, …, 6)
MPF forward	PF	ω_i, Γ_i (i = 5, 6)
MPF backward	PF	ω_i, Γ_i,$\bar{\theta}_i$ (i = 5, 6)
BCF turning	Body + caudal fin	ω_i, Γ_i,$\bar{\theta}_i$ (i = 1, 2, …, 4)
MPF turning	PF and/or pelvic fin	ω_i, Γ_i,$\bar{\theta}_i$ (i = 5, 6, 7)
Hybrid turning	All involved	ω_i, Γ_i,$\bar{\theta}_i$ (i = 1, 2, …, 7)
Sideward swimming	Pelvic fin	ω_i, Γ_i,$\bar{\theta}_i$ (i = 7)
Diving/surfacing	PF or BCF + PF	ω_i, Γ_i,$\bar{\theta}_i$ (i = 1, 2, …, 6)
Braking	All involved	ω_i, Γ_i,$\bar{\theta}_i$ (i = 1, 2, …, 4)

where θ_m^i is the driving signal fed to the motor, γ_i is the transformation coefficient, $u_{\max}^i$ is the upper bound for the driving signal of the ith oscillator, and $u_{\rm b}^i$ is the potential bias when the ith control surface stays at its reference position. For simplicity, the reference positions of the body, and the caudal fin, and the pelvic fin are in the sagittal plane, and those of the pectoral fins are in the horizontal position.

Since the fishlike swimming is regarded as sinusoidal motions, the interactions of different control surfaces can be attributed to coupled factors among CPG units. According to our proposed CPG parameter determination method in accordance with the traveling body wave [27], a set of feature parameters for the fish CPG network can be sought. Thus, through the coordinated control of body CPG, caudal CPG, pectoral CPG, and pelvic CPG, a diversity of swimming gaits such as coordinated forward swimming, backward swimming, hybrid turning, sideward swimming, diving/surfacing, and braking, are implemented. Table 1 summarizes the relationships between the swimming gaits and the involved control surfaces. For all these gaits, the swimming speed can be altered by adjusting the frequency ω_i and/or the amplitude Γ_i. Meanwhile, the angular bias $\bar{\theta}_i$ is used for turning maneuvers and 3-D swimming. Typically, $15° \le \overline{\theta_1} = \overline{\theta_2} \le 45°$ is set for the BCF turning; $\overline{\theta_5} = \overline{\theta_6} = 180°$ for the MPF backward; $0 < \overline{\theta_5} = \overline{\theta_6} < 90°$ for the PF-based diving; and $-90° < \overline{\theta_5} = \overline{\theta_6} < 0$ for the PF-based surfacing. It is critical to find appropriate CPG patterns that are closely associated with viable swimming gaits. In this study, a dynamic model of robotic fish swimming using Kane's method is first developed to guide the primary parameter search. A particle swarm optimization algorithm is further utilized to yield a relatively optimal parameter set [28]. The main control parameters that are empirically determined with the help of dynamic simulations are listed in Table 2. Note that an assumption is made in simplifying the CPG couplings that the descending weights are identical, implying that the ith CPG receives the influence from the (i–1)th one. Namely, $a_{i,i-1} = a_1$ and $b_{i,i-1} = b_1$, where i = 2, 3, …, n; so do the ascending coupling weights, holding

Table 2 CPG parameter values applied to robotic fish

Parameters	Value
ω	$\omega_i \in [0, 52)$
Γ	$\Gamma_i \in [0, 60]$
a	$a_1 = 0.3$, $a_2 = -0.2$
b	$b_1 = 0.2$, $b_2 = -0.1$
γ_i	0.0
u_b^i	30

$a_{i,i+1} = a_2$ and $b_{i,i+1} = b_2$. Note also that vision, infrared and depth sensing can be included in the control loop, serving as sensory feedback for online modulation.

5 Experiments and Results

5.1 Development of Robotic Fish

To evaluate the conceived design ideas and control framework, we try to build different physical robots serving as a repeatable testbed. A conceptual design of robotic fish with multiple control surfaces entirely consists of several elements: a rigid head, a multilink soft body, a peduncle and caudal fin, a pair of pectoral fins and a pelvic fin. Notice that the pelvic fin is used instead of the anal fin for sideward maneuverability and for ease of installation and waterproof protection. More specifically, the multilink soft body consists of four servomotors connected in series with aluminum links, providing four independent joints (J_1–J_4) around the yaw axis. The outside of the multilink soft body is wrapped by a compliant, crinkled rubber tube functioning as fish skin. It is worthwhile to note that the output torque of servomotors should increase from tail to head, so the driving motor for the first link has a bigger torque. Considering the caudal fin, in its final lash, may contribute as much as 40 % of the forward thrust in carangiform and thunniform swimming; a crescent-shaped caudal fin is connected to the last link via a slim peduncle made of polyvinyl chloride. The caudal fin is made of partly compliant material, polyurethane, for generating more thrust.

Concerning the accessory fins, as illustrated in Fig. 9, the pectoral fins are two oscillating foils laterally mounted at the rear lower position of the rigid shell, providing two independent joints (J_5 and J_6) around the pitch axis. Note that each pectoral fin, capable of 0°–360° rotation via a set of custom-built gears, can be controlled synchronously or independently. The pelvic fin located at the posterior bottom of the fish shell provides a single DOF joint (J_7) around the yaw axis. The imported pelvic fin plays a role like a rudder on a ship. For the purpose of waterproof protection, the rotations of the servomotors in these accessory fins are transmitted to the outside through a dynamic waterproof sealing. Currently, the oscillatory angles of J_1–J_4 are restricted to ±60° that of J_7 is limited to ±90°,

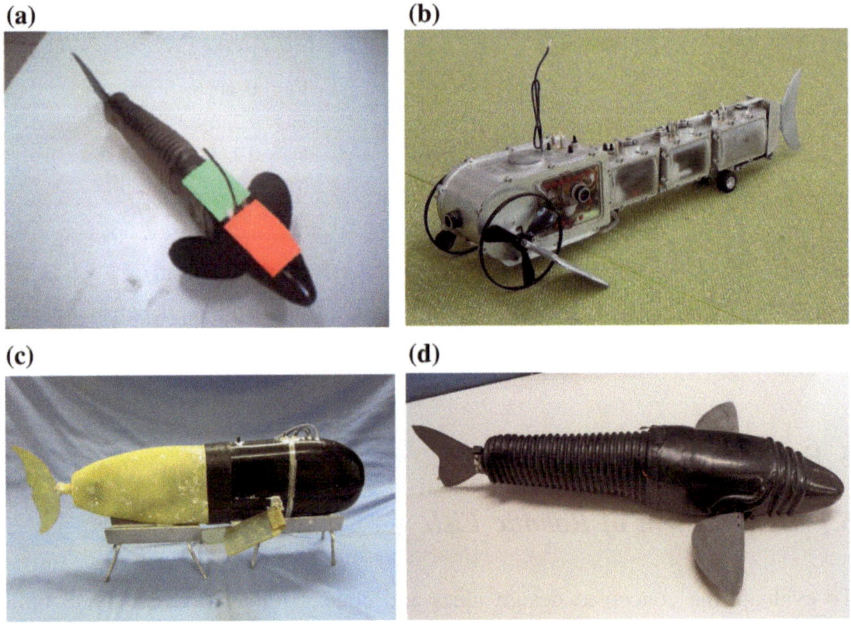

Fig. 10 Prototypes of different robotic fishes. **a** A three-joint robotic fish for the Robofish Water Polo (~380 mm × 40 mm × 78 mm, $L \times W \times H$). **b** A three-link amphibious robot (~700 mm × 320 mm × 150 mm). **c** A multi-fin robotic fish with embedded vision (~680 mm × 260 mm × 220 mm). **d** An *Esox lucius* inspired multimodal robotic fish (~614 mm × 83 mm × 81 mm)

whereas those of J_5 and J_6 are expanded to ±180° allowing for both forward and backward swimming. Such an adequately large oscillation range permits large-amplitude and fast-swimming modes.

To date, as shown in Fig. 10, a series of robotic fish prototypes have been developed in our laboratory. These robots share the same multilink structure but serve different purposes. Through extensive simulations and experiments, robotic fishes have achieved vivid swimming and even performed some simple mobile sensing tasks. We remark that two control methods: body wave-based control and CPG-based control, have been adopted in our robotic fish capable of multimodal swimming. For the former, the swimming control data is derived from the DFS. But the latter CPG controller is solved online with an embedded system.

5.2 Experimental Results

In order to evaluate the proposed design and control methods for multi-joint robotic fish, extensive experiments on multi-fin robotic fish (see Fig. 10c) were conducted in an indoor swim tank with clear water. During experiments, unless otherwise specified, the data points and error bars shown in subsequent figures were the averages and standard deviations of three runs. Currently, the robotic fish have successfully achieved autonomous obstacle avoidance and 3-D swimming. Available propulsive modes include the BCF-type swimming, the MPF-type swimming, and their combination. More quantitative results will be provided below.

The first experiment concerned the stability comparison of the CPG-based control method against the boy wave-based control method in the BCF-type swimming. Figure 11 shows the comparative oscillatory signals of J_1–J_4 calculated from (2) and (7), respectively. The amplitudes of two signals increase monotonically from head to tail. It is worth noting that the caudal fin functions as a principal propulsor in carangiform swimming. Since the caudal fin is rigidly attached to J_1 via the peduncle, the driving signal of J_1 should be carefully devised for highlighting the propulsive effect of the caudal fin. To this end, a minor phase lag is maintained between J_1 and J_2 (see Fig. 11b). With the oscillatory signals as the joint inputs of the robotic fish, Fig. 12 plots the measured pitch, yaw, and roll angles in the world reference frame. During testing, the oscillatory amplitude suddenly increased at 7 s, whereas the amplitude climbed steadily after 12 s. In contrast with the fish body wave method, the CPG-based control method obtained increased pitch and roll stability, yet without loss of yaw stability. Thanks to the intrinsic limit cycle characteristic of the CPGs that is insusceptible to small perturbations, the CPG-based control method provides the possibility to abruptly vary control parameters while ensuring smooth variations of swimming gaits. Just benefited by this enhanced swimming stability derived from the CPGs, for the embedded vision guided robotic fish, image dithering accompanied by the low-amplitude oscillation of the fish head could be effectively weakened so that steady image capture and visual tracking were available. Note that the CPG-based control method was applied in the following tests.

A second experiment was performed to evaluate the propulsive speeds among different swimming modes. Fish modulate their speeds depending on a hybrid of oscillatory frequency and amplitude of the involved control surfaces. Similar methods can be adopted in the speed control of the robotic fish performing the BCF, the PF, or the BCF + PF swimming. As a demonstration, four swimming cases with various amplitudes were examined. Notice that length-specific speeds expressed as body lengths per second (BL/s) are applied in order to be comparable with other fish robots. The following relationship was also maintained during testing: Amplitude_III = k_a Amplitude_II = k_a^2 Amplitude_I, where $k_a = 1.33$. Figure 13 shows the obtained results in the BCF-type swimming: the forward speed directly increased with the oscillatory amplitude and frequency till the actuators peaked their speed limits. Compared with the BCF swimming termed "Amplitude I", speeds were increased by 22.83 ± 9.10 % and 26.58 ± 7.73 % in "Amplitude II" and "Amplitude III", respectively. Remarkably, a combination of the BCF-type and the PF-type termed "Amplitude III + PF" in Fig. 13 attained an enhanced speed over

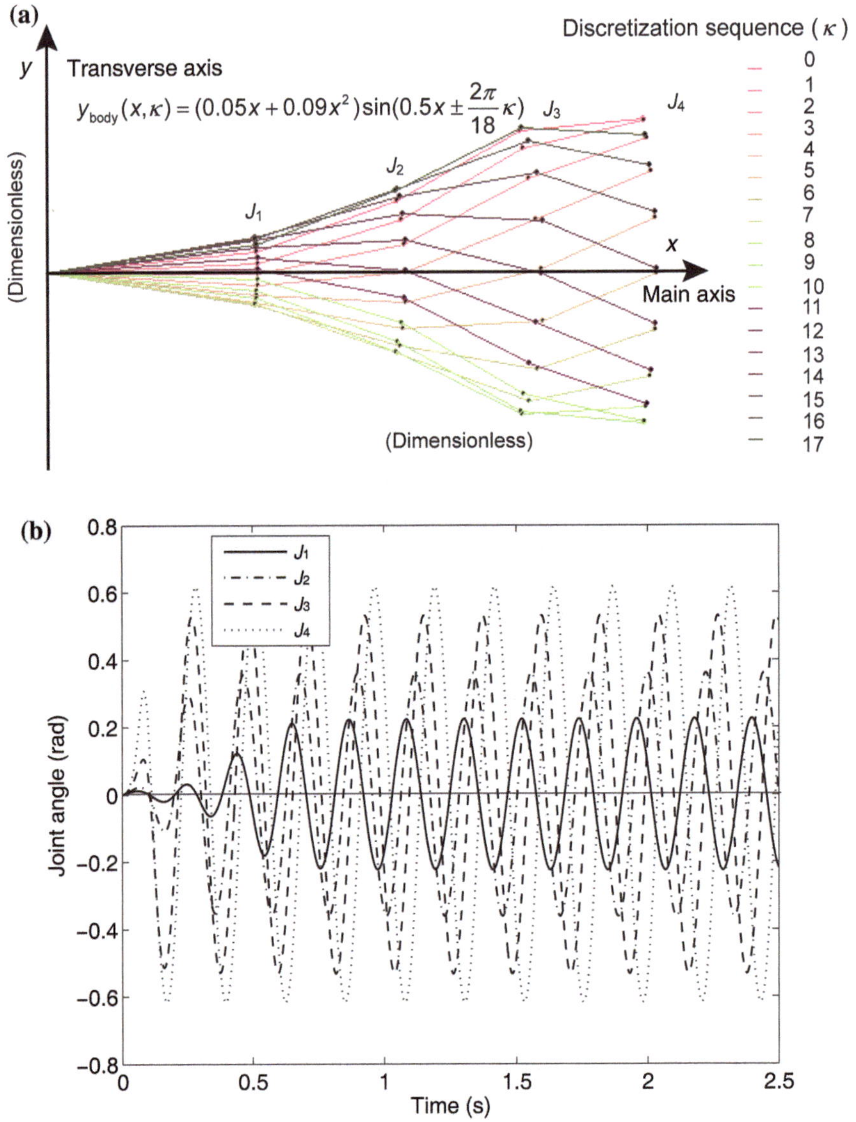

Fig. 11 Comparative oscillatory signals for the BCF-type swimming. **a** Oscillatory sequence of J_1–J_4 in the body wave-based control. **b** Oscillatory angles of J_1–J_4 in the CPG-based control

the entire range of oscillatory frequency. It is estimated that a speed increase of 33.34 ± 11.18 % was obtained in the "Amplitude III + PF" by contrast with the "Amplitude I", while a speed profit of 5.22 ± 2.85 % was even earned relative to the "Amplitude III". The induced speed increase in the "Amplitude III + PF" implies the cooperative capability of control surfaces in achieving a higher speed. The

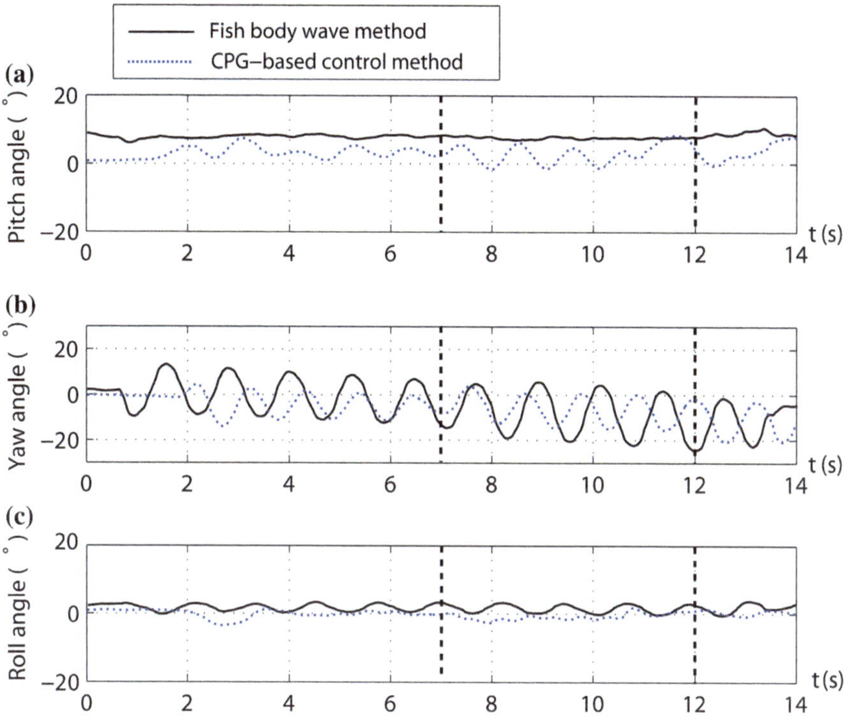

Fig. 12 Stability comparison of the CPG-based control method against the fish body wave method in the BCF-type swimming

Fig. 13 Speed test of the BCF swimming by changing the oscillatory frequency and amplitude

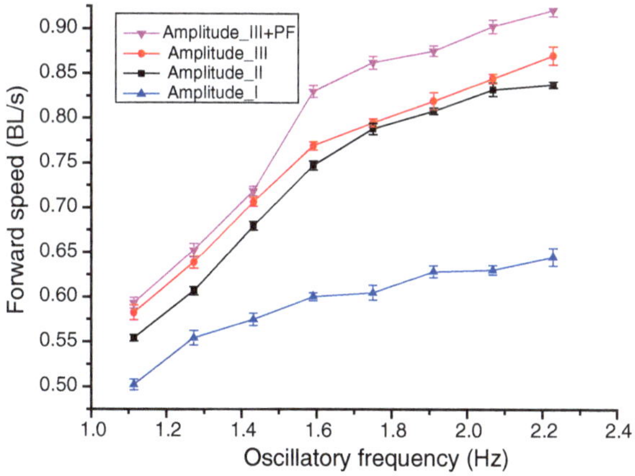

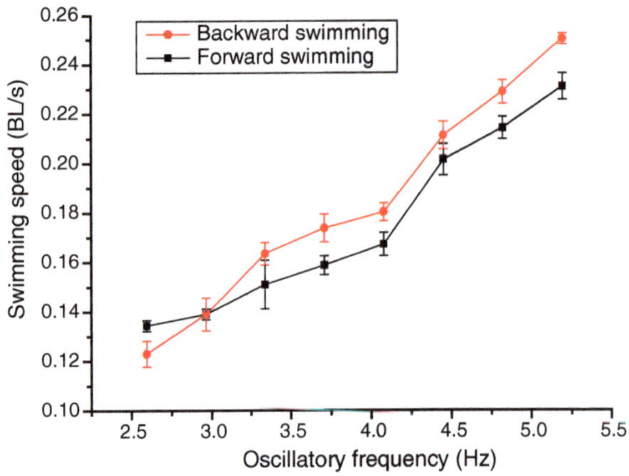

Fig. 14 Speed test of the PF swimming by changing the oscillatory frequency

maximum swimming speed of 0.71 m/s (equivalent to 1.04 BL/s) was obtained at a higher frequency of 3.5 Hz.

At the same time, the robotic fish could swim forward or backward by flapping the pectoral fins reciprocally and fleetingly, i.e., perform labriform swimming. As shown in Fig. 14, the speed of the PF-type swimming also rose over the plotted frequency region, but with a relatively low value compared to the BCF-type swimming. This suggests the body shape of the developed robotic fish is more appropriate to carangiform swimming rather than labriform swimming as far as speed is concerned. Notice also that some speed difference (positive or negative) could be detected in the same oscillatory frequency for the PF-type forward and backward swimming, but the difference was not so significant. Despite the same oscillatory frequencies and amplitudes being adopted, entirely consistent results were hard to obtain primarily due to differences of anterior–posterior body shape.

The third experiment concerned the exploration of the turning maneuvers. The dependence of the obtained turning radius and angular speed on the angular bias is summarized in Table 3, where the angular bias was superposed on the rhythmically oscillated body joints (J_1 and J_2) to induce turns. According to the experimental data, increasing the angular bias will shorten the turning radius (represented by body length independent of body size, BL for short) and achieve a larger angular speed (represented by rad/s) simultaneously. Statistical analysis showed that, the turning radius was decreased by 17.13 ± 4.95 % whereas the angular speed was

Table 3 Turning performance characterized by angular bias in different gaits at 2 Hz

	Angular bias (°)	15	30	45
BCF	Turning radius (BL)	0.84 ± 0.03	0.62 ± 0.03	0.23 ± 0.04
	Angular speed (rad/s)	0.65 ± 0.04	0.87 ± 0.02	1.58 ± 0.07
BCF + PF	Turning radius (BL)	0.74 ± 0.03	0.51 ± 0.03	0.18 ± 0.02
	Angular speed (rad/s)	0.87 ± 0.04	0.97 ± 0.04	1.78 ± 0.05

increased by 19.33 ± 12.58 % in the coordinated BCF + PF turns (where the left and right pectoral fins oscillate out of phase), compared with the BCF turns. Overall, the coordinated BCF + PF turning mode performed better than the BCF turning mode.

In another braking testing with an initial speed of around 0.4 m/s (equivalent to 0.59 BL), the robotic fish stopped in 2.3 s with a cooperative operation of the pectoral fins, the dorsal fin, and the pelvic fin (typically holding the fins vertical to the main axis plane), whereas a longer 5.2 s were needed to halt the fish without any braking measure. This experiment partly highlighted the need for coordinated control surfaces in rapid brake.

5.3 Discussion

As a specific biomimetics branch, robotic fish has grown rapidly in recent years. As yet, the obtained swimming performance is inferior to that of the biological counterpart. There are a variety of reasons for this discrepancy. First, although biological system is a useful source of inspiration, how to transfer the advantages of biological fish to robotic fish still remains unsolved. Second, how to integrate hydrodynamic advantages of free swimming into a fishlike mechatronic system needs to be synthetically tackled. Third, using soft robotics-related single joint or multi-joint to better replicate fishlike swimming necessitates innovative mechanism and actuator design [29]. Fourth, CPGs are a suitable paradigm to solve the problem of locomotion control of robots with multiple joints or DOFs. However, in the absence of higher control commands or sensory feedback, the CPG controller will be unable to adapt itself to uncertain and ever-changing environments. So how to include rich sensory feedback in the control loop of the CPGs so as to realize reactive locomotor patterns is a potential topic for future research. In addition, the applicability of robotic fish as a mobile sensing platform or a specific vehicle should be better demonstrated via common marine tasks.

6 Conclusions and Future Work

This chapter has described an overall design and control for fish-inspired swimming simulation and robotic implementation. Within the systematic framework taking account of hydrodynamic characteristics, mechatronic constraints, and biological control model, the detailed design procedures for multi-joint robotic fish have been summed up. Both body wave-based control and CPG-based control are proposed and compared on a robotic fish with multiple control surfaces, in terms of oscillatory signals and swimming stability. In particular, a bidirectional swimming data exchange has been well integrated into the DFS, enabling fish swimming data generation and testing. Accompanying with this software platform, various robotic fish and applicable control framework have been developed and tested. The obtained results have demonstrated relatively satisfactory swimming performance.

In the future work, much more effort should be devoted to breaking through the performance bottlenecks and tuning characteristic parameters in a relatively optimum fashion. To this end, we should continue to learn from fish and advance comprehensive hydrodynamic design and control for high-performance aquatic robots.

Acknowledgments This work is supported by the National Natural Science Foundation of China (nos. 61375102, 61333016, 61421004), the Beijing Natural Science Foundation (no. 3141002), and the Project-Based Personnel Exchange Program with CSC and DAAD (2013).

References

1. Bar-Cohen Y (ed) (2005) Biomimetics—biologically inspired technologies. CRC Press, Boca Raton
2. George A (ed) (2011) Advances in biomimetics. InTech, Rijeka
3. Lepora NF, Verschure P, Prescott TJ (2013) The state of the art in biomimetics. Bioinspir Biomim 8(1), 013001 (11 pp)
4. Triantafyllou MS, Triantafyllou GS (1995) An efficient swimming machine. Sci Amer 272 (3):64–70
5. Hu H, Low KH (2006) Guest editorial: special issue on biologically inspired robotic fish. Int J Autom Comput 3(4):323–324
6. Roper DT, Sharma S, Sutton R, Culverhouse P (2011) A review of developments towards biologically inspired propulsion systems for autonomous underwater vehicles. Proc IMechE Part M: J Eng Marit Environ 255(2):77–96
7. Liang J, Wang T, Wen L (2011) Development of a two-joint robotic fish for real-world exploration. J Field Robot 28(1):70–79
8. Tan X (2011) Autonomous robotic fish as mobile sensor platforms: challenges and potential solutions. Mar Technol Soc J 45(4):31–40
9. Sfakiotakis M, Lane DM, Davies JBC (1999) Review of fish swimming modes for aquatic locomotion. IEEE J Ocean Eng 24(2):237–252
10. Lauder GV, Drucker EG (2004) Morphology and experimental hydrodynamics of fish fin control surfaces. IEEE J Ocean Eng 29(3):556–571
11. Helfman GF, Collette BB, Facey DE, Bowen BW (2009) The diversity of fishes: biology, evolution, and ecology, 2nd edn. Wiley-Blackwell, New York
12. Lachat D, Crespi A, Ijspeert AJ (2006) Boxybot: a swimming and crawling fish robot controlled by a central pattern generator. In: Proceedings first IEEE/RAS-EMBS international conference biomedicine robot biomechatron, Pisa, Italy, pp 643–648
13. Kodati P, Hinkle J, Winn A, Deng X (2008) Microautonomous robotic ostraciiform (MARCO): hydrodynamics, design, and fabrication. IEEE Trans Robot 24(1):105–117
14. Yu J, Tan M, Wang S, Chen E (2004) Development of a biomimetic robotic fish and its control algorithm. IEEE Trans Syst Man Cybern B Cybern 34(4):1798–1810
15. Yu J, Wang L, Tan M (2007) Geometric optimization of relative link lengths for biomimetic robotic fish. IEEE Trans Robot 23(2):382–386
16. Yu J, Liu L, Wang L, Tan M, Xu D (2008) Turning control of a multilink biomimetic robotic fish. IEEE Trans Robot 24(1):201–206
17. Yu J, Wang K, Tan M, Zhang J (2014) Design and control an embedded vision guided robotic fish with multiple control surfaces. Scientific World J 631296:13

18. McIsaac KA, Ostrowski JP (2003) Motion planning for anguilliform locomotion. IEEE Trans Robot Autom 19(4):637–651
19. Lauder GV, Madden PGA (2007) Fish locomotion: kinematics and hydrodynamics of flexible foil-like fins. Exp Fluids 43(5):641–653
20. Schultz K (2004) Ken Schultz's field guide to freshwater fish. Wiley, New York
21. Grillner S, Kozlov A, Dario P, Stefanini C, Menciassi A, Lansner A, Kotaleski JH (2007) Modeling a vertebrate motor system: pattern generation, steering and control of body orientation. Prog Brain Res 165:221–234
22. Ijspeert AJ (2008) Central pattern generators for locomotion control in animals and robots: a review. Neural Netw 21(4):642–653
23. Yu J, Tan M, Chen J, Zhang J (2014) A survey on CPG-inspired control models and system implementation. IEEE Trans Neural Netw Learn Syst 25(3):441–456
24. Videler JJ, Hess F (1984) Fast continuous swimming of two pelagic predators, saithe (Pollachiusvirens) and mackerel (Scomberscombrus): a kinematic analysis. J Exp Biol 109 (1):209–228
25. Yu J, Liu L, Tan M (2008) Three-dimensional dynamic modelling of robotic fish: simulations and experiments. T I Meas Control 30(3–4):239–258
26. Yu J, Wang M, Tan M, Zhang J (2011) Three-dimensional swimming. IEEE Robot Autom Mag 18(4):47–58
27. Wang M, Yu J, Tan M (2008) Parameter design for a central pattern generator based locomotion controller. In: Proceedings 1st international conference intelligent robotic applications (LNAI 5314), Wuhan, China, Part I, pp 352–361
28. Wu Z, Yu J, Tan M (2012) CPG parameter search for a biomimetic robotic fish based on particle swarm optimization. In: Proceedings IEEE international conference robot biomim, Guangzhou, China, pp 563–568
29. Chu W-S, Lee K-T, Song S-H, Han M-W, Lee J-Y, Kim H-S, Kim M-S, Park Y-J, Cho K-J, Ahn S-H (2012) Review of biomimetic underwater robots using smart actuators. Int J Precis Eng Manuf 13(7):1281–1292

Research on Robotic Fish Propelled by Oscillating Pectoral Fins

Yueri Cai, Shusheng Bi and Hongwei Ma

Abstract Swimming mode utilizing oscillating pectoral fins possess characteristics of high efficiency, high stability, high maneuverability, and higher swimming velocity compared with other swimming modes of fish. Broad application prospect and important research value are shown by robotic fish propelled by oscillating pectoral fins. The research on bionic fish of this kind has become a hot pot. In this chapter, structure characteristics and pectoral fin motion deformation during oscillation of the nature sample, cownose ray, are analyzed. Main structure parameters of the sample cownose ray and simple mathematical model of oscillating movement of the pectoral fin are obtained. Finally, a robotic fish propelled by oscillating pectoral foils featuring with organic combination of form bionic and function bionic is developed.

Keywords Biomimetics · Robotic fish · Pectoral fin · Oscillating propulsion · Cownose ray

1 Introduction

The ocean covers most areas of the planet and possesses abundant mineral resources and species resources. Plenty of attention has been paid on the development and utilization of ocean resources. Underwater robots attract much attention because of their great potential value in marine military strategy, marine mineral exploration, marine biological research, and other civilian areas.

Y. Cai · S. Bi (✉) · H. Ma
Robotics Institute, Beihang University, Beijing, People's Republic of China
e-mail: ssbi@buaa.edu.cn

Y. Cai
e-mail: caiyueri@me.buaa.edu.cn

H. Ma
e-mail: mahongwei@buaa.edu.cn

© Springer-Verlag Berlin Heidelberg 2015

R. Du et al. (eds.), *Robot Fish*, Springer Tracts in Mechanical Engineering,
DOI 10.1007/978-3-662-46870-8_5

As the key technology of underwater robots, underwater propulsion technology has been the focus of research institutions around the world. Most of the traditional underwater robots with rudder steering control are driven by propeller. The larger thrust and faster speed can be obtained. But there are also many disadvantages. For example, the effect of fluid disturbance is significant. The maneuverability of underwater robot is poor at low speed. It is difficult to further improve the propulsion efficiency. Recently, the research attention has been shifted from cruising speed to autonomy, stability, and concealment in the complex environment [1, 2]. Given the functional limitations of traditional underwater robots, the exploration of new underwater propulsion is badly needed.

With millions of years' nature selection and evolution, aquatic animals have developed prefect body shapes and efficient modes for underwater locomotion [3–7], which inspires researchers to design different novel mechanical structures, especially the ability that fish obtain thrust by utilizing fluid. Fish owns advantages in speed, efficiency, and maneuverability, which make fish ideal objects in the studies of improving propulsion performance of the existing underwater vehicles and the development of new underwater vehicle with high efficiency, high maneuverability, and less environmental disturbance [8–12].

Fish locomotion can be generally classified into two categories based on their physiological structures and propulsion mechanisms: Body and/or Caudal Fins (BCF) mode and Median and/or Paired Fins (MPF) mode [13–16]. The BCF mode performs better in acceleration and velocity, while the MPF mode has better maneuverability and can achieve more stable movement with low noise [15]. In the future, the robotic fish belonging to the MPF mode can be expected to be used in underwater detection, water monitoring, investigation, etc., especially in complex water conditions.

2 Literature Review

This part contains two parts: the research on oscillating pectoral fins and the development of bionic robotic fish propelled by pectoral fins. For the researches on oscillating pectoral fins, the innovation points are focused on the structure and new materials applications. The research object is to explore the pectoral fin propulsion mechanism by the construction and experimental study of various bionic pectoral fins. The design optimizations of bionic pectoral fins are carried out. Finally, the feasibility of bionic fish based on nature sample's structure and motion features is verified.

In the aspect of the construction of bionic fish, the robotic prototypes developed have achieved the basic swimming functions. The research emphasis is still on the measurements of the related function parameters. Consequently, the simple remote control mode is generally adopted. Several prototypes are installed sensors in order to realize the autonomous swimming. The swimming performance of these prototypes should further be improved.

2.1 Research State of Flexible Oscillating Pectoral Fin Units

Fish belonging to MPF mode use the paired flexible pectoral fins as the main thrust source for their driving. The constructions and tests of oscillating pectoral fins are foundation for the research on bionic fish. Researchers from several countries have carried out related studies on the construction and propulsion performance of oscillating pectoral fins. The research achievements can be references for the development of robotic fish.

In 2002, Renee Boileau from University of British Columbia built a two-dimensional oscillating pectoral fin test platform to explore the applications of Rajiform motion mode on the underwater propulsion. The platform was built by the structure and motion parameters of *Gymnura micrura*. The SMA (Shape Memory Alloy) was used as the actuator [17]. From Fig. 1, the pectoral fin made of polyester film consisted of 11 SMA rods skeleton. The length and width are 460 and 420 mm, respectively. The SMA rods were driven according to a certain phase difference in order to form kinematic wave along the pectoral fin. But no obvious thrust was obtained in the test. The reason is that the SMA rods are not connected with the pectoral fin reasonably. The bionic pectoral fin could not achieve desired deformation as *G. micrura*.

In 2005, Clark from Princeton University extracted the main structural features of cownose ray's pectoral fin. The experimental apparatus was developed to study the thrust and vortex structures generated by oscillating pectoral fins, as shown in Fig. 2 [18]. The tridimensional bionic pectoral fin was fabricated by the integral pouring. Four rigid rods distributed along the chordwise direction were driven by crank–rocker mechanism with a certain phase difference. The kinematic wave was transmitted along the bionic pectoral fin. The experiments for thrust tests were

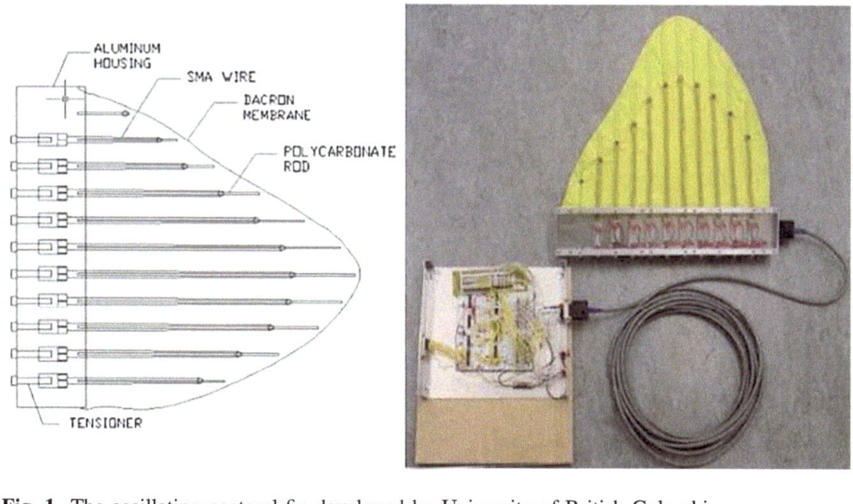

Fig. 1 The oscillating pectoral fin developed by University of British Columbia

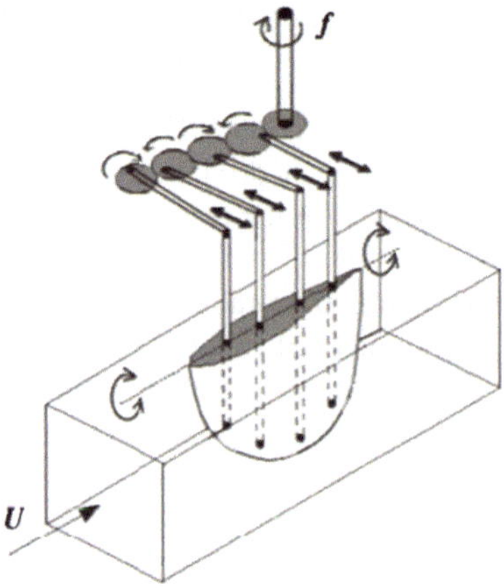

Fig. 2 The experimental apparatus of bionic pectoral fin developed by Princeton University

carried out in water tank. The flow velocity was 0.11 m/s and corresponding Reynolds number (based on the average chord length of bionic pectoral fin) was 11,400. The thrust coefficient increased with the rise of the Strouhal number. The maximum propulsion efficiency tested was about 40 %. The way of vortex observation played an important role in the study on the bionic pectoral fin. The feasibility of bionic propulsion utilizing the oscillating pectoral fin mode was verified by the test results.

In 2006, Brower from TUFTS University built an integrated pectoral fin structure according to the structure and motion characteristics of manta ray's pectoral fin [19]. The metal rigid skeleton was embedded into the bionic pectoral fin. The pectoral fin was made of PVC plastic and driven by pneumatic components, as shown in Fig. 3. As a result of the high density and larger stiffness of PVC material, the deformation generated by pneumatic mechanism was so little that the

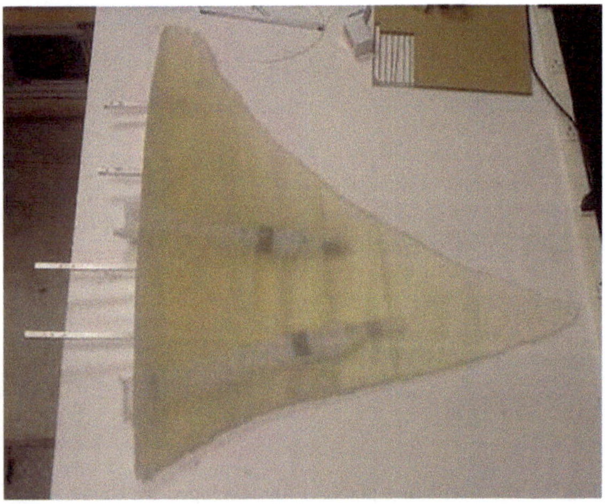

Fig. 3 The bionic pectoral fin developed by Tufts University

corresponding thrust did not drive the bionic pectoral fin. However, the way of making the integrated pectoral fin could be a reference for the construction of bionic pectoral fin.

In the same year, Moored from University of Virginia designed the bionic pectoral fin based on tensegrity structure. Then the structure of pectoral fin was optimized. The simulation results showed that the bionic pectoral fin utilizing in-tensegrity structure could achieve similar deformation as manta ray's pectoral rays [20]. The biomimetic pectoral fin was fabricated. The servo was chosen as actuators, as shown in Fig. 4.

Since 2006, researchers from Robotics Institute of Beihang University used flexible rubber sheet as fin surface material. The carbon fiber rods were chosen as the leading edge of pectoral fin. The bionic pectoral fin was fabricated. The hydrostatic thrust tests were carried out through the designed platform, as shown in Fig. 5 [21, 22]. The results showed that the thrust can be generated by the pectoral fin utilizing the sinusoidal motion. The thrust increased with the rise of oscillating frequency and amplitude. The maximum thrust was 0.85 N when oscillating frequency is 0.8 Hz and the oscillating amplitude is 40°. A biomimetic fish utilizing this kind of bionic fin was developed, and the swimming experiments proved that the bionic pectoral fin could generate sufficient thrust for locomotion of the robotic fish.

In 2009, Tomokazu Nakamura from Kitakyushu University built the test device of bionic pectoral fin, as shown in Fig. 6. The fluid simulation for flexible oscillating pectoral was carried out through the finite element method [23]. Hydrostatic thrust experimental results showed that the pectoral fin with width of 0.24 m could generate thrust with the maximum value of 0.38 N and lift with the maximum value of 20 N. The twisting torque with the value of 1.6–1.7 Nm was generated through differential motion of paired pectoral fin. So the drive mode with pectoral fin was proved to have high maneuverability.

Fig. 4 The bionic pectoral fin based on tensegrity structure

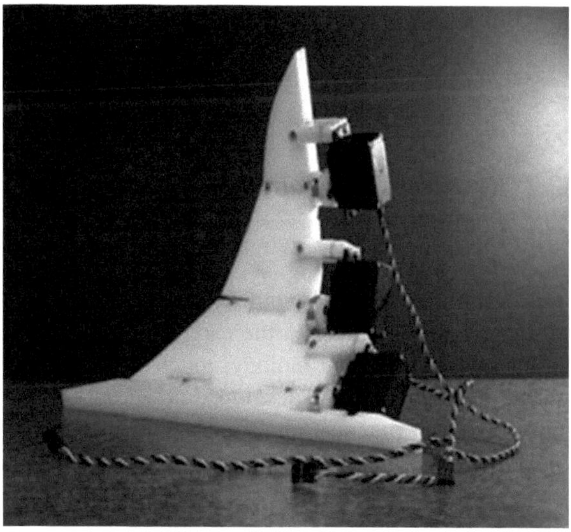

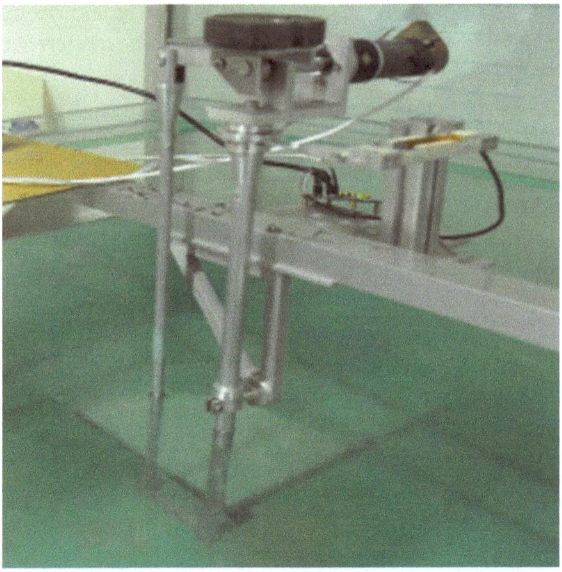

Fig. 5 The bionic pectoral fin imitating manta ray

Fig. 6 The test platform of the bionic pectoral fin imitating manta ray

2.2 Research State of Robotic Fish Propelled by Oscillating Pectoral Fins

Pectoral fins were usually utilized to balance body, control direction, and improve turning maneuverability in the early research stage of bionic fish. With the development of bionics and biology, much more attention had been paid to the MPF mode. Most robotic fish developed could only realize swimming function, but not equipped with practical application ability. According to the degree of structural flexibility, the bionic fish developed could be divided into the following three categories:

(1) The rigid body combined with simplified flexible pectoral fins. Such type of bionic fish simplified the body structure of nature sample to a great extent, especially the structure of pectoral fin. The motion features were only retained. The complex three-dimensional structure of pectoral fin was ignored in order to achieve the most important goal, the underwater propulsion. The middle body of prototype was rigid, and the fin surface was usually designed by membrane or platy flexible material. The leading edge of pectoral fin was usually made of the rigid or flexible material and it played a driving role.

Typical prototype was developed by Japanese researchers in 2004. This prototype was the first robotic fish propelled by oscillating pectoral fin, as shown in Fig. 7 [24]. The flexible material was used in fabricating the pectoral fins. The fin bones were made of hardened steel belt. The rigid dual four-bar linkage drove the leading edge of the pectoral fin and the oscillating motion of the pectoral fin was obtained. The length of the prototype was 0.65 m, the wingspan was 0.5 m, and the weight was 0.64 kg. The motion direction of the prototype was controlled by the tail rudder. The diving depth was about 1.5 m, the duration was half an hour, and its leakproofness was good. The maximum swimming velocity was about 0.6 m/s. The right yaw phenomenon in a straight line movement was shown because the thrust generated on each side of the pectoral fins were slightly different.

From 2005, the researchers from Robotics Institute, Beihang University have been devoting their efforts to the research on the robotic fish propelled by oscillating pectoral fins. Cownose ray was chosen as the bionic object. Based on the biology researches and movement observation of cownose ray, five generations of robotic fish named Robo-ray I-V were developed from the simple functional bionic to combination of the functional bionic and morphological bionic, as shown in Fig. 8.

The first generation of Robo-ray I is shown in Fig. 8a. The length of the prototype was 0.5 m, the wingspan was 0.6 m, and the displacement was 3.4 kg. The designed operation depth was 3 m, and the maximum swimming speed was 0.7 m/s. The material of the middle flat body was fiberglass. The material of pectoral fin's leading edge was carbon fiber rod. It was bonded with flexible silicone rubber sheet to constitute the oscillating pectoral fin. The two pectoral fins were driven by only one motor. The tail rudder was used to achieve heaving and turning motion [25].

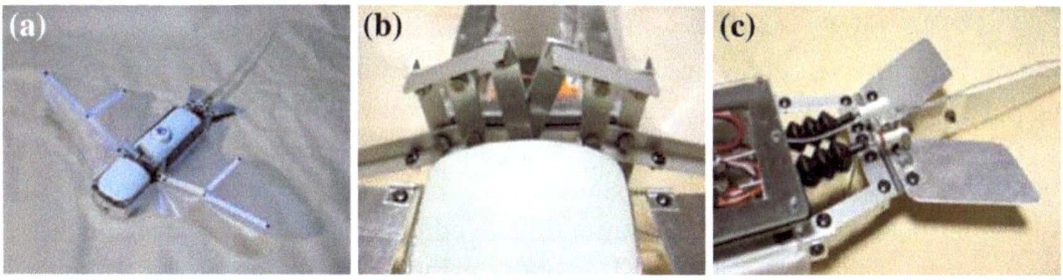

Fig. 7 The first robotic fish propelled by oscillating pectoral fins. **a** Prototype, **b** drive mechanism, **c** tail unit

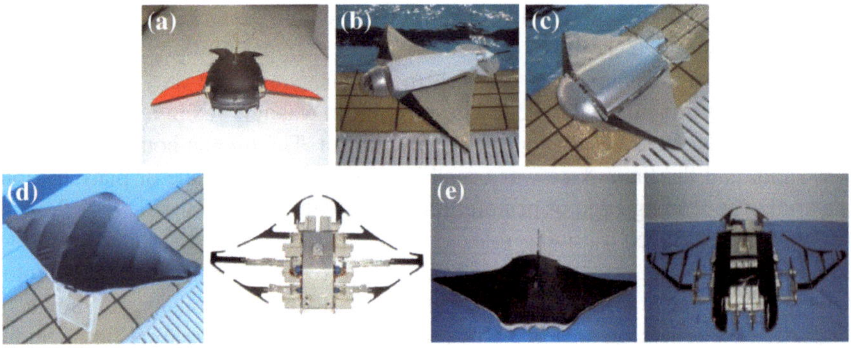

Fig. 8 The bionic fish imitating cownose ray developed by Beihang University. **a** Robo-ray I, **b** Robo-ray II, **c** Robo-ray III, **d** Robo-ray IV, **e** Robo-ray V

The second generation of Robo-ray II is shown in Fig. 8b. The length of the prototype was 0.7 m, the wingspan was 0.98 m, and the displacement was 7 kg. The designed operation depth was 8 m and the maximum swimming speed was 0.9 m/s. The structure of the pectoral fin was similar to Robo-ray I. However, the leading edge of the pectoral fin was designed more reasonably. The fin surface was more similar to the projective plane of cownose ray. Moreover, each of the pectoral fin was driven by an independent DC motor, which improved the controllability and turning mobility of the prototype [26].

The third generation of Robo-ray III is shown in Fig. 8c. The length of the prototype was 0.4 m, the wingspan was 0.62 m, and the displacement was 3.2 kg. The designed operation depth was 8 m, and the maximum swimming speed was 0.6 m/s. The Robo-ray III was designed with the original design method. However, it was integrated with the attitude sensor, depth sensor, and obstacle avoidance sensor. These sensors were mainly used to carry out the related study, including the autonomous movement of the robotic fish, the attitude control, autonomous obstacle avoidance, and path planning [21].

The fourth generation of Robo-ray IV is shown in Fig. 8d. The length of the prototype was 0.5 m, the wingspan was 0.67 m, the maximum swimming speed was about 0.4 m/s, and the maximum operation depth was 8 m. Compared with the previous generations utilizing the single fin ray structure and passive deformation of pectoral fins, the three-dimensional shape design and multi-fin rays structure were used in Robo-ray IV. Moreover, the deformation of pectoral fin could be controlled by multi-fin drive mechanism. The combination of functional bionic and morphological bionic was achieved in the fourth-generation robotic fish [22, 27].

The fifth generation of Robo-ray V is shown in Fig. 8e. The length of the prototype was 0.46 m, the wingspan was 0.83 m, the maximum swimming speed was about 0.48 m/s, and its maximum dive depth was 8 m. The controllable active deformation of the pectoral fin and passive flexible deformation were combined in the Robo-ray V. The combination of functional bionics and morphological bionics was achieved better.

In 2010, the robotic fish imitating manta ray was developed by Harbin Institute of Technology. The robotic fish was propelled by SMA (Shape Memory Alloy), as shown in Fig. 9 [28]. Based on the characteristics of SMA, the high stability and the propulsion without noise were achieved. The maximum speed was 0.079 m/s, and the minimum turning radius was 0.118 m.

In 2011, the robotic fish imitating manta ray, respectively, developed by Princeton University (USA) and University of Virginia (USA) were reported in Science [29]. The oscillating pectoral fins with tensegrity mechanism were used in the prototype developed by University of Virginia (Fig. 10). The prototype developed by Princeton University is shown in Fig. 11. The body was made by 3D printing, and both sides of pectoral fins were driven by dual fin rays.

(2) Rigid skeleton combined with complex flexible deformation pectoral fin. The internal driving skeleton was made of rigid material or material with flexible

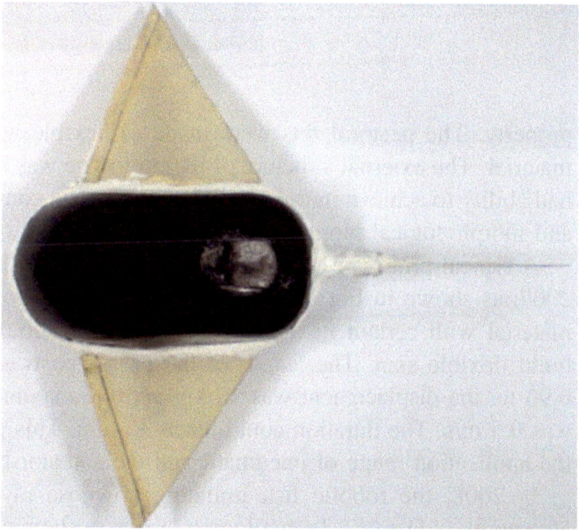

Fig. 9 The bionic fish propelled by SMA developed by Harbin Institute of Technology

Fig. 10 The bionic fish developed by University of Virginia

Fig. 11 The bionic fish developed by Princeton University

property. The pectoral fins were made of flexible waterproof or water-permeable material. The external structure of the prototype was similar to nature sample and it had ability to achieve more complex motion. The combination of functional bionics and morphological bionics was partially achieved.

A typical prototype was Aqua_ray developed by Festo Company (Germany) in 2007, as shown in Fig. 12 [30]. The internal skeleton was built by the composite material with certain flexibility. The plastic material with fiberglass was used to build flexible skin. The length of the prototype was 0.615 m, the wingspan was 0.96 m, the displacement was 10 kg, and the maximum forward swimming speed was 0.5 m/s. The duration could reach 30 min. This prototype was used to expand the application range of pneumatic muscle actuator belonging to Festo Company.

In 2008, the robotic fish imitating cownose ray was developed in National University of Defense Technology (China), as shown in Fig. 13 [31]. The prototype structure was flat. The structure was divided into two parts: middle cabin and pectoral fins. The eight actuators and the corresponding transmission mechanism

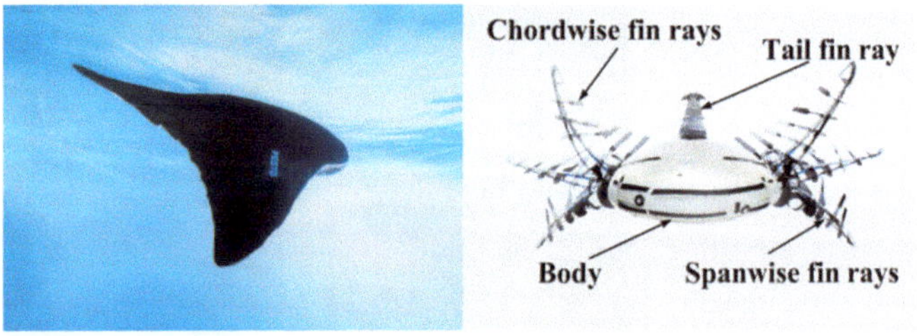

Fig. 12 Aqua_ray

Fig. 13 The cownose ray-I developed by NUDT (China)

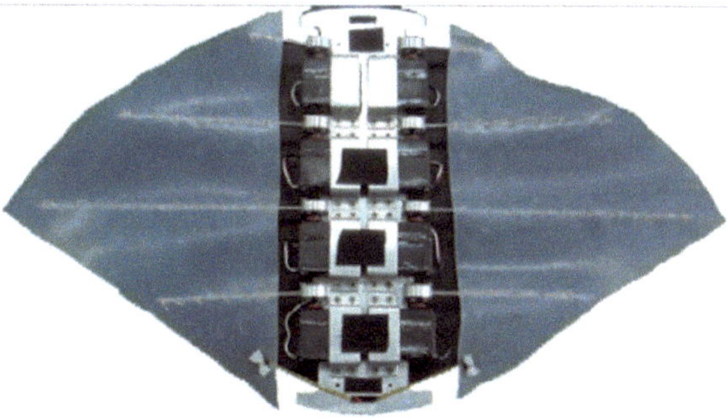

were mounted in middle cabin which was rigid. The triangular pectoral fin was composed of multi-fin rays and flexible fin surface. The weight of the prototype was 1 kg, the length was 0.3 m, and the wingspan was 0.5 m. The actuators could control the pectoral fins to oscillate in a certain phase difference. The kinematic wave was transmitted along the pectoral fin. The number of kinematic wave was 0.4. Forward thrust could be obtained in this mode. The maximum swimming speed of the robotic fish was 0.13 m/s, the backward speed was 0.15 m/s, and the pivot turning could be achieved with 8 s.

Since 2009, Nanyang Technological University (Singapore) had cooperated with National University of Defense Technology (China), and carried out the related study on the robotic fish imitating manta ray. The latest prototype RoMan-II is shown in Fig. 14 [32]. Both sides of its pectoral fins were driven by three independent groups of rigid fin rays. Bionic swimming bladder was used to achieve the depth control of the robotic fish. The maximum swimming speed of the prototype was 0.3 m/s. The minimum turning radius was 0.1 m, and 4 kg load could be carried.

In 2012, based on the research of skeletal structure of cownose ray's pectoral fin and the numerical calculations, ten groups of fin rays were used to build an oscillating pectoral fin skeleton by the University of Virginia (USA), as shown in Fig. 15b [33]. Elastic O-rings were used to connect the fin rays, which imitated the conjunctive tissue between the fin rays of cownose ray's pectoral fin. Each fin ray was composed of different number of artificial segments. The segments were

Fig. 14 The robotic fish imitating manta ray

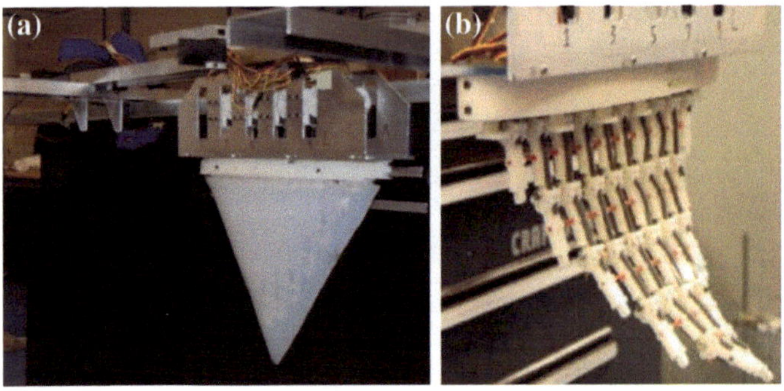

Fig. 15 The bionic pectoral fin developed by University of Virginia (USA). **a** Underwater test, **b** structure of internal skeleton

designed to be hinge-jointed together in series using stainless steel axle bolts, and all artificial segments were made by 3D printing technology. The oscillating pectoral fins' skeleton was embedded in the silicone rubber casting structure, which constituted a three-dimensional flexible oscillating pectoral fin (Fig. 15a). The oscillating pectoral fin was driven by ten groups of servos, and had chordwise and spanwise flexible deformation property. The structure of the oscillating pectoral fin was closer to the pectoral fin structure of cownose ray. However, there were many servos. The complex control strategy was needed.

(3) A full flexible body meant that the whole body was made of full flexible materials. According to the deformation demands of oscillating pectoral fin and structure characteristics of cownose ray, the flexible distribution of the body was designed. The design of the prototype was close to the combination of the functional bionics and morphological bionics. It was difficult to achieve self-propelled function of this kind of prototype. But it was more effective to explore the influence of the structure characteristics of cownose ray on the propulsion performance.

Typical prototype was Manta Robot developed by Osaka University (Japan), as shown in Fig. 16 [34, 35]. Since 2005, Osaka University had studied comparatively the pectoral fins with symmetric and asymmetric stiffness. The differences were analyzed in the propulsion performance. On the basis of studying on the passive and active flexible pectoral fin, a pneumatic rubber actuator was successfully designed in 2007. The actuator was applied to the fabrication of a bionic fish, the Manta Robot. It was driven by pneumatic cavity and designed with the silicone rubber. The prototype was propelled by an external gas source. The length of the prototype was 0.15 m, the wingspan was 0.17 m, and the maximum swimming speed was 0.1 m/s. The prototype can achieve oscillating motion similar to manta ray. However, the driving source of the prototype was installed externally. Meanwhile, there was no load space of the prototype. It was difficult to realize the self-control and long-distance swimming.

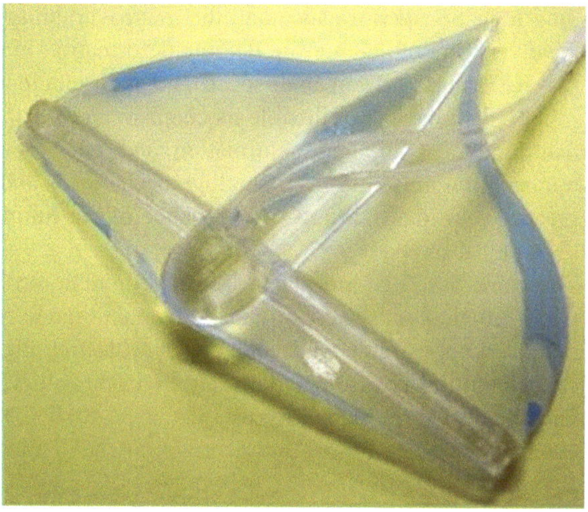

Fig. 16 Manta Robot

2.3 *Analysis of Research State on the Robotic Fish Propelled by Oscillating Pectoral Fins*

The researches on the robotic fish propelled by oscillating pectoral fins are carried out later than other kinds of robotic fish. The oscillating motion and the structure of nature sample were simplified in most of the researches on the oscillating pectoral fin units and bionic prototype. The goal is to verify the self-driven or basic swimming functions of the bionic fish developed. However, the related researches on the key parameters which can affect propulsion performance are less. There is a certain gap between the robotic fish utilizing oscillating pectoral fins and its nature sample in speed, maneuverability, and efficiency. Research status of pectoral fin units and bionic fish are analyzed, as follows:

(1) Bionic degree

Functional bionics is given priority. The robotic fish propelled by oscillating pectoral fins are designed with simplification of the complex structure features and locomotion features of nature samples. Simplified sheet-like pectoral fins with passive flexible fin surface's deformation only realize propulsion function. The pectoral fins with controllable active deformation are significant for the propulsion performance [36]. Bionic pectoral fins with simplified structures only achieve a lower degree of imitation.

(2) Drive form

In addition to utilizing SMA or pneumatic artificial muscle as drive source, the robotic fish are mostly propelled by servo motors. This leads to higher motion noise. The complex transmission mechanism constrains the performance of the robotic fish. The driving and transmission mechanisms are rigid and isolated. Therefore, the desired combination of rigidity and flexibility mechanism are not realized. It results in flexible movement lack of pectoral fins and reducing of the drag reduction performance and the maneuverability of the robotic fish [37]. Fish have very smooth composite flexible deformation in the swimming process, which plays a crucial role in high-speed and high-efficient propulsion. In the study of robotic fish, the use of flexible mechanism and flexible material allows the robotic fish produce compliant deformation and effectively reduce the resistance and transmission noise [38, 39]. Flexible mechanisms and flexible materials will play a great role in improving the performance of the future robotic fish.

(3) Control method

The cable-based control and wireless remote control are two main methods. The two methods limit the controllable movement range of the robotic fish. Meanwhile, the autonomous underwater movement with multiple sensors is not stable because of the complexity of the underwater environment. As for the future research work, the robotic fish propelled by oscillating pectoral fins must be guaranteed to have enough space for sensors and autonomous navigation system in order to meet the actual needs of autonomous swimming.

(4) Research on propulsion mechanism

The research on propulsion mechanism of swimming mode utilizing caudal fin has become increasingly mature. There are better mathematical models and analysis results of related hydrodynamic observation. But for swimming mode utilizing oscillating pectoral fins, the related theory is still lacking.

(5) Analysis of typical characteristics

Researches on typical characteristics of pectoral fin utilizing oscillating motion are few, including the general relationship between the structure and motion parameters of oscillating pectoral fins and propulsion performance, the relationship between distribution of spanwise and chordwise flexibility and propulsion performance, and the ground effect which can improve the propulsive efficiency of the kind of robotic fish. The empirical formulas derived from such typical characteristics can provide references for designing the robotic fish propelled by pectoral fin.

3 Analysis of the Bionic Structure and Motions of Cownose Ray

3.1 Introduction

The analysis of the structure and the motion characteristics of cownose ray is the basis to realize the combination of function bionic and the structure bionic. Based on the structure characteristics from the previous research of the biology and anatomy results of Myliobatidae, the structure characteristics of cownose ray are obtained, including the cross section shape of pectoral fin and internal skeleton structure. Meanwhile, the pectoral fin's motion law and the key parameters of the amplitude, frequency, and kinematic wave passing on the pectoral fin are obtained, which are derived from video analysis of free-swimming cownose ray in an aquarium. Finally, the structure characteristics and the simplified motion model of cownose ray guiding the bionic design of robotic fish are obtained.

3.2 The Structure Characteristics of the Cownose Ray

The structure characteristics of the cownose ray mainly include the shape, size, internal skeleton, and muscle. These characteristics can provide guidance for the design of robotic fish. Cownose ray is a kind of cartilaginous fish, which is characterized by streamlined flat body, high aspect ratio pectoral fins, and wing-like planform at top view, as shown in Fig. 17. According to the structure characteristics and the motion function, the body of the cownose ray can be mainly divided into the following parts: the left pectoral fin, the right pectoral fin, the tail fins, and the central body. The left and right pectoral fins produce the main thrust. The tail fins play a major role in adjusting stability and mobility. The central body mainly remains rigid, but it bends a little in high maneuvering movements.

The pectoral fin of the cownose ray mainly consists of the internal skeleton, dorsal abductor, ventral abductor epidermis, etc. [41–43]. The special bone structure of the pectoral fin is the key factor to achieve large amplitude, strongly deformed oscillating movements. The skeleton of the pectoral fin consists of the fin base and the fin rays, as shown in Fig. 18.

Fin base includes the propterygium, metapterygium, and mesopterygium, which supports the body. The combination of the fin base and scapula constitute the head cavity and abdominal cavity. So the pectoral fins cannot rotate around the scapula. All fin rays of pectoral fin are radiated from the fin base to the pectoral fin edge and the angles between the fin base and fin rays are about 10°–90°. Each fin ray is

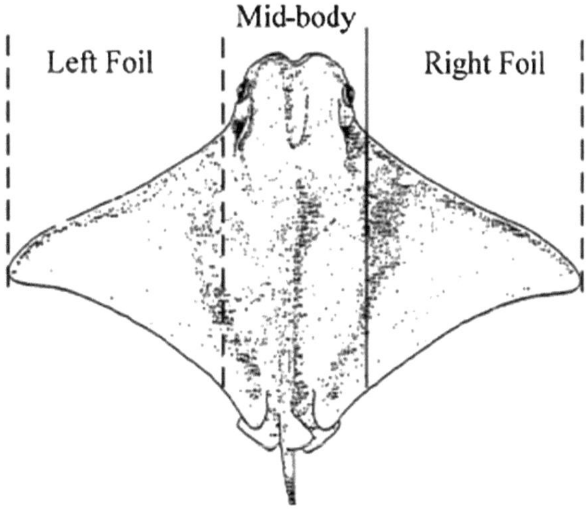

Fig. 17 Top view of cownose ray (modified from Ref. [40])

Fig. 18 Skeletal structure of cownose ray's pectoral fin

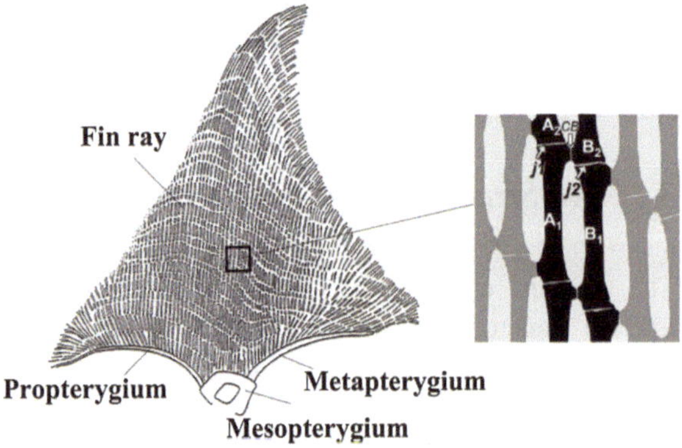

comprised of a number of radials, which are connected by tendons and joints. This unique configuration allows great flexibility in the radial direction of the whole pectoral fin. During the oscillation, the dorsal abductor and ventral abductor alternate contraction and stretch to drive the radials to form the oscillations of the entire pectoral fin [42].

The bone structures have a lot in common among the fish propelled by oscillating pectoral fins. There are cross bracings between the fin rays, which connect the adjacent radials along the chordwise direction (see the red triangle zone in Fig. 19). At the edge of the pectoral fin, there are no links between the fin rays, but there are gaps between them. As shown in Fig. 18, fin ray A is joined to fin ray B by a cartilaginous extension (CB) and the radial A_1 is joined to the radial A_2 by the calcification j_1. This makes sure a close connection between the fin rays, which can enhance the stiffness and reduce the flexibility in the chordwise direction. But Myliobatidae propelled by undulating pectoral fins has no such organization between the adjacent radials, which ensures better flexibility and larger deformation in the chordwise direction.

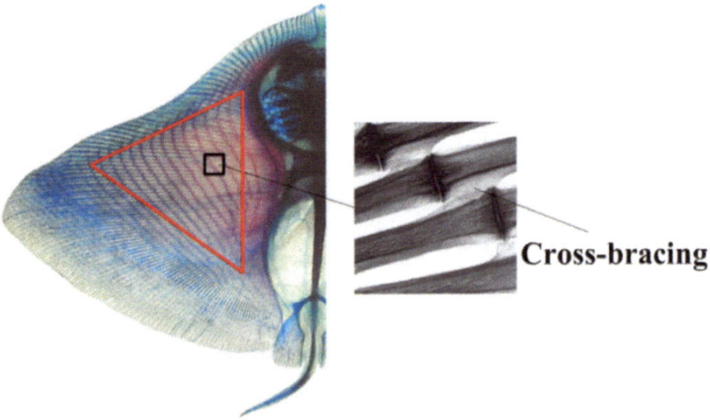

Fig. 19 Skelton flexibilty distribution of *Gymnura marmorata* (modified from Ref. [44])

According to the biological structure researches on the oscillating pectoral fins of Myliobatidae, the following characteristics of the oscillating pectoral fins can be summed up:

(1) The pectoral fin consists of a lot of radials, which connect one to another into a fin ray along the spanwise direction and are radially distributed from the fin base to the pectoral fin edge. So the oscillating pectoral fin has good spanwise deformation capacity.
(2) There are cross connections between the adjacent fin rays, which enhance the stiffness of the pectoral fin and reduce the deformation ability along the chordwise direction. So there is no more than 0.5 propulsion wave passing on the pectoral fin in normal forward swimming. This feature is the most fundamental structure difference between the oscillating pectoral fins and the undulating pectoral fins.
(3) Near the front and root of the pectoral fin, the calcification of the radial junctions is the most obvious and the rigidity is the largest. It makes sure that the pectoral fin can withstand the fluid load and actively control the deformation along the chordwise and spanwise direction driven by the muscle. Near the pectoral fin trailing edge and the tip, there are gaps between the adjacent fin rays. The rigidity is the smallest and the flexibility is the largest in this area. It ensures that the pectoral fin can form passive deformation obviously under the action of the fluid, in order to facilitate the transmission of the kinematic waves. Thus the complex deformation in the oscillating pectoral fin is the result of the active and passive deformation together.

In summary, the design of the robotic fish should base on the biological structure of the prototype, and the previous researches are significant to guide the design of the oscillating pectoral fins.

3.3 The Kinematic Analysis of the Cownose Ray

The cownose ray swims in flexible ways in different environments, such as cruise, acceleration, pitching, and yawing. It swims in the normal cruise state with better repeatability and stability for most of time. Therefore, the cruise state is chosen for the kinematic analysis in this section. The kinematic analysis of the cownose ray focuses on the pectoral fin motion, which can be equivalent to the pass of kinematic waves along the pectoral fin [44]. The kinematic analysis includes fin's flap motion procedure, the shape of kinematic waves, the wave's frequency, amplitude, and the wave number in the pectoral fin during cruise. A simplified mathematical model is established according to the kinematic analysis of the cownose ray's pectoral fin. Tail fins also play a regulatory role on the pitching stability in the swimming process of the cownose ray. So the motion of the tail fin is also analyzed.

3.3.1 The Analysis Method

First, the locomotion videos of cownose ray were obtained in aquarium. Second, the leading edge and trailing edge were extracted from different views of the videos. Then, the procedure of fin's flapping and the locomotion of specific points on the edges is analyzed.

The coordinate system set for the analysis in this section is shown in Fig. 20. The origin is set on the root point of the leading edge where there is no obvious deformation during linear cruise motion, the *x*-axis is perpendicular to the middle chordwise cross section along spanwise direction, the *y*-axis is with the body axis, and the *z*-axis is determined by the right-hand rule. The maximum vertical displacement of the fin tip in all of the video frames analyzed is set as the unit 1, which makes the motion displacement in different video frames comparable, even if the amplitudes are different.

3.3.2 Flapping Procedure of the Pectoral Fin

Leading edge and trailing edge deformations during flapping of the pectoral fin are extracted from some proper video frames, such as the typical frames of a linear swimming cownose ray seen from the lateral view, as shown in Fig. 21. The leading edge and trailing edge motion curve seen from lateral view at different moment at time intervals of 1/5 s is shown in Fig. 22.

During downstroke, the basal of leading edge begins to move downward first, and then the distal part is pulled after it. At the end of the stroke when the distal part is still moving downward toward its maximum position, the basal is already moving upward. Similarly, the basal pulls the distal part upward during upstroke. This case is not obvious for the trailing edge seen from Fig. 22. The trailing edges are not coincident at different moments. There are two reasons: the cownose ray pitches during the cruise, its trailing part of the pectoral fin is very flexible to form large passive deformation. It can be obtained from Fig. 22 that the upstroke amplitude of the pectoral fin can reach 0.5 times of the whole fin length and 2 times of the

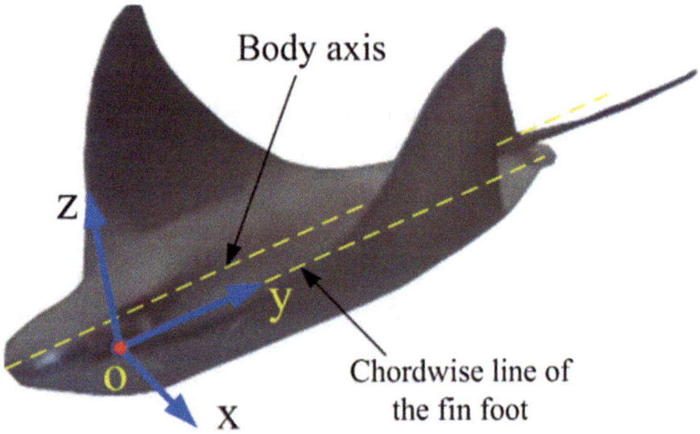

Fig. 20 Coordinate system definition

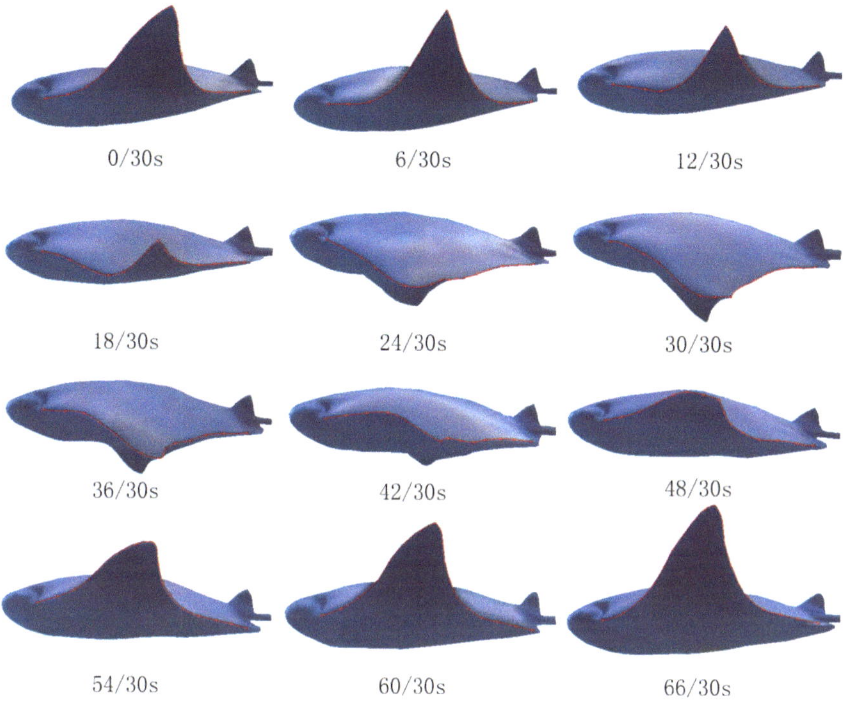

Fig. 21 Typical video frames of the cownose ray seen from the lateral view

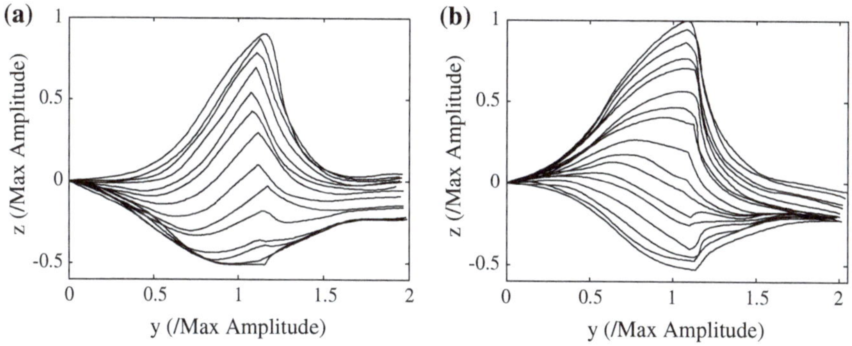

Fig. 22 Pectoral fin profiles of a cownose ray during the flapping in lateral view. **a** Downstroke, **b** upstroke

downstroke amplitude. It means that the pectoral fin flapping of cownose ray has the spatial asymmetry.

Trailing edge deformations during flapping of the pectoral fin are extracted from some proper video frames, such as the typical frames of a linear swimming cow-nosed ray seen from the back view, typically as shown in Fig. 23.

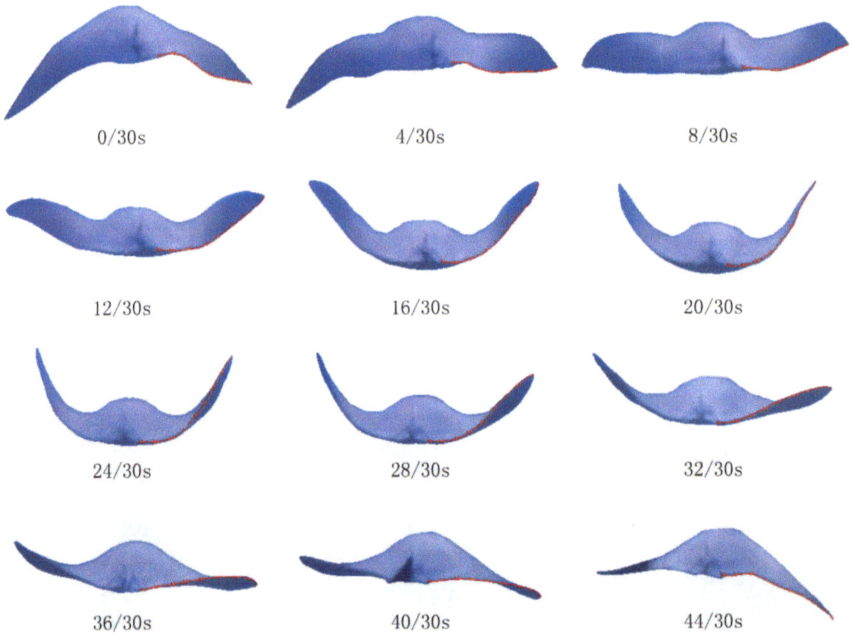

Fig. 23 Typical video frames of the cownose ray seen from the back view

The profiles of the extracted trailing edge contours are integrated on the same graph, as shown in Fig. 24. During the downstroke, the fin tip starts to move down, but the fin root remains in upward movement, as shown in Fig. 24a. The upstroke process is opposite, as shown in Fig. 24b. From Fig. 24, it can be seen that the basal of trailing edge always lags behind in phase during the flapping motion.

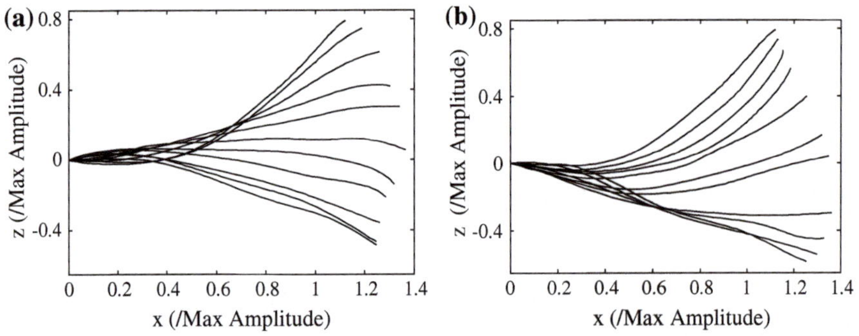

Fig. 24 Pectoral fin profiles of a cownose ray during the flapping in back view. **a** Downstroke, **b** upstroke

By analyzing cownose ray's flapping movement, the following results can be achieved: locomotion of the pectoral fins is equivalent to the pass of kinematic wave front to back, parallel to the chordwise direction; the vertical amplitude of upstroke is large, about half the body length; the flap motion is not symmetric by the horizontal plane, for the amplitude above horizontal plane is about twice as much as that below horizontal plane.

3.3.3 The Kinematic Wave Number

The number of kinematic waves presented on the pectoral fin is defined as the value of the wavelength divided by the length of the fin base. Videos that are seen from a direction perpendicular to the leading edge or trailing edge reflect their actual shape considering the refraction of water. The length of the leading edge and trailing edge is assumed to be constant in the analysis, if the video frames are properly chosen. Therefore, the phase lag between the middle points on the edges and the fin tip can be obtained. Furthermore, the number of kinematic waves can be calculated.

As shown in Fig. 25, the middle point on the leading edge has phase lead for 0.1 waves than fin tip; the middle point has phase lag for 0.07 waves, close to that in the leading edge. The phase difference is not obvious at maximum upper position in the trailing edge, which probably results from the lower velocity at the position and arc shape of the trailing edge. The data of the leading edge are valid for computing the number of waves. According to the definition, the number of kinematic waves in pectoral fins is about 0.4, coincident with the result in Ref. [31].

Finally, the conclusions are:

1. Pectoral fin's locomotion of the cownose ray is equivalent to the pass of oscillating motion wave from front to back.
2. The number of kinematic waves presented on pectoral fins is about 0.4.

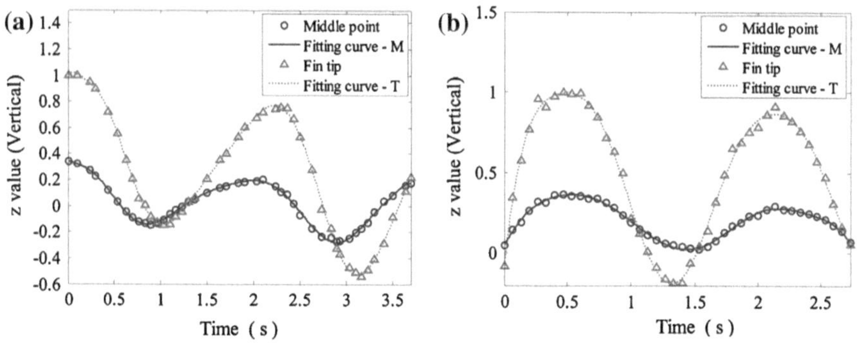

Fig. 25 Vertical displacement curve of the middle points on the pectoral fin leading edge and trailing edge and the fin tip of a sample cownose ray [45]. **a** Leading edge condition, **b** trailing edge condition

4 Design and Optimization of Drive Mechanism

4.1 Introduction

Based on the structure bionic, the mechanical system of robotic fish is designed to the combination of internal skeleton (similar to the fin rays of cownose ray) and external flexible skin (similar to the epidermal tissue of cownose ray). Then the appropriate actuator is chosen to drive the internal skeleton. The internal skeleton includes eight drive mechanisms. The deformation of the flexible skin is controlled by the drive mechanisms in order to realize the flexible movement of the robotic fish. Lastly, the drive mechanisms are designed and optimized according to the results of bionic research.

4.2 Driving Mechanism Design

Analyzing locomotion of cownose ray, it is concluded that locomotion of pectoral fins is equivalent to the oscillating wave passing from front to back. Shape of the profile can be approximated by cubic function $y = a(t)x^3$. To realize the locomotion, mechanism with three revolution joints is set as shown in Fig. 26a. However, the mechanism needs three actuating motors making it difficult to control; the mechanism is heavy; the worst point, the mechanism needs inner space of pectoral fins severely. To conquer these disadvantages, two-stage rocker–slide mechanism is designed, as shown in Fig. 26b. The novel mechanism needs only one actuating motor. Although it cannot perform as good as the mechanism driven by three motors in mimicking the profile of cownose ray, it is enough for the design of robotic fish. The structure of this novel mechanism is shown in Fig. 27a. All the link rods are made of aluminum except the distal rod, which is made of carbon fiber to provide elasticity. The mechanism in Fig. 27b was tested initially, in which the mechanism oscillated smoothly.

Several fin rays are required to mimic the kinematic wave passing on pectoral fins. Three groups of the fin ray mechanism are enough for a flapping pectoral fin, as there are only 0.4 waves presented on the pectoral fin. The pectoral fin of

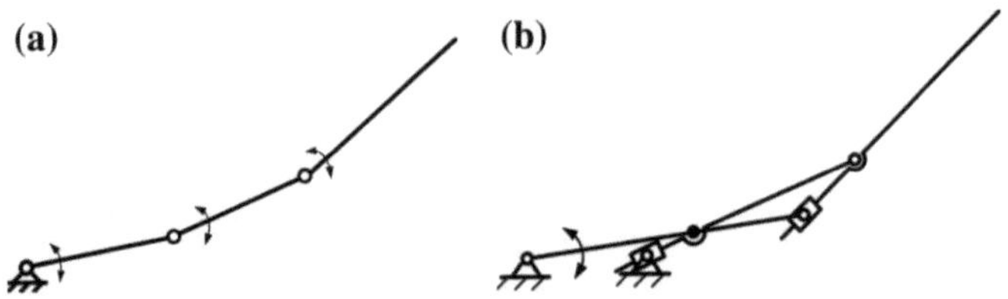

Fig. 26 Drive mechanism of multi-fin (where *arrows* are required actuators) [45]. **a** Mechanism with three revolution joints, **b** two-stage rocker–slide mechanism

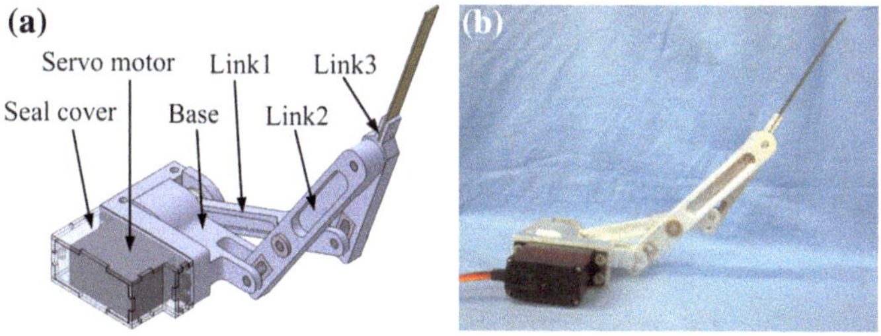

Fig. 27 Structure of the two-stage rocker–slide mechanism. **a** Model in CAD, **b** actual mechanism

Fig. 28 Fin rays with different joint number are designed

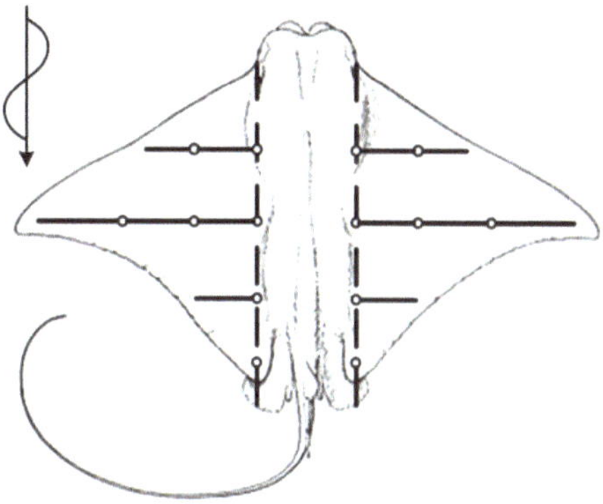

cownose ray is approximate to a triangle, so the joint number of each fin ray is different. There are 2, 3, 1 joints for the fin rays from front to back, respectively, and driving mechanisms for tail fins are designed on both sides symmetrically, as shown in Fig. 28.

4.3 Optimization Design

Key dimensions of drive mechanisms are required to be optimized to mimic the locomotion of cownose ray to higher extent. The optimization parameters include the distance between fin rays and dimensions of the two-stage rocker–slide mechanism.

4.3.1 Optimization of Distance Between Fin Rays

The distance between fin rays is an important factor that affects the extent of mimicking the locomotion of cownose ray. So it is necessary to optimize the distance to minimize the error. The number of waves is about 0.4, obtained by

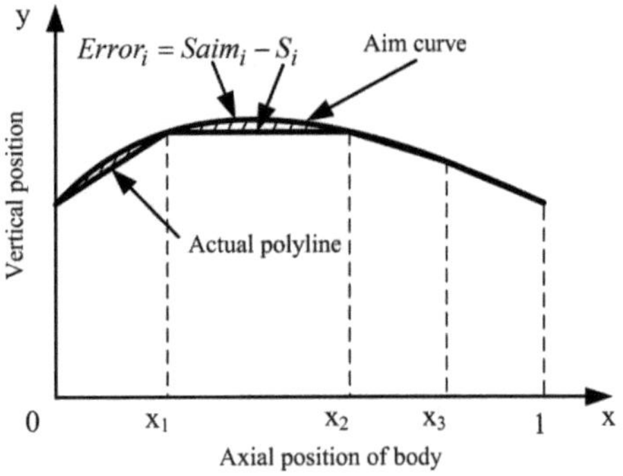

Fig. 29 *Dashed area* is used to estimate the fitting extent

dividing the length of the fin base by the wavelength. The number of control points is 5 besides 2 point on the fin base. $x_1,\ x_2,\ x_3$ illustrate the positions of fin rays. As shown in Fig. 29, the dashed area surrounded by the aim curve and the actual polyline is the target to estimate the fitting extent. The object function is the sum of dashed areas during a period which is discretized to 80 time points, that is:

$$\text{Error} = \sum_{i=1}^{80} |\text{Error}_i| = \sum_{i=1}^{80} |\text{Saim}_i - S_i| \tag{1}$$

where, Saim_i is the integral of the aim curve; S_i is the integral of the polyline. Mark $f(x, i)$ as the aim curve at time i, then:

$$f(x, i) = \sin(0.8\pi x + \frac{2\pi i}{80}) \tag{2}$$

$$\text{Saim}_i = \int_0^1 f(x, i)\text{d}x \tag{3}$$

$$\begin{aligned} S_i = \ & \tfrac{1}{2}(f(0,i) + f(x_1,i))y_1 + \tfrac{1}{2}(f(x_1,i) + f(x_2,i))(x_2 - x_1) \\ & + \tfrac{1}{2}(f(x_2,i) + f(x_3,i))(x_3 - x_2) + \tfrac{1}{2}(f(x_3,i) + f(1,i))(1 - x_3) \end{aligned} \tag{4}$$

The constraint condition is:

$$0 < x_1 < x_2 < x_3 < 1 \tag{5}$$

The object is to minimize the value of object function (1). Uniting (1–5), optimized result is obtained with the initial values 0.2, 0.4, 0.6 for $x_1,\ x_2,\ x_3:\ x_1 = 0.2683,\ x_2 = 0.5000,\ x_3 = 0.7317$.

4.3.2 Dimension Optimization of the Pectoral Fin Ray Mechanism

As discussed before, during cruise the locomotion of cownose ray can be approximated by cubic function $y = a(t)x^3$, and the real shape of span section can be approximated by the three link rods of the mechanism, as shown in Fig. 30.

Similar to the optimization of the distance between fin rays above, a period is discretized to 80 time points, the object function is the sum of the error between the aim curve and the actual shape of span section, that is:

$$\text{Error} = \sum_{i=1}^{80} |\text{Error}_i| = \sum_{i=1}^{80} |\text{Saim}_i - S_i| \tag{6}$$

where, Saim_i is the integral of the aim curve at time i; S_i is the integral of the actual polyline at time i shown in Fig. 30, then:

$$\text{Saim}_i = \int_0^{xt(i)} a(i)x^3 \text{d}x \tag{7}$$

$$S_i = \frac{1}{2}x_1(i)y_1(i) + \frac{1}{2}[y_1(i) + y_2(i)][x_2(i) - x_1(i)] + \frac{1}{2}[y_2(i) + y_3(i)][x_3(i) - x_2(i)] \tag{8}$$

where, $xt(i)$ is the horizontal coordinate of the tip of ideal curve; $(x_1(i), y_1(i))$, $(x_2(i), y_2(i))$, $(x_3(i), y_3(i))$ are coordinates of the three revolution joints. As shown in Fig. 30, l_1, l_2, l_3 are the length of the link rods, and $\theta_1(i)$, $\theta_2(i)$, $\theta_3(i)$ are the angle of the link rods at time i, so:

$$x_1(i) = l_1 \cos\theta_1(i) \tag{9}$$

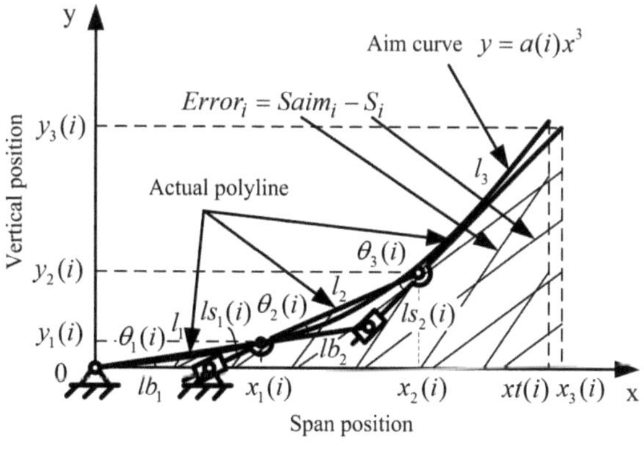

Fig. 30 Optimization of dimensions of fin rays

$$y_1(i) = l_1 \sin\theta_1(i) \tag{10}$$

$$x_2(i) = x_1(i) + l_2 \cos[\theta_1(i) + \theta_2(i)] \tag{11}$$

$$y_2(i) = y_1(i) + l_2 \sin[\theta_1(i) + \theta_2(i)] \tag{12}$$

$$x_3(i) = x_2(i) + l_3 \cos[\theta_1(i) + \theta_2(i) + \theta_3(i)] \tag{13}$$

$$y_3(i) = y_2(i) + l_3 \sin[\theta_1(i) + \theta_2(i) + \theta_3(i)] \tag{14}$$

It is assumed that the length of span profile during locomotion is a constant L:

$$\int_0^{xt(i)} \sqrt{1 + [3a(i)x^2]^2} \mathrm{d}x = L \tag{15}$$

To decrease the number of optimizing parameters, all the revolution joints are restricted in the aim curve when it is at the maximum upper position. Thus, the maximum angle of the three joints $\theta m_1,\ \theta m_2,\ \theta m_3$ can be obtained, and:

$$lb_1 = l_1 \sin\theta m_2 / \sin(\theta m_1 + \theta m_2) \tag{16}$$

$$lb_2 = l_2 \sin\theta m_3 / \sin(\theta m_2 + \theta m_3) \tag{17}$$

where, lb_1 and lb_2 were defined in Fig. 30. At time i, the first revolution joint is restricted in the aim curve to simplify the optimization, that is:

$$y_1(i) = a(i)x_1(i) \tag{18}$$

Uniting (9), (10), and (18), the angle of the first joint $\theta_1(i)$ can be obtained, then:

$$ls_1(i) = \sqrt{{l_1}^2 + {lb_1}^2 - 2l_1 lb_1 \cos\theta_1(i)} \tag{19}$$

$$\theta_2(i) = \arcsin[lb_1 \sin\theta_1(i)/ls_1] \tag{20}$$

$$ls_2(i) = \sqrt{{l_2}^2 + {lb_2}^2 - 2l_2 lb_2 \cos\theta_2(i)} \tag{21}$$

$$\theta_3(i) = \arcsin[lb_2 \sin\theta_2(i)/ls_2] \tag{22}$$

where, $ls_1(i)$ and $ls_2(i)$ are defined in Fig. 30.

Uniting (6–22), the value of object function can be acquired based on the length of the three link rods. The result of the optimization is $l_1 = 96.6667$, $l_2 = 96.6667$, $l_3 = 96.6667$ with the initial values 90, 90, 90 for $l_1,\ l_2,\ l_3$.

5 Design of the Bionic Fish

5.1 Introduction

The multi-fin ray mechanism and external flexible skin are the typical features of this robotic fish. After the design idea of driving mechanism is proposed, the engineering fulfillment of robotic prototype is completed, including its mechanical and electrical systems. The initial testing results show that some aspects of design are undesired, and then some improved design is implemented aiming at solving these problems raised.

5.2 Mechanical Design

The mechanical structure of the bionic fish with oscillating paired pectoral fins contains two main parts. One is the inside skeleton including the three fin rays on each lateral side and two at the back of the middle control case. The other part is the flexible soft body with distributed flexibility to meet the flapping motion requirements of the inside skeleton.

5.2.1 Internal Skeleton

The structure of fin rays is designed and machined according to the optimization results. Figure 31 shows the inside skeleton of all fin rays at maximum upper amplitude, which shows the ability that the pectoral fin can realize the desired oscillating movement.

The two-stage slide–rocker mechanism is shown in Fig. 32. The linkages III of all fin rays are made of carbon fiber plates featuring flexibility with thickness of 1 mm. For the fin rays with only one linkage, the whole fin ray is made of carbon fiber. Research results show that the moderate flexibility of the fish in nature greatly benefits its propulsion efficiency.

5.2.2 Flexible Skin

Research results show that the body flexibility of the fish in nature greatly benefits its propulsion, such as reducing resistance, lowering noise, and increasing efficiency. The flexible body is made of two-component silicon rubber with outstanding elasticity and anti-tear properties. Some small changes are made to meet the machining and assembly requirements, and to make the body a streamlined shape. The plane body shape, the body size including the length and span, and the shape of the chordwise sections are all designed according to the biology and

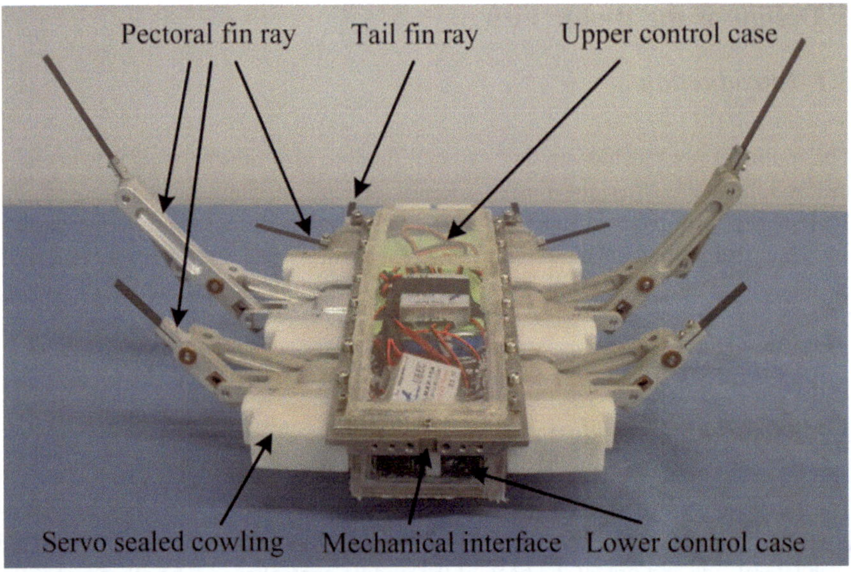

Fig. 31 Internal skeleton of robotic fish with control and driving units [45]

Fig. 32 Photo of middle fin ray optimized without the servo waterproof cover [45]

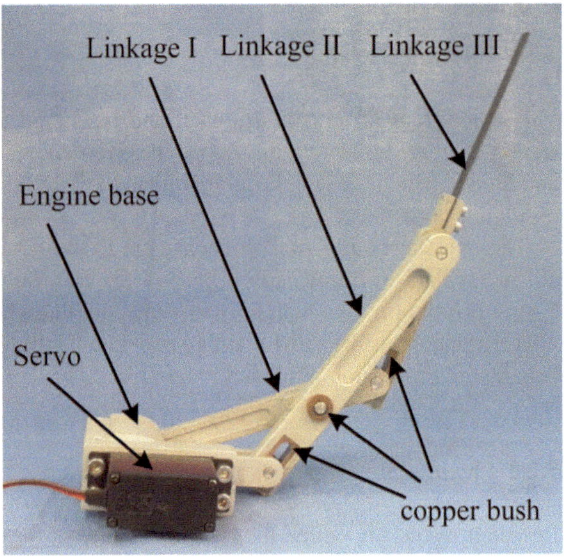

anatomy results. The mold and pouring process of the outer flexible body are shown in Fig. 33. Chordwise sections of the flexible body are simplified to series of airfoil shapes (NACA0015), as shown in Fig. 34. The average wall thickness changes gradually from 10 mm at the fin base to 3 mm at the fin tip, which is obtained from

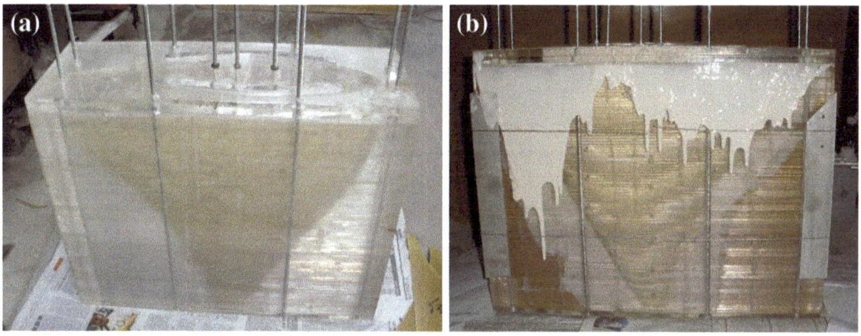

Fig. 33 Mold and pouring of the flexible body [45]. **a** before pouring, **b** after pouring

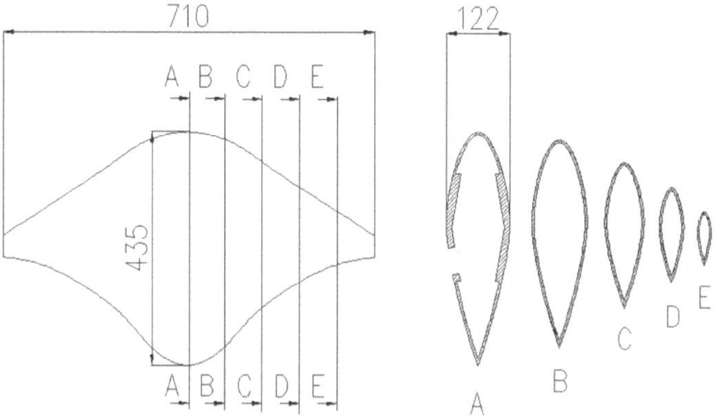

Fig. 34 The design of fish skin with chordwise NACA series sections from fin base to tip

pouring experiments, except for the much thicker middle part with a V-type inside polyline, which is designed for the fixation between the middle control case and the flexible body.

5.3 Control Method

A commercial wireless remote control method is employed. The control diagram is shown in Fig. 35. Actual tests indicate that when the water depth is no deeper than 2 m, the wireless product can provide effective control signals. Each pectoral fin ray is driven by a high-torque servo, with a typical peak current of 2 A. Considering stability, two lithium batteries are used. One with capacity of 20 Ah is utilized to supply the servos, and the other with capacity of 3 Ah is for the wireless receiver and control circuits. The prototype has at least 1 h endurance time based on the battery capacity.

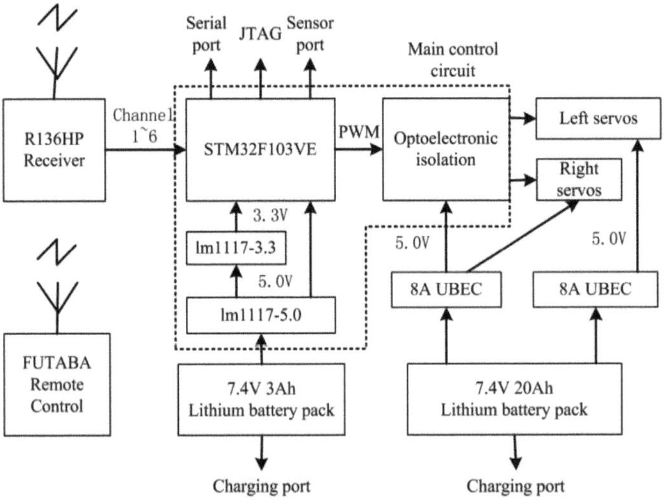

Fig. 35 Control diagram of the bionic fish

Servos used for the pectoral fins and the tail part are controlled through different methods. The two servos in the tail are controlled directly by the radio receiver. Each of the three servos for one pectoral fin is controlled by the microcontroller STM32F103VE. The phase difference between the fin rays is programmed ahead, aiming to realize 0.4 times the wave passing on the pectoral fin. The amplitude and frequency of the flapping motion are directed by signals from the wireless remote controller.

5.4 Bionic Prototype

All the electrical components, including the control circuits, batteries, and the wireless radio receiver, are sealed in the middle control case. The servos used are all sealed separately by a plastic shell and connected to the control case with waterproof cables. The inside skeleton was submerged in a 1.5 m depth water tank to verify its waterproof performance. For a 36 h test, there was no leakage condition.

Assembly of the bionic prototype is easy. The inside skeleton is put into one half of the flexible body, with the switch button and the LED light for the working state left outside through pinholes, as shown in Fig. 36a. In the same way, the other half of the flexible body is put on the inside skeleton, as shown in Fig. 36b. The gap between the two flexible half-bodied is bonded by the same liquid silicon rubber as the body itself. Some counterweight blocks are bonded with the waterproof shells of the servos to make the bionic fish suspend in water with a horizontal attitude when the bionic fish is in the stationary state.

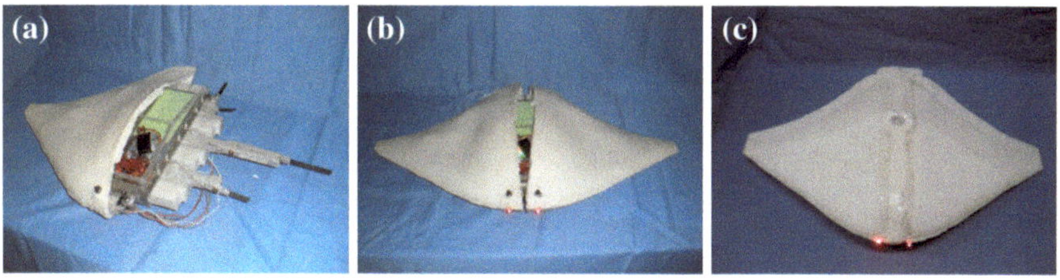

Fig. 36 Assembly process of the inside skeleton and the outer flexible body and the final bionic prototype [45]

A photo of the final bionic prototype filled with the inside skeleton and control parts is shown in Fig. 36c. It has a body length of 0.46 m, wingspan of 0.71 m, and weight of 5.31 kg after counterbalancing, which is similar to a mature cownose ray in nature.

6 Testing of the Bionic Fish

In order to verify whether the bionic fish can realize the desired flapping motion and kinematic wave transmission on its pectoral fin, we analyze the motion of bionic prototype. The preliminary experiments consist mainly of motion analysis and the underwater attitude observation of bionic fish. A FastCam 1280 PCI high-speed camera is used to record their motions. A sampling frequency of 125 frames and a shutter time of 1/500 s are employed.

The prototype is put on a stand and held stationary by a fixed frame, as shown Fig. 37. The water depth is 0.5 m. Parameter settings of the high-speed camera are the same as used in the observation of the inside skeleton. Profiles during the down-flapping period of the leading edge from the front view are obtained, as shown in

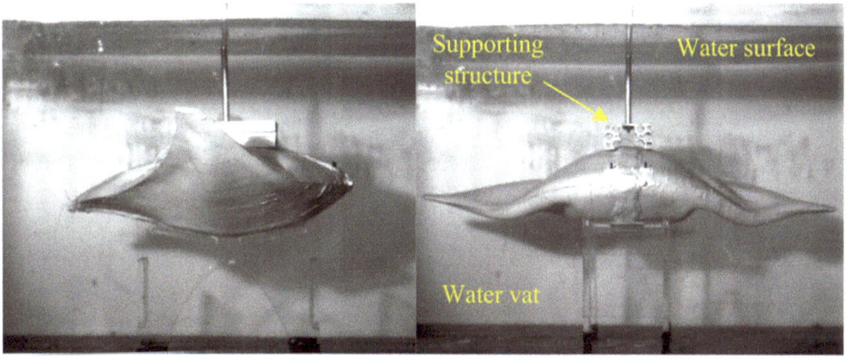

Fig. 37 Motion analysis of the bionic prototype using the high-speed camera [45]

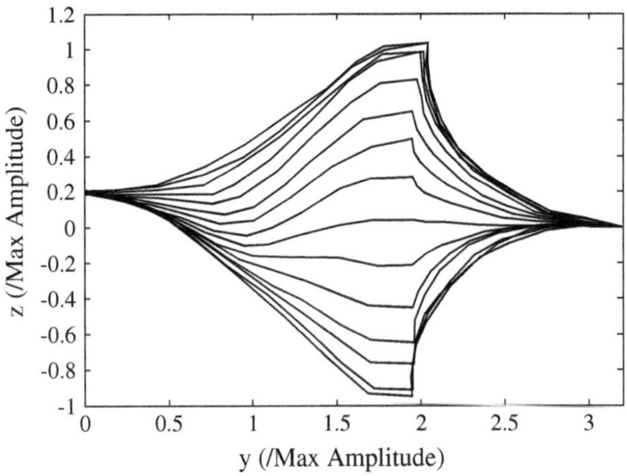

Fig. 38 Pectoral fin profiles of the bionic fish during the downstroke cycle in lateral view

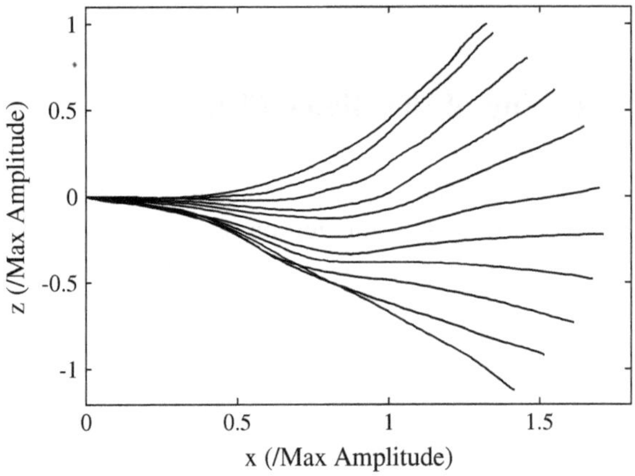

Fig. 39 Leading edge curve in front view during the downstroke of a flapping cycle

Figs. 38 and 39 separately. All the lengths utilized are nondimensionalized by dividing the maximum amplitude of the pectoral fin, employing the same method as in analyzing motion of the cownose ray.

According to the profiles, passing waves from the head to tail and from the middle chord to the fin tip are obtained by the bionic prototype. The flapping amplitude of the prototype is smaller than the cownose ray in nature, as shown in Figs. 22 and 24, which is mainly caused by the elasticity of flexible body and hydrodynamic resistance of water. Deformation of the trailing edge is not as much as expected. However, an effective and positive attack angle can be observed, attributed to the differential motion of the fin rays.

A preliminary swim test of the bionic fish is carried out in a water tank. The water tank is 3 m in length, 2 m in width, and the water depth is 0.5 m. The swimming velocity of bionic fish can reach 0.26 m/s, which is 0.55 times of body length per second. The effect of swimming is not good. The reason is that the practical oscillating amplitude of pectoral fin cannot reach the given amplitude, which is mainly

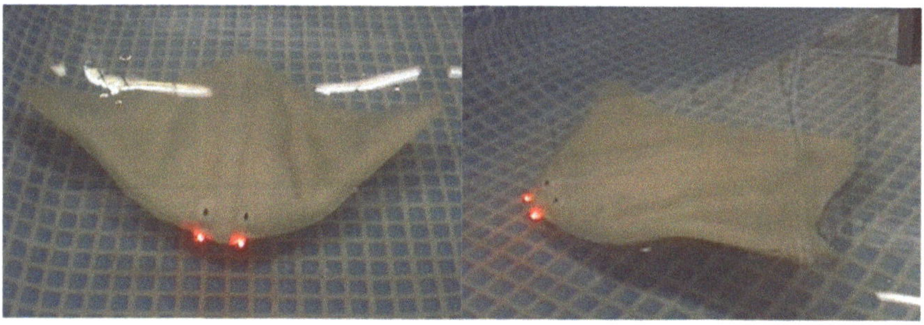

Fig. 40 Video snapshots of the fish robot swimming in a water tank [45]

caused by the elasticity of flexible body and hydrodynamic resistance of water. Images of the bionic fish swimming in the water tank are shown in Fig. 40.

7 The Improved Prototype

With regard to the improved prototype, the key aspect is to overcome the contradiction from separating the rigidity and flexibility of the skin. First, the flexibility of skin is increased to a large extent to release the motion resistance of fin rays caused by skin. Seen from the experiments, the skin made of poured silicon rubber cannot meet our needs. Polyurethane fiber is chosen as the skin material. The elastic deformation capability of this material is about 200 %, which means slighter resistance.

Second, chordwise supporting frames are designed, which is set on the mobile fin linkages inside the skin to preserve the shape of pectoral fin. The mold and pouring process of the supporting frames are shown in Fig. 41. The frames are utilized to resist the hydraulic pressure and realize smooth deformation of the skin.

Fig. 41 The mould of chordwise supporting silicon rubber rib

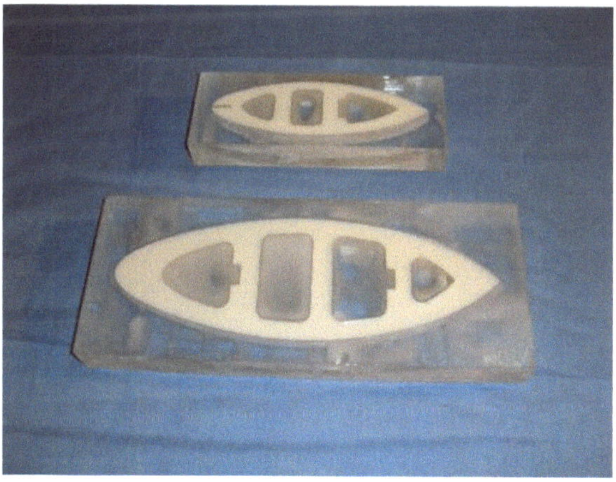

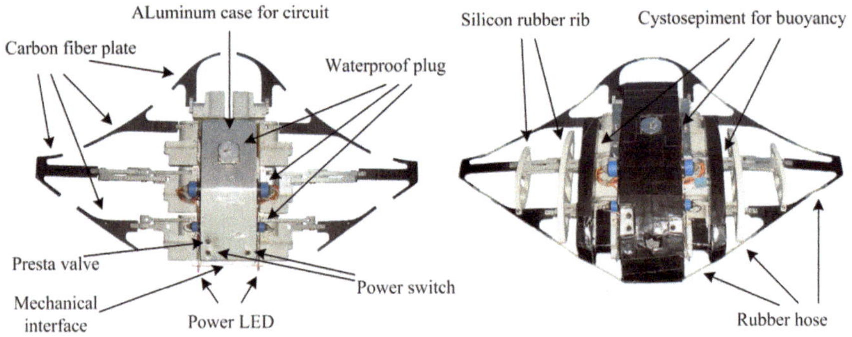

Fig. 42 The improved inside skeleton of robotic fish

One of the advantages of permeable skin is avoiding instability of robotic fish caused by the hydraulic pressure. The skin made of polyurethane fiber is just the right option. Meanwhile, it leads to the loss of the outer waterproof defense, so the waterproofness of electrical case and servo module must be considered carefully. As shown in Fig. 42, the improved middle housing is made of brazed aluminum frames with a thickness of 2.5 mm, which is utilized to improve the waterproofness.

Displacement of the robotic fish reduced to 2.5 kg in permeable design, whereas the weight of the prototype is 4.2 kg. We use cystosepiment with density of 20 kg/m^3 to increase buoyancy. They are incised according to the cross profile of cownose ray. The cystosepiment for buoyancy satisfies two requirements, increasing displacement and maintaining the body shape. The robotic fish prototype with permeable skin is shown in Fig. 43.

Fig. 43 The bionic robotic fish in external view

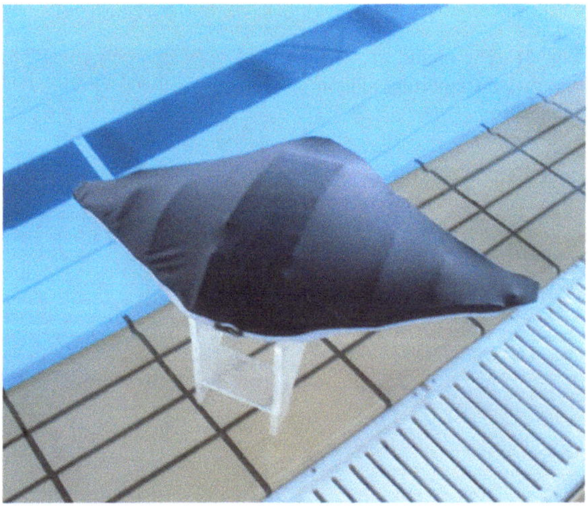

8 Underwater Experiments of the Robotic Fish

8.1 Introduction

A series of underwater experiments are carried out to verify the effectiveness of the robotic fish design, including swimming experiments and thrust tests. The purpose of the swimming experiments is to test the movement performance of the robotic fish and obtain the performance parameters, including the maximum swimming speed, the turning radius, and the endurance time. Then the effectiveness of the robotic fish design is verified through the movement state. The thrust tests are carried out to study the change rule of thrust produced by the robotic fish on the different motion parameters (frequency, amplitude, difference, etc.). The mechanism of thrust is also studied preliminarily to verify the validity of the design. The thrust experiments provide the theoretical basis for the performance improvement of robotic fish.

8.2 Swimming Experiments

Swimming experiments can intuitively show the performance of the robotic fish. The basic motions of the robotic fish are realized by remote control, including cruise, turning, diving and surfacing, and backward motion. The performance parameters of the robotic fish are measured, including the maximum speed, the minimum turning radius, the cruise duration. The tests of the swimming speed of the robotic fish are conducted in Beihang University's natatorium, as shown in Fig. 44.

The borders of the black ceramics at the bottom of swimming pool are used as references, and boundary distance is 225 mm. The time which the robotic fish takes

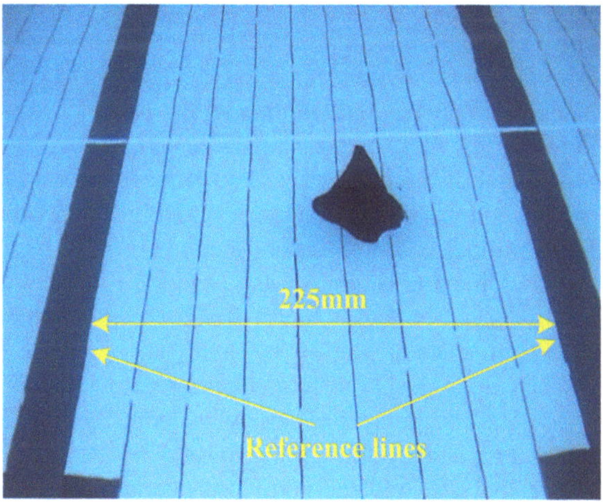

Fig. 44 The swimming speed test of the robotic fish

through the boundary distance is recorded. Then the actual swimming speed is obtained. According to the tests, the maximum speed of the robotic fish is obtained in which the frequency is 0.6 Hz, the oscillating amplitude is 20°, and the phase difference is 1/10 wave length. The maximum swimming speed is 0.3 m/s.

Using symmetrical flapping motion, the turning radius of the robotic fish is zero. In this case, the kinematic wave of the lateral pectoral fin passes backward, the phase difference of the adjacent fin rays is 1/10 wave length, and the forward thrust is produced. Meanwhile, the kinematic wave of the interior pectoral fin passes forward, the phase difference of the adjacent fin rays is 1/10 wave length, and the backward thrust is produced. It is concluded that the robotic fish is equipped with good mobility when the shapes of the pectoral fins are actively controlled by the multi-fin ray driving mechanism.

The backward thrust is produced when the kinematic waves presented on the pectoral fins on both sides pass forward, and the robotic fish will swim backward. According to the tests, the maximum backward speed of the robotic fish is obtained when the frequency is 0.6 Hz, the oscillating amplitude is 20°, and the phase difference is 1/10 wave length. The maximum speed is 0.2 m/s. In nature, the cownose ray has no backward motion. Thus the robotic fish is equipped with the backward motion when the shapes of the pectoral fins are actively controlled by the multi-fin ray driving mechanism.

Experiments show that the various kinds of motion including the cruise, turning, diving and surfacing, and backward motion are realized by the robotic fish. It is concluded that the robotic fish is equipped with good performance capability. In the swimming process, the robotic fish performs good pitching maneuverability too. The diving and surfacing motions of the robotic fish are realized by controlling the angle of caudal fin.

8.3 Thrust Experiments

The purpose of this experiment is to get the driving forces generated by the robotic fish in different motion state (cruise, turning, etc.) and different motion parameters (flapping frequency and amplitude, the phase difference between the fin rays). As shown in Fig. 45, the experiments are carried out in a 2 m × 3 m water tank, and the water depth is 0.6 m. The distance between the extreme position of the pectoral fin and the bottom of water tank is ensured as 0.1 m when the robotic fish is installed. The coordinate system is established in Fig. 45. The thrust direction is along x-axis, the yaw direction is along y-axis, and the lift direction is along z-axis.

Six-axis force sensor Smart 300 is the core component in the experiment. The main parameters are shown in Table 1.

The forces closely affect performances of the robotic fish are the thrust F_x and torque M_z. The thrust F_x is closely related to the swimming speed, and the torque M_z is closely related to the turning movement. Therefore, the thrust F_x and the torque M_z are analyzed in detail here. In the experiments, the testing time of each motion

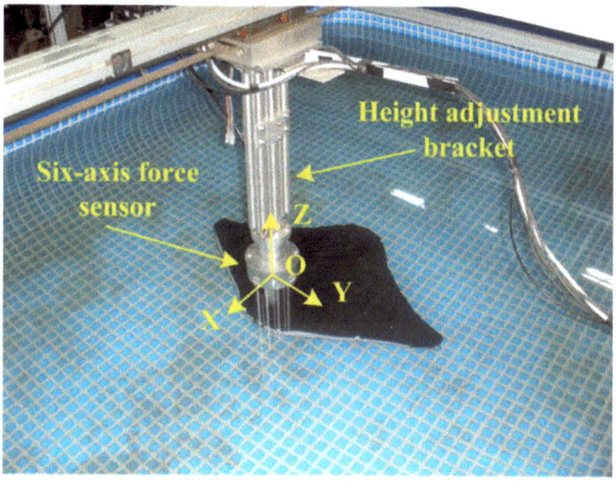

Fig. 45 Experiment device and the definition of the coordinate system

Table 1 Main parameters of six-axis force sensor

Parameters	Range
F_z	±50 N
F_x	±10 N
F_y	±10 N
M_x	±3 Nm
M_y	±3 Nm
M_z	±1 Nm

modes is longer than five full flapping cycles. In the analysis process, three full flapping cycles in the middle are chosen as the object. The up–down symmetric flapping motion is employed in the experiments.

1. Cruising

The cruising movements with different frequency and amplitude are carried out to study the performance of the robotic fish with different control parameters. The frequency utilized varies from 0.3 to 0.6 Hz, at 0.1 Hz intervals. The amplitude utilized varies from 5° to 20°, at 3° intervals. The phase difference is fixed at 1/10 wavelength. Analysis of the experimental data is focused on the thrust F_x, and the experimental data is shown in Table 2.

Seen from Table 2, the average thrust increases with the increment of the frequency and the amplitude. The change rule of the maximum thrust is similar to the change rule of average thrust. The maximum thrust is obtained in frequency 0.6 Hz, amplitude 20°. The change curve between thrust F_x and time is shown in Fig. 46. In the symmetric flapping motion, the value of the average thrust is 3.21 N, and the value of the maximum thrust is 20.48 N. In the swimming experiments, the maximum speed is also obtained in these motion parameters.

From Fig. 46, it can be seen that the change period of the thrust is the same as the flapping period. The negative thrust is produced during the flapping, which reduces the propulsive efficiency. The reason might be that the shape of the pectoral fin is

Table 2 The thrust data in the cruising (unit: N)

Frequency (Hz)	Amplitude					
	5°		8°		11°	
	Average	Maximum	Average	Maximum	Average	Maximum
0.3	−0.10	1.63	−0.03	2.53	0.39	4.06
0.4	0.11	2.96	0.38	5.15	0.70	7.65
0.5	0.25	4.85	0.73	9.38	1.15	13.06
0.6	0.30	8.63	0.99	13.09	1.42	15.08
Frequency (Hz)	Amplitude					
	14°		17°		20°	
	Average	Maximum	Average	Maximum	Average	Maximum
0.3	0.90	6.83	0.96	8.65	1.36	11.11
0.4	1.05	9.94	1.78	13.25	1.95	14.02
0.5	1.93	13.42	2.29	15.22	2.96	16.92
0.6	2.07	17.28	2.87	18.64	3.21	20.48

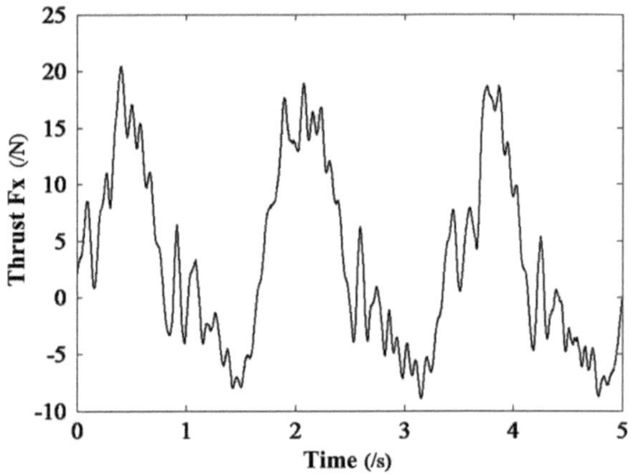

Fig. 46 The change curve of thrust with time-varying

actively controlled by three fin rays without considering the role of water on the deformation of the pectoral fin. The robotic fish should be improved in the future.

In the study on the relationship between phase difference and thrust, frequency is set as 0.5 Hz, and the amplitude is set as 20°. The wavelength utilized varies from 6/80 to 10/80, at 2/80 of wavelength intervals, as shown as in Table 3. From Table 3, it can be seen that the robotic fish obtains the maximum thrust in which the 10/80 of wavelength is used, that is the number of kinematic wave is 0.4.

2. Turning

In the turning condition, the lateral pectoral fin with the maximum amplitude is controlled by the remote control, and the kinematic wave of the lateral pectoral fin passes backward. The turn mode can be divided into four modes according to the

Table 3 The thrust data with different phase difference

Phase difference (wavelength)	6/80		8/80		10/80	
	Average	Maximum	Average	Maximum	Average	Maximum
Thrust (N)	2.91	15.58	2.96	16.92	2.31	14.61

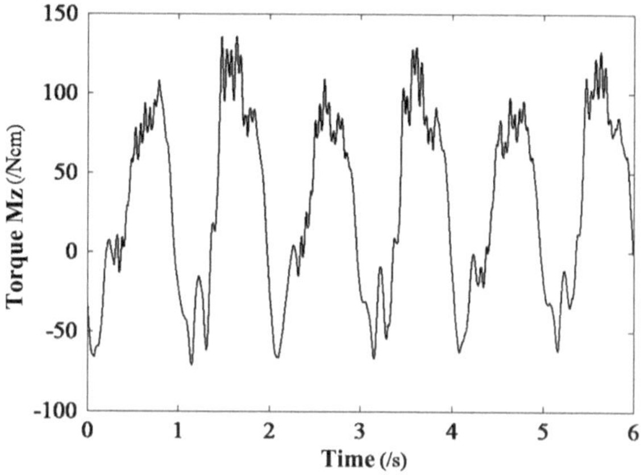

Fig. 47 The change curve of torque with time-varying in mode 4

motion of the interior pectoral fin. The flapping amplitude is half of the lateral pectoral fin, and the kinematic wave of the lateral pectoral fin passes backward (mode 1). The flapping amplitude of the interior pectoral fin is zero (mode 2). The flapping amplitude is half of the lateral pectoral fin, and the kinematic wave of the lateral pectoral fin passes forward (mode 3). The flapping amplitude is the same with the lateral pectoral fin, and the kinematic wave of the lateral pectoral fin passes forward (mode 4). In the mode 4, the robotic fish is equipped with the minimum turning radius. The turning radius is zero in the swimming experiment.

In the hydrostatic thrust test, the turning torque is measured in mode 4. In the test process, the frequency is set as 0.5 Hz, the amplitude is set as 20°, and the phase difference is 1/10 of wavelength. The measured value of torque M_z is 0.3 Nm, the maximum value is 1.36 Nm. The change curve between torque M_z and time is shown in Fig. 47. It can be seen that the change curve of the torque has two peaks in each flapping period. The negative torque is produced when the pectoral fin is in the extreme position. Thus, improving the pectoral fin motion in the extreme position is significant for the performance improvement of the robotic fish.

9 Conclusion

Based on the research on cownose ray, the robotic fish propelled by oscillating pectoral fins is designed according to the requirements of functional bionics and morphology bionics in this chapter. The swimming experiments show that the motion effect of the robotic fish fulfills the design objectives. The key features of the bionic underwater propulsion with pectoral fin can be realized, and the mechanical design is rational. The conclusions are as follows:

1. The motion analysis of the pectoral fin of the robotic fish shows that it can achieve the approximate motion as cownose ray, which also proves that the simplified kinematic model is practical.
2. Multi-fin ray driving mechanism realizes the spanwise section flapping motion of the cownose ray. The kinematic wave length and direction in the robotic fish's pectoral fin can be actively controlled in order to ensure the maneuverability (especially in the excellent turning performance). The functional bionics is truly realized.
3. The design of the permeable flexible skin and the silicon rubber supporting ribs solve the problems caused by the deformation resistance of the pouring skin. The similar streamline shape as the cownose ray is obtained. The morphology bionics is realized to some extent. Pitching stability is a key problem in the pectoral fin propulsion mode, but the developed robotic fish can swim with no obvious pitching motion in the experiments, which is closely benefited from the three-dimensional shape design.

But there are still some problems need to be solved about the robotic fish, the follow-up studies need to be explored further:

1. The thrust tests show that the robotic fish produces negative thrust in about half a cycle time. So finding the primary cause of this phenomenon is an important way to improve the propulsion performance of the kind of robotic fish.
2. The control parameters of the robotic fish's fin rays are given in advance. But it is not like cownose ray that can realize the real-time control of pectoral fin motions according to the variation of the surrounding environments. So realizing the same bionic closed-loop control is a deeper question for the robotic fish development in future.

References

1. Yuru X, Pengchao L (2011) Development trend of underwater robot. Chin J Nat 33 (3):125–132
2. Xuelun P (2004) Research state and development trend of underwater robot. Robot Tech Appl 4:43–47
3. Roper DT, Sharma S, Sutton R et al (2011) A review of developments towards biologically inspired propulsion systems for autonomous underwater vehicles. Proc Inst Mech Eng Part M J Eng Marit Environ 225(2):77–96
4. Naeem W, Sutton R, Ahmad SM et al (2003) A review of guidance laws applicable to unmanned underwater vehicles. J Navig 56(1):15–29
5. Yuh J (2000) Design and control of autonomous underwater robots: a survey. Auton Robot 8 (1):7–24

6. Triantafyllou MS, Techet AH, Hover FS (2004) Review of experimental work in biomimetic foils. IEEE J Ocean Eng 29(3):585–594
7. Bandyopadhyay PR (2005) Trends in biorobotic autonomous undersea vehicles. IEEE J Ocean Eng 30(1):109–139
8. MacIver MA, Fontaine E, Burdick JW (2004) Designing future underwater vehicles: principles and mechanisms of the weakly electric fish. IEEE J Ocean Eng 29(3):651–659
9. Altringham JD, Ellerby DJ (1999) Fish swimming: patterns in muscle function. J Exp Biol 202 (23):3397–3403
10. Lauder GV, Drucker EG (2002) Forces, fishes, and fluids: hydrodynamic mechanisms of aquatic locomotion. News Physiol Sci 17:235–240
11. Blake RW (1983) Swimming in the electric eels and knifefishes. Can J Zool 61(6):1432–1441
12. Yueri C, Shusheng B (2011) Research advances of bionic fish propelled by oscillating paired pectoral foils. J Mech Eng 47(19):30–37
13. Blake RW (2004) Fish functional design and swimming performance. J Fish Biol 65 (5):1193–1222
14. Sfakiotakis M, Lane DM, Davies JBC (1999) Review of fish swimming modes for aquatic locomotion. IEEE J Ocean Eng 24(2):237–252
15. Triantafyllou MS, Triantafyllou GS, Yue D (2000) Hydrodynamics of fishlike swimming. Annu Rev Fluid Mech 32(1):33–53
16. Colgate JE, Lynch KM (2004) Mechanics and control of swimming: a review. IEEE J Ocean Eng 29(3):660–673
17. Bioleau R, Fan L, Moore T (2002) Mechanization of rajiform swimming motion: the making of Robo-ray. Engineering Physics Project Laboratory, University of British Columbia, Vancouver
18. Clark RP, Smits AJ (2006) Thrust production and wake structure of a batoid-inspired oscillating fin. J Fluid Mech 562:415–429
19. Brower T (2006) Design of a manta ray inspired underwater propulsive mechanism for long range, low power operation. Tufts University
20. Moored KW, Taylor SA, Bliss TK et al (2006) Optimization of a tensegrity wing for biomimetic applications. In: Proceedings of 45th IEEE conference on decision and control, San Diego, pp 2288–2293
21. Jun G, Shusheng B, Ji L, Yueri Cai (2011) Design and hydrodynamic experiments on robotic fish with oscillation pectoral fins. J Beijing Univ Aeronaut Astronaut 37(3):344–350
22. Zheng L, Bi S, Cai Y et al (2010) Design and optimization of a robotic fish mimicking cownose ray. In: IEEE international conference on robotics and biomimetics, ROBIO 2010, Tianjin. IEEE computer society, pp 1075–1080
23. Nakamura T, Yamamoto I (2009) Research on fluid analysis simulator for elastic oscillating fin of biomimetic underwater robots. In: ICROS-SICE international joint conference 2009. Fukuoka international congress center, Japan, pp 3886–3889
24. http://www.imae-kagaku.com/mech.htm. Accessed 10 Oct 2014
25. Xu Y, Zong G, Bi S et al (2007) Initial development of a flapping propelled unmanned underwater vehicle (UUV). In: Proceedings of the 2007 IEEE international conference on robotics and biomimetics, Sanya, pp 524–529
26. Gao J, Bi S, Xu Y et al (2007) Development and design of a robotic manta ray featuring flexible pectoral fins. In: IEEE international conference on robotics and biomimetics, Sanya, pp 519–523
27. Cai YR, Bi SS, Zheng LC (2010) Design and experiments of a robotic fish imitating cownose ray. J Bionic Eng 7(2):120–126
28. Wang Z, Wang Y, Li J et al (2009) A micro biomimetic manta ray robot fish actuated by SMA. In: IEEE international conference on robotics and biomimetics, Guilin, pp 1809–1813
29. Elizabeth P (2011) Bio-inspired engineering: manta machines. Science 232:1028–1029
30. http://www.festo.com/cms/en_corp/9786.htm. Accessed 10 Oct 2014
31. Yang SB, Qiu J, Han X (2009) Kinematics modeling and experiments of pectoral oscillating propulsion robotic fish. J Bionic Eng 6:174–179

32. Zhou CL, Low KH (2010) Better endurance and load capacity: an improved design of manta ray robot (RoMan-II). J Bionic Eng 7(Supplement 1):137–144
33. Russo RS (2012) Biomechanical modeling of ray pectoral fins to inform the design of AUV propulsion systems. University of Virginia, Charlottesville
34. Suzumori K, Endo S, Kanda T, Kato N et al (2007) A bending pneumatic rubber actuator realizing soft-bodied manta swimming robot. In: IEEE international conference on robotics and automation, Angelicum University, Rome, pp 4975–4980
35. Ando Y, Kato N, Suzuki H, Suzumori K (2006) Elastic pectoral fin actuators for biomimetic underwater vehicles. In: Proceedings of the sixteenth international offshore and polar engineering conference, San Francisco. ISOPE, Cupertino, pp 260–270
36. Alben S, Madden PG, Lauder GV (2007) The mechanics of active fin-shape control in ray-finned fishes. J Royal Soc Interf V4:243–256
37. Walker JA (2000) Does a rigid body limit maneuverability? J Exp Biol V203:3391–3396
38. Deyuan Z, Jun C, Xiang L et al (2010) Bioforming methods of bionic manufacturing. J Mech Eng 46(5):88–92
39. Cooper AJ, Carpenter PW (1997) The stability of rotating disc boundary-layer flow over a compliant wall. Part I. Types I and II instabilities. J Fluid Mech 350:231–259
40. Heine C (1992) Mechanics of flapping fin locomotion in the cownose ray, Rhinoptera bonasus (elasmobranchii, myliobatidae). Duke University
41. Summers AP, Koob-Emunds MM, Kajiura SM et al (2003) A novel fibrocartilaginous tendon from an elasmobranch fish (Rhinoptera bonasus). Cell Tissue Res 312(2):221–227
42. Schaefer JT, Summers AP (2005) Batoid wing skeletal structure: novel morphologies, mechanical implications, and phylogenetic patterns. J Morphol 264(3):298–313
43. Rosenberger LJ, Westneat MW (1999) Functional morphology of undulatory pectoral fin locomotion in the stingray Taeniura lymma (Chondrichthyes: Dasyatidae). J Exp Biol 202 (24):3523–3539
44. Rosenberger LJ (2001) Pectoral fin locomotion in batoid fishes: undulation versus oscillation. J Exp Biol 204(2):379–394
45. Cai YR, Bi SS, Zheng LC (2012) Design optimization of a bionic fish with multi-joint fin rays. Advanced Robotics 26(1–2):177–196

Soft-Body Robot Fish

Pablo Valdivia y Alvarado and Kamal Youcef-Toumi

Abstract New mechanism paradigms are needed to improve robot performance and resilience in harsh environments. Soft continuous mechanisms can be simpler and more robust than the stiff-discrete assemblies traditionally found in robotics. Herein, a methodology to design underactuated soft robots is presented and its application to fish-like robots is outlined. The principle is simple, a flexible viscoelastic continuous body is designed with the appropriate material heterogeneity such that dominant modes of vibration match desired body motions. Design examples are given for different fish-like robots and actuation units. Fabrication techniques and testing procedures are presented along with locomotion performance. The approach is general and can be applied to bodies with more complex target kinematics and offers a promising alternative for mobile robots.

1 The Need for New Robot Bodies

Many field applications would benefit from unmanned systems that can function and survive for extended periods of time in harsh and unstructured liquid environments. Oceans are a prime example, marine environments are extremely unforgiving on hardware. All marine-bound equipment requires periodic maintenance which limits deployment cycle times. Environment complexity and deployment scale add further challenges. As a result, autonomous marine robots require excellent locomotion efficiency, mechanical robustness, and maneuverability to survive and successfully accomplish mission goals.

P. Valdivia y Alvarado (✉) · K. Youcef-Toumi
Mechanical Engineering Department, Massachusetts Institute of Technology, Cambridge, MA, USA
e-mail: pablov@mit.edu

K. Youcef-Toumi
e-mail: youcef@mit.edu

© Springer-Verlag Berlin Heidelberg 2015

R. Du et al. (eds.), *Robot Fish*, Springer Tracts in Mechanical Engineering,
DOI 10.1007/978-3-662-46870-8_6

Biology is inspiring a wide range of promising locomotion approaches that could potentially address these requirements. However, new hardware paradigms are needed to take full advantage of these approaches. Bio-inspired locomotion requires mechanisms with large numbers of degrees of freedom (DOF) to efficiently transfer momentum to a liquid environment. This poses significant problems as the added mechanical complexity can prevent long-term deployments. If mechanical transmissions are not properly sealed and protected from the environment, corrosion and biofouling can quickly degrade components and prevent long-term deployments. Simplicity and mechanical robustness are key for prolonging hardware survival.

1.1 Rigid and Discrete Versus Soft and Continuous

Traditional robot bodies are rigid and discrete, mechanisms consist of assemblies of gears, pulleys, cables, linkages, etc. As a result, the complexity of mechanical transmissions increases with required DOFs. This can be detrimental to the mechanical robustness of a robot as the probabilities of failure increase with mechanism part count.

In contrast, the natural dynamics of a soft flexible body can be exploited to reduce mechanism complexity. The modes of vibration of a structure depend on geometry, material properties, and excitation (i.e., forces being applied). Hence, a structure can be designed such that the dominant modes of vibration match desired body motions when a reduced number of actuation inputs is applied. If successful, such a mechanism is in principle simple and robust as it only consists of a flexible body and an excitation source. To illustrate this idea, Fig. 1 shows the differences between the traditional approach of designing a fish robot and the proposed

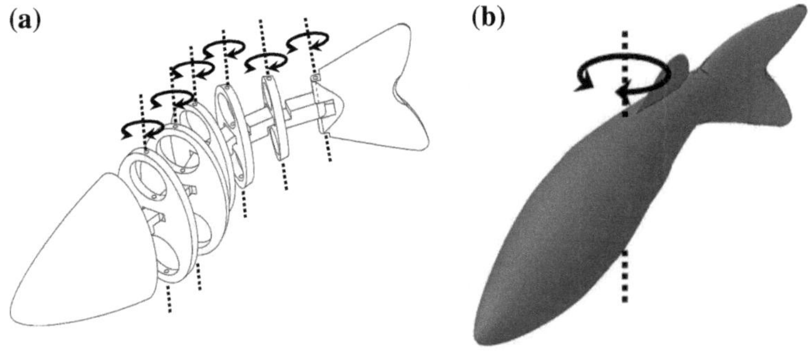

Fig. 1 Traditional discrete-stiff mechanism versus underactuated soft body continuous mechanism approach for a robotic fish: **a** The body undulations needed for locomotion are traditionally replicated using manipulator-like mechanisms with actuators driving each DOF directly or through pulleys, linkages, or geared transmissions; **b** in contrast, a soft body can be manufactured with the appropriate material heterogeneity such that its dominant modes of vibration match the required body motions when excited by a reduced number of actuators

underactuated soft body approach. The body undulations required for fish-like locomotion are traditionally replicated using manipulator-like mechanisms with actuators driving each DOF directly or through transmissions involving pulleys, cables, linkages, and gears. Alternatively, a soft body can be manufactured with a material heterogeneity such that its natural dynamic response matches desired body motions. In addition, a continuous soft body provides better protection from the environment for delicate components as all mechanisms are encapsulated and protected inside the body.

1.2 Previous Work

The basic principles of the underactuated soft body approach, applied to fish-like robots, were first proposed by Valdivia y Alvarado and Youcef-Toumi in [28, 29], and subsequently formalized in [21, 30]. The approach relies on two principles: using soft bodies instead of traditional stiff mechanisms, and allowing passive mechanism dynamics achieve target motions. Earlier work in the areas of manufacturing, passive robotic mechanisms, and biomimetics enabled crucial tools and principles for developing the approach.

Fabricating heterogeneous mechanisms traditionally involves assembling different discrete components and addressing issues related to fastening, sealing, and wear due to friction. Additive prototyping ideas such as shape deposition manufacturing [18] enabled simultaneous fabrication and assembly of components and mechanisms with complex geometries and heterogeneous materials. The tools and techniques developed allowed the manufacture of complex continuum structures and marked a shift away from the lumped-parameter type mechanisms prevalent in automation.

The control of dynamic systems benefits from the knowledge of the physics governing the problem. The idea of exploiting mechanism and environment dynamics to achieve control goals was famously explored by Mcgeer on passive dynamic walkers [17]. Kubow and Full [14] studied similar concepts on cockroach legs. Having the dynamic system play a role in control is a very powerful tool that enables simplification of both mechanism and overall control requirements.

Finally, animal locomotion has provided inspiration for mobile robot design and control in many different areas. Applications in liquid environments sparked interest on fish-like locomotion given its perceived high efficiency. The work by Barrett et al. [4] inspired the development of fish-like robots to understand the hydrodynamics involved in fish locomotion and to develop new mobile robot platforms. Work on fish-like robots has benefited from many different approaches, and as a result they represent an ideal platform to test underactuated soft robot ideas.

Herein, the main points of the underactuated soft body approach are presented along with representative performance measurements. Several practical considerations are discussed and suggestions for further work are presented. With advances in the fields of material synthesis and actuation new and exciting applications of these ideas are emerging.

2 Design Methodology

The main goal of the design methodology is to identify the proper material heterogeneity needed in a soft body to achieve modes of vibration that match desired body motions. The methodology consists of five steps:

(i) Model desired body or appendage kinematics. Derive models of the motions required to achieve the desired locomotion performance. As seen in Fig. 2, these motions can involve only a section of the body, in the case of an appendage, or the entire body.
(ii) Model soft body dynamics. Derive continuum mechanics models of the dynamics (vibrations) in a flexible viscoelastic body. The models usually involve a system of partial differential equations and their complexity is determined by the assumptions regarding geometry and deformation.
(iii) Plug in desired kinematics into dynamic models. The approach does not attempt to find the system response by solving the dynamic equations. The desired solutions, the motions of interest, are the kinematic equations found in (i). Kinematic models can be plugged into the dynamic equations and boundary conditions transforming a set of partial differential equations into a set of algebraic and trigonometric equations.
(iv) Use appropriate constraints to solve the system of equations. For locomotion in fluid environments maintaining neutral buoyancy is important and characteristic geometrical features must also be conserved to achieve the hydrodynamics of interest. These constraints narrow the choices on the density of the materials used and body geometry.

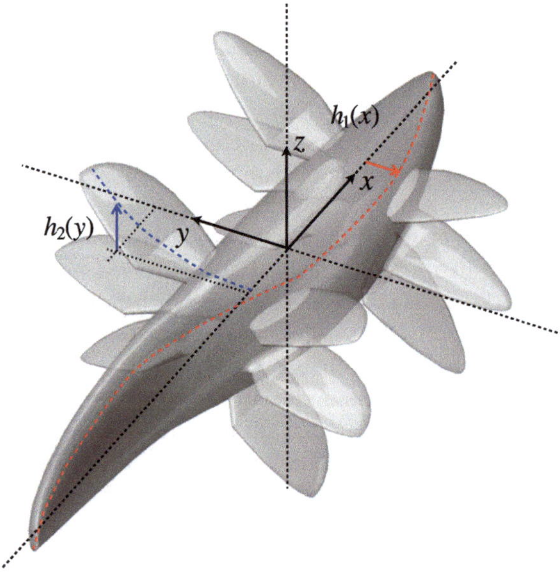

Fig. 2 Design methodology: the body or appendage dynamics of a biological creature is of interest to improve the locomotion performance of man-made vehicles, the kinematics of interest and the dynamics of a flexible body are modeled, the desired motions (e.g., $h_1(x)$, $h_2(y)$) are inserted into a dynamic model to solve for the required material distributions

(v) Solve for model parameters. Finally, an actuation distribution (number, location, magnitude, and frequency) is prescribed and the system of equations is solved using the desired constraints. The model parameters solved for are the material properties of the continuum: modulus and viscosity distributions.

The design methodology is set up as an inverse problem. Rather than solving for the response of a dynamic system (e.g., soft flexible body), model parameters that can enable desired system responses are identified.

To illustrate the approach, in the remainder of this section the design methodology is applied to the design of carangiform (e.g., bass) and thunniform (e.g., tunas) robotic fish. According to the classification of fish swimming modes by Breder [5], carangiform and thunniform fish use mostly tail and caudal fin motions for propulsion. Earlier studies focused on these swimming modes due to their presumed high hydrodynamic efficiencies [4].

2.1 Modeling Body Kinematics

Figure 3 shows the outline of a fish-like viscoelastic body of length l moving at an average speed U. Body motions are described with respect to a Cartesian (x, y, z) reference frame attached to the body frontal end. Carangiform and thunniform fish swim by propagating traveling waves along the length of their bodies. The body traveling wave moves down the body at a speed V and has a wavelength λ. Barrett

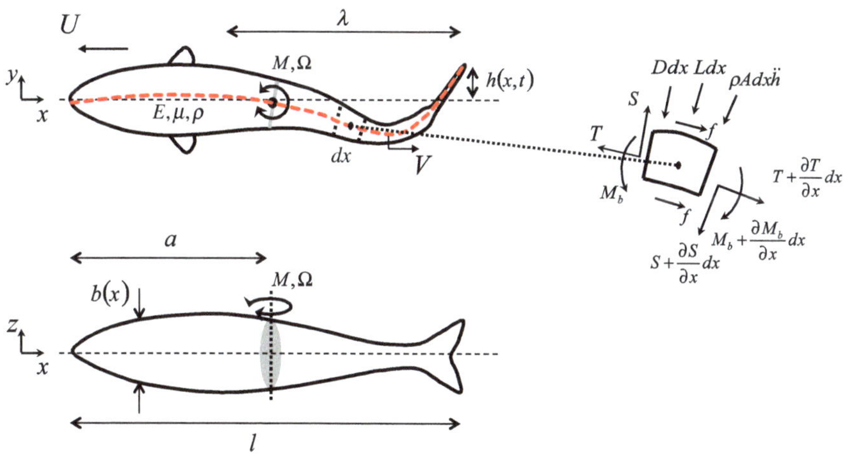

Fig. 3 Fish model: *Top* and *side views* of a fish-like body of length l moving forward at an average speed U. Lateral body motions $h(x, t)$ in the body frame of reference take the form of traveling waves with a constant wavelength λ. A concentrated moment of magnitude M and actuation frequency Ω located at a distance $x = a$ excites body motions. The free body diagram of a section of the body shows relevant forces

et al. [4] proposed a simple model for the body center line lateral deflections h of the form: $h(x,t) = H(x)\sin(\omega t - \kappa x)$, where $H(x) = c_1 x + c_2 x^2$ is an amplitude envelope, ω is the caudal fin flapping frequency, and $\kappa = 2\pi/\lambda$ is the wave number. Valdivia y Alvarado [21] proposed a slightly different model using $H(x) = a_1 + a_2 x + a_3 x^2$ instead, which could be used to describe both carangiform and thunnifrom swimming modes (values for a_1, a_2, and a_3 depend on fish swimming mode). These models are reasonable approximations of fish kinematics in the body frame of reference. For analysis in an inertial reference frame rigid body motions (rotations $h_r(t)x$ and translations $h_t(t)$) must be added [15]. The kinematics can then be modeled as,

$$h(x,t) = H(x)\sin(\omega t - \kappa x) + h_t(t) + h_r(t)x, \tag{1}$$

h_t and h_r can be found using the problem boundary conditions. To facilitate calculations, in this study the definition of the amplitude envelope used is modified to $H(x) = c_1 \cos(c_2 x) + c_3$ (values for c_1, c_2, and c_3 also depend on fish swimming mode). All relevant model parameters are listed in Table 1 along with their definitions. Values for the parameters in Eq. (1) for both fish swimming modes are listed in Table 2 (c_i values are given only for carangiform swimming mode). Figure 4 shows the center line body planar kinematics described by Eq. (1) at four different times, $t = \{\tau/4, \tau/2, 3\tau/4, \tau\}$, during the tail flapping period τ. The amplitude envelope $H(x)$ is shown as dashed black lines. Outlines of the body during the same times are shown in the upper portion of the figure.

2.2 Modeling Body Dynamics

Figure 3 also shows the forces acting on a section of the fish-like viscoelastic body as it is actuated to move by a concentrated moment M_a. The body is assumed to have an elliptical cross section with major and minor radii $R(x)$ and $r(x)$, respectively. For simplicity, bending is assumed to be the dominant deformation mode. Hence, when the concentrated moment is located at a distance $x = a$ from the body front end, $M_a = \delta'(x-a)\, M\sin(\Omega t)$, the dynamics governing body lateral vibrations, $h(x,t)$, are given by [21],

$$\frac{\partial^2}{\partial x^2}\left(EI\frac{\partial^2 h}{\partial x^2} + \mu I\frac{\partial^2 h}{\partial t \partial x^2}\right) = -L - \rho A\frac{\partial^2 h}{\partial t^2} + \delta'(x-a)M\sin(\Omega t)\,, \tag{2}$$

where E, μ, and ρ are the modulus of elasticity, the viscosity, and the density of the materials along the flexible body. $A = \pi R(x)r(x)$ and $I = \frac{\pi}{4}R(x)r(x)^3$ are the local cross-sectional area and the second moment of area of the fish-like body. In this analysis, E and μ are assumed to be functions of x only, and ρ is assumed to be constant throughout the body length. The hydrodynamic force L acting on the body surface is approximated by the added mass effects [21],

Table 1 Model parameters

Parameter	Definition	Units
A	Cross-sectional area at x	(m^2)
a	Location of concentrated actuation moment	(m)
a_1, a_2, a_3	Variables defining the amplitude envelope	(m), (−), (m^{-1})
b	Body depth at x	(m)
β	Added mass coefficient	(−)
C_d	Body drag coefficient	(−)
c_1, c_2, c_3	Alternative variables to define amplitude envelope	(m), (rad/m), (m)
E	Body modulus of elasticity (a function of x only)	(N/m^2)
ε	Body strain	(−)
η	Forward propulsive efficiency	(−)
H	Amplitude enveloped	(m)
H_l	Caudal fin peak-to-peak flapping amplitude	(m)
h	Body lateral deflections (a function of x and t)	(m)
I	Second moment of area at x	(m^4)
i	Current consumed by all actuators	(A)
κ	Wave number, $\kappa = 2\pi/\lambda$	(rad/m)
L	Hydrodynamic force normal to the body surface	(N)
l	Body length	(m)
λ	Wavelength of body traveling wave	(m)
M_a	Concentrated actuation moment	(Nm)
M	Magnitude of concentrated actuation moment	(Nm)
m	Added mass (a function of x only)	(kg)
μ	Body viscosity (a function of x only)	(Pa s)
Ω	Frequency of actuation moment	(rad/s)
ω	Caudal fin flapping frequency	(rad/s)
R	Body cross section major radius (a function of x only)	(m)
r	Body cross section minor radius (a function of x only)	(m)
ρ	Body density (assumed to be constant)	(kg/m^3)
ρ_f	Density of liquid medium	(kg/m^3)
T	Forward propulsive force (thrust)	(N)
t	Time	(s)
τ	Caudal fin flapping period	(s)
U	Average forward swimming speed	(m/s)
V	Backward propagating speed of traveling wave	(m/s)
v	Voltage input to all actuators	(V)

Table 2 Parameters for the two proposed kinematic models

Swimming mode	κ	a_1	a_2	a_3	c_1	c_2	c_3
Carangiform	$\frac{7}{l}$	$\frac{0.004\omega l}{2\pi}$	$-\frac{0.02\omega}{2\pi}$	$\frac{0.04\omega}{2\pi l}$	$\frac{\cos^{-1}(0.13l-1)-\pi}{7l/10}$	$\pi - c_1 0.3l$	$1+0.2l$
Thunniform	$\frac{9}{l}$	$0.02l$	-0.12	$\frac{0.2}{l}$			

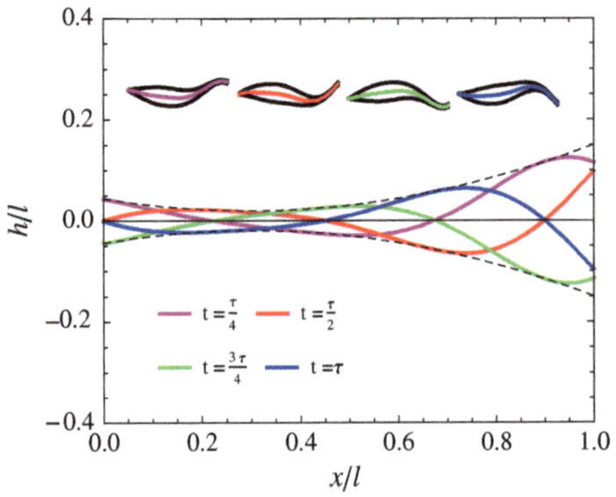

Fig. 4 Fish-like body kinematics as described in Eq. (1). Body center line is shown at four different times during the swimming cycle. *Dashed black lines* show the amplitude envelope $H(x)$. Outlines of a fish-like body during the same cycle times are also shown

$$L \approx m\frac{\partial^2 h}{\partial t^2} \tag{3}$$

where $m = \frac{1}{4}\beta\pi b^2\rho_f$ is the added mass of a cross section at x, β is a geometry dependent constant, $b = 2R(x)$ is the body depth at x, and ρ_f is the density of the fluid medium. Equation (2) assumes that the constitutive relations for the visco-elastic body materials are of the form,

$$\sigma = E\varepsilon + \mu\frac{\partial \varepsilon}{\partial t}. \tag{4}$$

where σ and ε are stress and corresponding strain of the materials. Since the fish-like body is moving freely in a liquid medium both shear and moments at the body ends ($x = 0$ and $x = l$) must be zero. The corresponding boundary conditions can be defined as [21],

$$\begin{aligned}
\left[EI\frac{\partial^2 h}{\partial x^2} + \mu I\frac{\partial^2 h}{\partial t \partial x^2} - M_a\right]_{x=0} &= 0\\
\left[EI\frac{\partial^2 h}{\partial x^2} + \mu I\frac{\partial^2 h}{\partial t \partial x^2} - M_a\right]_{x=l} &= 0\\
\left[\frac{\partial}{\partial x}\left(EI\frac{\partial^2 h}{\partial x^2} + \mu I\frac{\partial^2 h}{\partial t \partial x^2} - M_a\right)\right]_{x=0} &= 0\\
\left[\frac{\partial}{\partial x}\left(EI\frac{\partial^2 h}{\partial x^2} + \mu I\frac{\partial^2 h}{\partial t \partial x^2} - M_a\right)\right]_{x=l} &= 0
\end{aligned} \tag{5}$$

Internal components (batteries, actuators, sensors, etc.) have different densities. To account for the effects of embedded hardware a simple modification in the main dynamic equations is needed. In the case of a fish-like body, Eq. (2) can be modified to,

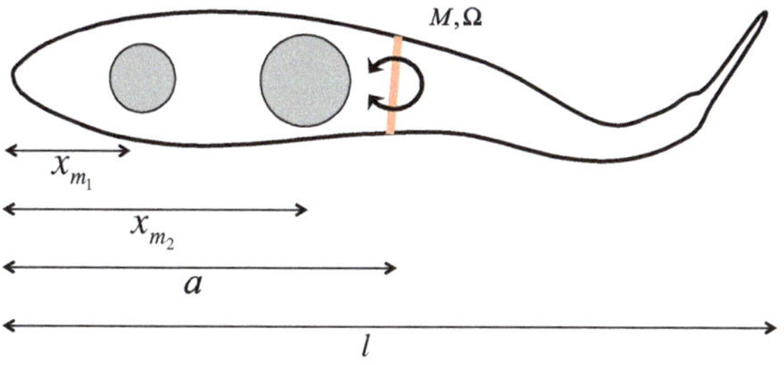

Fig. 5 Batteries and other components can be modeled as lumped masses inside a flexible body. The example in the figure shows two regions of concentrated mass as *two gray circles*. Their contributions to the dynamic behavior is described by Eq. (6)

$$\frac{\partial^2}{\partial x^2}\left(EI\frac{\partial^2 h}{\partial x^2}+\mu I\frac{\partial^2 h}{\partial t\partial x^2}\right)=-L-\left(\rho A+\sum_{i=0}^{n}\delta(x-x_{m_i})m_i\right)\frac{\partial^2 h}{\partial t^2}+\delta'(x-a)M\sin(\Omega t) \tag{6}$$

where the summation term accounts for the lumped masses, m_i, of n embedded component, located at distances x_{m_i} along the body. Figure 5 shows a modified body model example with two concentrated masses; m_1 and m_2.

The dynamic models in Eqs. (2) and (6) account for bending deformations only. However, shear and inertia terms can have large contributions and should be included when implementing body motions involving large curvature variations. Timoshenko and Cosserat beam models are examples of more detailed dynamic descriptions that account for these terms and can enhance the analysis accuracy [2, 20]. Cosserat theory can also be used for more complex three-dimensional kinematics.

2.3 Problem Constraints

Material and geometrical constraints must also be imposed. To facilitate depth control, the body should have a constant density ρ close to the density of the liquid medium ρ_f,

$$\rho \approx \rho_f. \tag{7}$$

It is also important for the body geometry to maintain particular features that enable characteristic fish hydrodynamics. In particular, carangiform body features include a streamlined front end, a large concentration of volume and mass toward the front of the body immediately behind the nose, a tapering peduncle region, and a high aspect ratio caudal fin. Thunniform bodies present a more streamlined body

Table 3 Body geometry parameters

Swimming mode	R_1	R_2	R_3	R_4	r_1	r_2	r_3	r_4
Carangiform	$0.14l$	$\frac{2\pi}{1.6l}$	$0.00008l$	$\frac{2\pi}{1.1l}$	$0.045l$	$\frac{2\pi}{1.25l}$	$0.06l$	$\frac{2}{l}$
Thunniform	$0.1l$	$\frac{2\pi}{1.57l}$	$0.00008l$	$\frac{2\pi}{0.81l}$	$0.055l$	$\frac{2\pi}{1.25l}$	$0.08l$	$\frac{2}{l}$

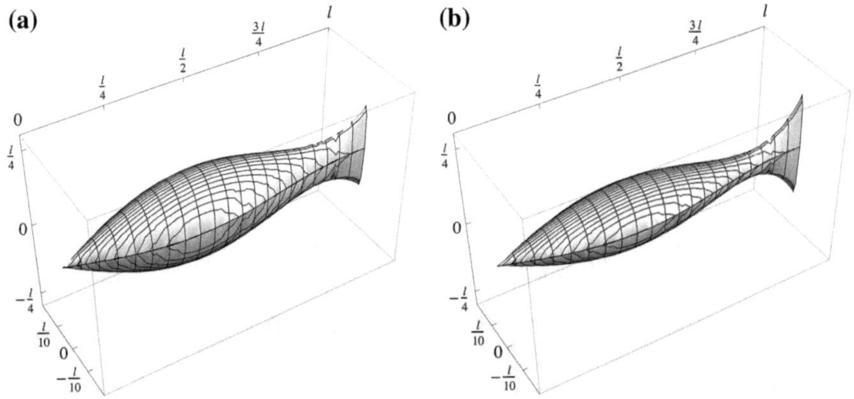

Fig. 6 **a** Fish body geometry for a carangiform model based on Eq. (8). **b** Fish body geometry for a thunniform model based on Eq. (8). Thunniform geometry has a more streamlined profile with a thinner peduncle and a higher aspect ratio caudal fin

with a thinner peduncle and a higher aspect ratio caudal fin with a lunate shape. For carangiform and thunniform swimmers, geometrical features can be modeled using the following expressions for the major and minor body cross-sectional radii [21],

$$\begin{aligned} R(x) &= R_1 \sin(R_2 x) + R_3(e^{R_4 x} - 1) \\ r(x) &= r_1 \sin(r_2 x) + r_3 \sin(r_4 x) \end{aligned} \tag{8}$$

Values for the parameters R_i and r_i, for both carangiform and tunniform bodies, are listed in Table 3. Example geometries for both body types, as defined by Eq. (8), are shown in Fig. 6. Pectoral, dorsal, and anal fins are omitted for now as the analysis concerns forward motion which is achieved primarily through body and caudal fin motions.

2.4 Solving the Inverse Problem

The design objective is to find the material distributions along a viscoelastic body such that its forced vibrations match required body motions. In the case presented here, the desired forward swimming fish kinematics, described in Eq. (1), are plugged in Eqs. (2), (3), and (5), which describe the dynamics governing the motions of the body. Equation (2) is used rather than Eq. (6) for simplicity but the approach to solving the problem is the same.

Assuming the system is linear, the frequency of body oscillations should match the actuation frequency, thus $\Omega = \omega$. After integrating Eq. (2) twice, it can be rewritten as,

$$\begin{aligned} &\cos(\omega t)[EI\gamma_3 + \mu I\omega\gamma_4] + \sin(\omega t)[EI\gamma_4 - \mu I\omega\gamma_3 - u(x-a)M] \\ &\quad = \cos(\omega t)S_1 + \sin(\omega t)S_2 \end{aligned} \tag{9}$$

where $u(x - a)$ is a unit step function and γ_3, γ_4, S_1, and S_2 are given by,

$$\begin{aligned} \gamma_3 &= \left(\frac{\partial^2 H}{\partial x^2} - H\kappa^2\right)\cos(\kappa x) - 2\frac{\partial H}{\partial x}\kappa\sin(\kappa x) \\ \gamma_4 &= \left(\frac{\partial^2 H}{\partial x^2} - H\kappa^2\right)\sin(\kappa x) - 2\frac{\partial H}{\partial x}\kappa\cos(\kappa x) \\ S_1 &= \int_0^l\int_0^l [(\rho A + m)H\omega^2\cos(\kappa x)]\mathrm{d}x\mathrm{d}x \\ S_2 &= \int_0^l\int_0^l [(\rho A + m)H\omega^2\sin(\kappa x)]\mathrm{d}x\mathrm{d}x \end{aligned} \tag{10}$$

Expressions for $E(x)$ and $\mu(x)$ are found for a particular design frequency, ω_d, by solving the trigonometric relation in Eq. (9),

$$\begin{aligned} E(x) &= \frac{S_1\gamma_3 + S_2\gamma_4 + u(x-a)M\gamma_4}{I(\gamma_3^2 + \gamma_4^2)} \\ \mu(x) &= \frac{S_1\gamma_4 - S_2\gamma_3 - u(x-a)M\gamma_3}{I\omega(\gamma_3^2 + \gamma_4^2)} \end{aligned} \tag{11}$$

Solution examples for a carangiform body shape of length $l = 1$ m and forward swimming kinematics as defined in Table 3 (with $\omega = \omega_d = 17\,\mathrm{rad\,s^{-1}}$) are shown in Fig. 7. Both $E(x)$ and $\mu(x)$ are affected by changes in the amplitude M and location a of the actuation and need not always have positive values. When the actuation is a single concentrated moment, the required modulus of elasticity reaches a maximum close to the caudal peduncle region ($x \sim 0.75l$). For a given magnitude M, moving the actuation location a toward the front of the body ensures that $E(x)$ remains positive until $x \sim 0.8l$. Increasing a creates a negative-valued region around $x \sim 0.35l$ (see Fig. 7a). For a given actuation location, increasing the moment amplitude increases peak values (see Fig. 7b). In all scenarios, $E(x)$ displays a negative region at the caudal end, $x > 0.85l$.

With the same actuation, the required viscosity is negative for most of the frontal part of the body and reaches a positive maximum in the caudal fin region at $x \sim 0.9l$. Moving the actuation location toward the front of the body increases the

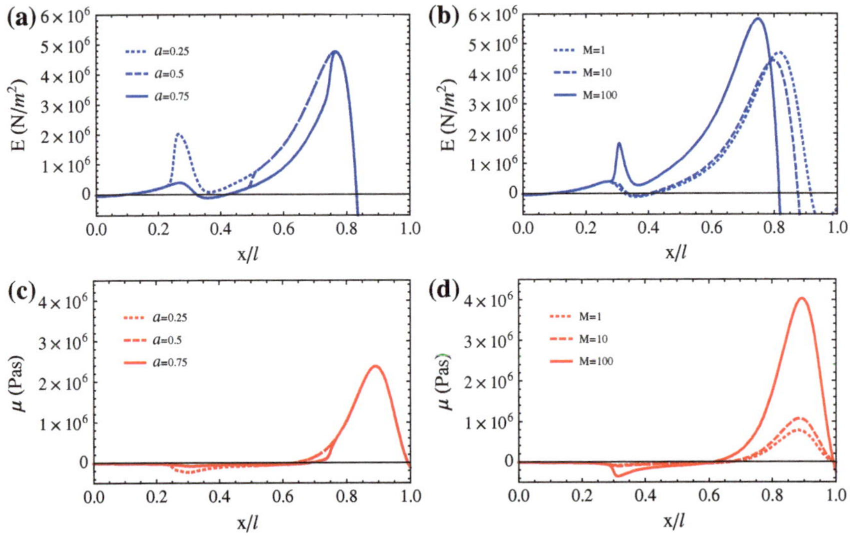

Fig. 7 Required modulus E and viscosity μ distributions along the length of a carangiform body of length $l = 1$ m undergoing forward swimming motions as defined by Eq. (1) and the parameters in Table 3 ($\omega = \omega_d = 17\,\mathrm{rad\,s^{-1}}$). Distributions are continuous and differentiable. **a** Effects of actuator location $a[m]$ on $E(x)$. **b** Effects of actuation magnitude $M[N]$ on $E(x)$. **c** Effects of actuator location $a[m]$ on $\mu(x)$. **d** Effects of actuation magnitude $M[N]$ on $\mu(x)$

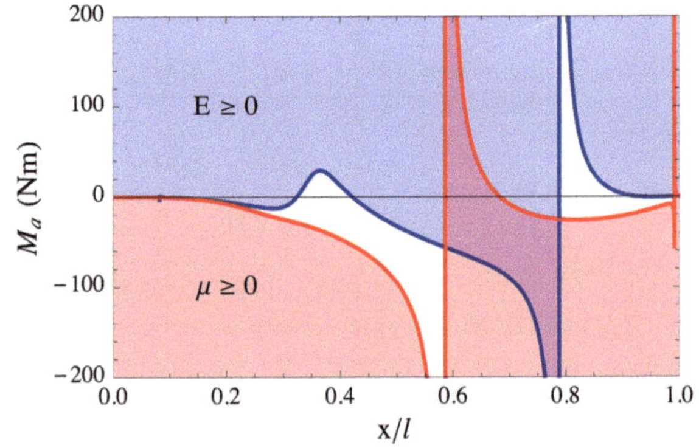

Fig. 8 Bounds for actuation distribution to achieve positive material distributions. *Blue* region satisfies the condition $E \geq 0$, and *red* region satisfies the condition $\mu \geq 0$. Overlapping regions satisfy both conditions and can be used to guide the selection of an actuation distribution $M_a(x)$

negative region (see Fig. 7c). Increasing actuation amplitude increases the values of positive ($x \sim 0.9l$) and negative ($x \sim 0.3l$) peaks (see Fig. 7d).

Materials displaying negative modulus of elasticity and negative viscosity are physically realizable, but are more challenging to manufacture. Instead, the analysis of Eq. (11) can provide guidelines for the actuation distributions that would yield positive values for both $E(x)$ and $\mu(x)$. Figure 8 shows actuation bounds that would guarantee both $E(x)$ and $\mu(x)$ are positive. The bounds are found by solving the system of inequalities $E \geq 0$, $\mu \geq 0$. The region in blue represents the actuation distribution that would ensure $E \geq 0$ for $x \in [0, l]$ and the region in red represents the actuation distribution that would ensure $\mu \geq 0$ for $x \in [0, l]$. The overlapping regions satisfy both conditions and can guide the construction of an actuation distribution.

This example shows a simple implementation of the soft underactuated robot design methodology. Practical issues regarding proper manufacturing are addressed in Sect. 4.

2.5 Optimization

The problem of solving for material and actuation properties can also be set as a constrained optimization. Minimizing the number of actuators, actuation amplitude, or energy consumption, and maximizing efficiency, swimming speed, or thrust are all feasible. Initial work in this area of optimization can be found in [21]. In this study details of this approach are omitted, but the performance models presented in Sect. 5 are linked to body kinematics and can hence be used to solve for material and actuation properties. This particular approach will be presented in more detail in future studies.

3 Soft Body Fish Designs

Prototyping and actuating a soft viscoelastic fish-like body with the required material property distributions $E(x)$ and $\mu(x)$ described by Eq. (11) present several challenges. A mechanical implementation of concentrated moments and forces can be achieved with standard actuation. While the mechanisms used at this stage are not entirely soft, they allow the feasibility of the proposed approach to be tested. Fabrication of the required continuous material distributions is more challenging and a simple casting approach that can approximate the required distributions is discussed.

Figure 9 shows two examples of mechanisms used to approximate a concentrated moment inside flexible viscoelastic bodies. The torque of a servomotor is transferred to a plate through a four-bar-linkage type flexure (see Fig. 9a) or a cable connection (see Fig. 9b). The servo mounting base and the plate are embedded at different locations inside the flexible body and are made of materials at least one order of magnitude stiffer than the flexible body continuum. As a result the actuation creates bending in the viscoelastic body about an axis located in between the servo and the plate. In both cases, the flexure and cable regions need to move independently of the surrounding body which requires a low shear modulus material layer (lubrication) surrounding the moving components.

Figures 10 through 13 show designs of soft robot fish inspired by carangiform and thunniform swimmers. Internal components are fabricated using fusion

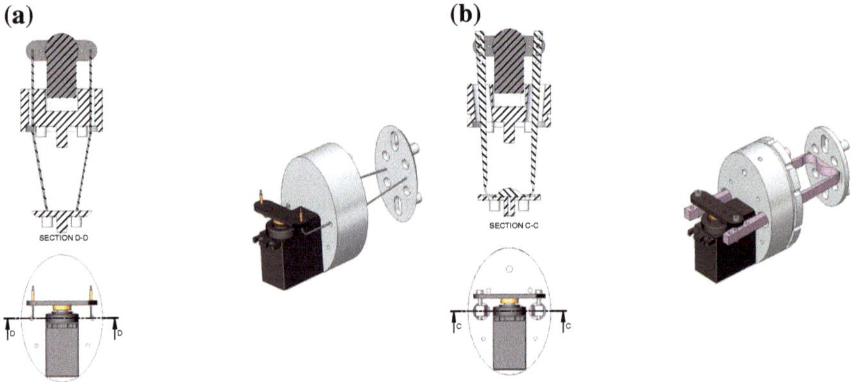

Fig. 9 Concentrated moment actuation. A servomotor is mounted to a stiff base embedded within a viscoelastic body (body not shown in drawings). The motor shaft is connected to a stiff plate, also embedded inside the flexible body at a distance away from the stiff base. Two designs allow force transmission from the servomotor to the stiff plate: **a** Cable/tendon drive. **b** Four-bar linkage flexure

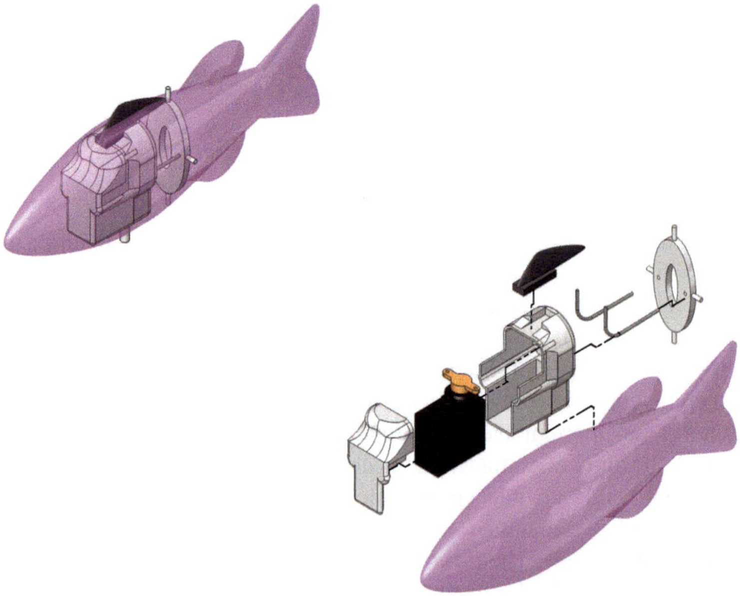

Fig. 10 Design for a soft robot fish inspired by a bass. Carangiform-like body with a caudal fin, two dorsal fins, and one anal fin. Body is powered by a single servo. The design consists of eight individual components including the viscoelastic body

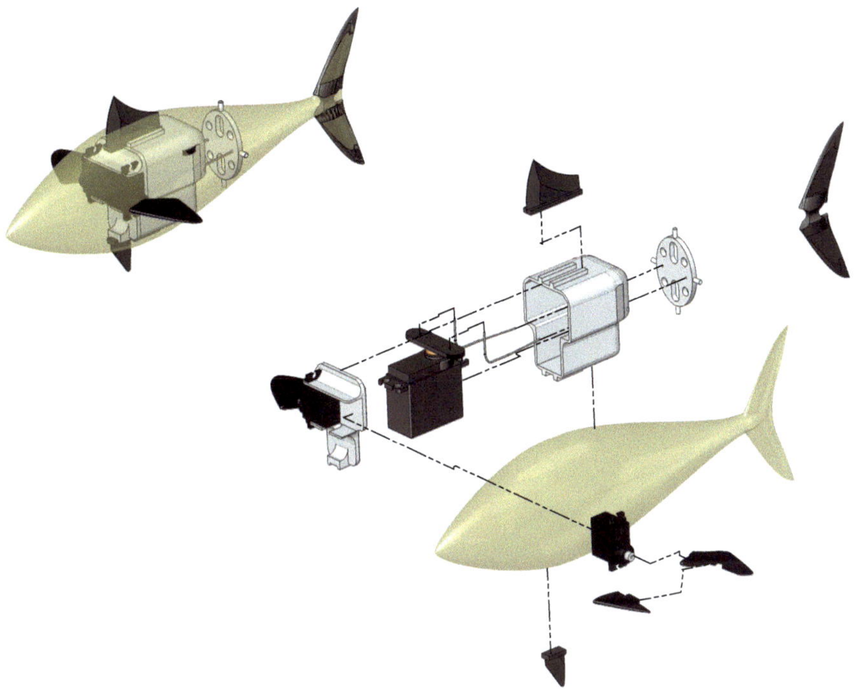

Fig. 11 Design for a soft robot fish inspired by a mackerel. Thunniform-like body with two individually controlled lateral fins, one dorsal fin, one ventral fin, and a high aspect ratio caudal fin. Body motions are powered by a single servo. The design consists of 16 individual components including the viscoelastic body

deposition manufacturing with ABS and polycarbonate plastics. Designs are simple, robust, and involve small part numbers compared to traditional bio-inspired robots. One actuator is sufficient to excite modes of vibration required for forward locomotion. Figure 10 shows the design of a carangiform type swimmer inspired by a bass. The design uses a cable-driven actuation unit and only 8 components are needed including the flexible body. Figure 11 shows the design of a thunniform type swimmer inspired by a mackerel. The design uses a cable-driven actuation unit and has two independently driven pectoral fins. The design consists of 16 components including the flexible body. Figure 12 shows the design of a thunniform type swimmer inspired by a shark. The design uses a flexure-driven actuation unit and has two independently driven pectoral fins. The prototype is fully autonomous and has power (lithium-ion cells) and a control unit (single board computer) embedded within the body. The design consists of 16 components including the flexible body. Figure 13 shows the design of a thunniform type swimmer inspired by a tuna. The design uses a cable-driven actuation unit and has two independently

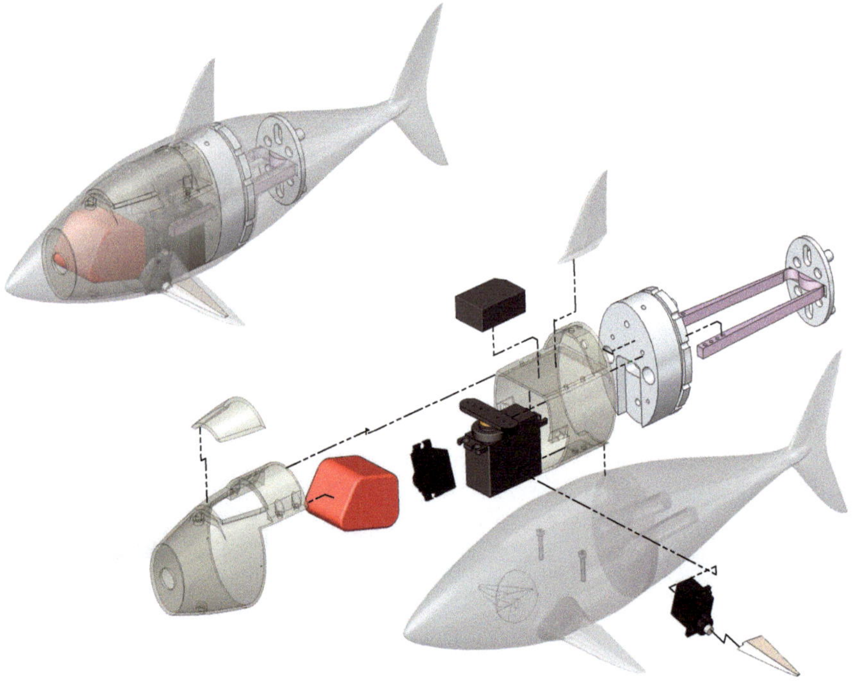

Fig. 12 Design for a soft robot fish inspired by a shark. Thunniform-like body with two individually controlled lateral fins, one dorsal fin, and a high aspect ratio caudal fin. Body is powered by a single servo. The design consists of 16 individual components including the viscoelastic body. Batteries and control unit are embedded inside the body

driven pectoral fins. The prototype is fully autonomous and has power (lithium-ion cells) and a control unit (single board computer) embedded within the body. The design consists of 21 components including the flexible body. The robotic mackerel, shark, and tuna designs have more streamlined peduncles and higher aspect ratio caudal fins with inserts to increase flexural rigidity. All prototypes are neutrally buoyant and do not use buoyancy control units. The mackerel, shark, and tuna designs control depth by changing lift forces as they swim forward using individually controlled pectoral fins. The fins also enable yaw and pitch turns as well as limited roll motions. Standard servomotors are used for simplicity but the approach can also be used with more sophisticated actuation such as electroactive polymers [3]. Wireless communications are used to update and control autonomous units.

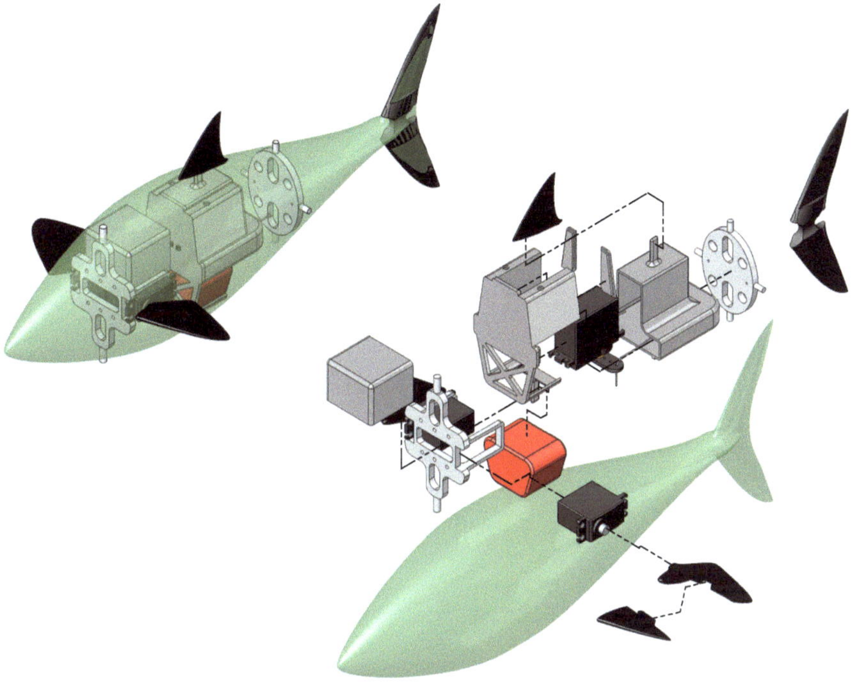

Fig. 13 Design for a larger scale soft robot fish inspired by a tuna. Thunniform-like body with two individually controlled lateral fins, one dorsal fin, and a high aspect ratio caudal fin. Body is powered by a single servo. The design consists of 21 individual components including the viscoelastic body. Batteries and control unit are embedded inside the body

4 Fabrication

Continuous material distributions as in Eq. (11) are needed to achieve a flexible body with the proper modes of vibration. At present, these types of distributions cannot be easily manufactured via standard or rapid prototyping techniques. The required material distributions are instead discretized and cast sequentially using silicone polymers. Given the symmetry of carangiform and thunniform geometries, a two part mold is often sufficient to cast the body. Pectoral fins can be added through a secondary casting process. The concept and the steps involved in the casting process are shown in Fig. 14. A two part mold of the fish-like volume is machined from a wax block. The molds can also be printed using fused deposition manufacturing (FDM) or other three-dimensional printing technologies. The advantage of wax and the plastics used in FDM is that they naturally do not form a chemical bond to silicone polymers and the application of a thin mold release layer is enough to facilitate demolding.

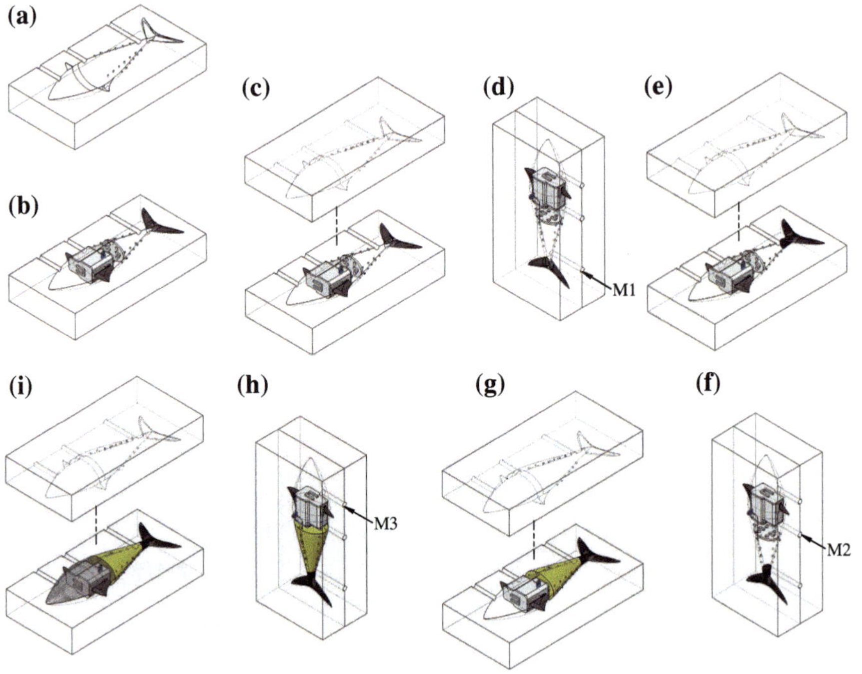

Fig. 14 A two part mold is used for fish-like bodies. Mold components are 3D printed or machined from wax blocks and material distributions are approximated using casting techniques. The original modulus and viscosity distributions are discretized in n sections and each section is casted consecutively using different viscoelastic polymers with matching material properties. The steps shown are for a thunniform prototype where the flexible body is cast with three different material regions. Manufacturing steps: **a** A mold half with three molding ports is prepared for casting; **b** internal components are positioned inside the mold half; **c** opposite mold half is aligned and the mold is closed; **d** first material (M1) is poured in and mold is oriented to constrain material location during curing; **e** after curing mold is opened to remove any material excess; **f** mold is closed, a second material (M2) is poured in and the mold is reoriented for curing; **g** after curing mold is opened to remove material excess; **h** mold is closed, third and last material (M3) is poured in, and the mold is reoriented for curing; **i** after final curing the mold is opened and the prototype is ready

Figure 14a shows a mold half with three molding ports. Previously printed internal mechanisms and housing components are positioned inside the mold half as shown in Fig. 14b. The opposite mold half is aligned and the mold is closed (Fig. 14c). The mold is reoriented to allow proper leveling during the casting process (Fig. 14d). The discretized material distribution is cast by sections starting from one end of the body. An appropriate number of ports should be added in a mold accordingly. Section lengths are controlled through the volume of silicone polymer poured inside the mold and any excess can be removed after demolding. Figure 14e shows the mold being opened for material excess removal after casting the tail section. After excess removal the mold is closed and reoriented to enable

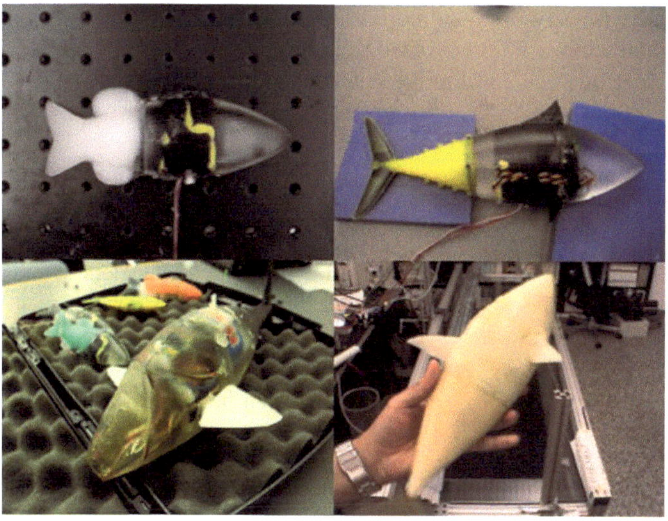

Fig. 15 Finished prototypes. Clockwise from *top left* small carangiform body (*bass*) resulting from design in Fig. 10, medium-sized thunniform body (*mackerel*) resulting from design in Fig. 11, medium-sized thunniform body (*shark*) resulting from design in Fig. 12, larger sized thunniform body (*tuna*) resulting from design in Fig. 13. Different viscoelastic materials appear with different *colors*

casting of the following section as shown in Fig. 14f. Newly cast silicone sections form a chemical bond with previously cast sections provided common surfaces are not contaminated during the process. The casting process continues in the same manner, section by section until the body is complete (Fig. 14g–i). The example body shown in Fig. 14 only has three discrete sections.

Platinum cure silicone rubbers are used for casting body sections due to the compatibility of their material properties with required modulus and viscosity ranges. Silicone rubbers cure at room temperature but the process can also be accelerated using an oven. Different silicone compounds were characterized using a dynamic mechanical analyzer [21]. The viscoelastic behavior of silicone polymers is highly nonlinear but for moderate strains, the constitutive relations governing stress–strain relations can be reasonably approximated by Eq. (4).

Figure 15 shows four completed prototypes of the designs in Figs. 10 through 13. Different materials can be distinguished by their color and surface appearance. The body surface in all prototypes is continuous and discontinuity-free as a result of the casting process. All delicate components are encapsulated inside the bodies and protected from external environment conditions.

Unfortunately, discretization of the continuous material distributions can lead to errors in body motions and a decrease in locomotion performance. In addition, within a discrete section, a single material often cannot satisfy both required modulus of elasticity and viscosity values. A simple solution is to attempt to match the required viscosity first, and correct modulus discrepancies with purely elastic embedded structures. Future work on material synthesis is crucial as is the capability to tailor material constitutive relations.

5 Locomotion Performance

Locomotion performance is characterized using three parameters: average forward swimming speed U, propulsive thrust T, and locomotion efficiency η. Based on Lighthill's elongated body theory [15], simple locomotion performance models can be derived.

The average forward propulsive thrust, T, can be approximated by,

$$T \sim m_l H_l^2 \left(\omega^2 - U^2 \left(\kappa^2 + \frac{1}{l^2} \right) \right) \tag{12}$$

where H_l and m_l are the peak-to-peak oscillation amplitude and the added mass at the caudal fin tip ($x = l$), respectively. The average swimming speed, U, can be approximated by,

$$U \sim \omega H_l \sqrt{\frac{m_l}{\rho_f C_d l^2 + m_l H_l^2 \left(\kappa^2 + \frac{1}{l^2} \right)}} \tag{13}$$

where C_d is the estimated body drag coefficient. The total propulsive efficiency η is given by,

$$\eta = \frac{UT}{vi} \tag{14}$$

where v and i are the voltage supplied and the current consumed by the prototype's actuators. The body kinematics can also be approximated by simplifying Eq. (2) using an order of magnitude analysis. The caudal fin peak-to-peak oscillation amplitude H_l can be approximated as,

$$H_l \sim \frac{M}{(l-a)^2 \sqrt{\left(\frac{EI}{(l-a)^4} - (\rho A + m_l)\omega^2 \right) + \left(\frac{\mu I \omega}{(l-a)^4} \right)}} \tag{15}$$

and the wave number κ is given by,

$$\kappa = \frac{1}{l} \tan^{-1} \left(\frac{\frac{\mu I \omega}{(l-a)^4}}{\frac{EI}{(l-a)^4} - (\rho A + m_l)\omega^2} \right) \tag{16}$$

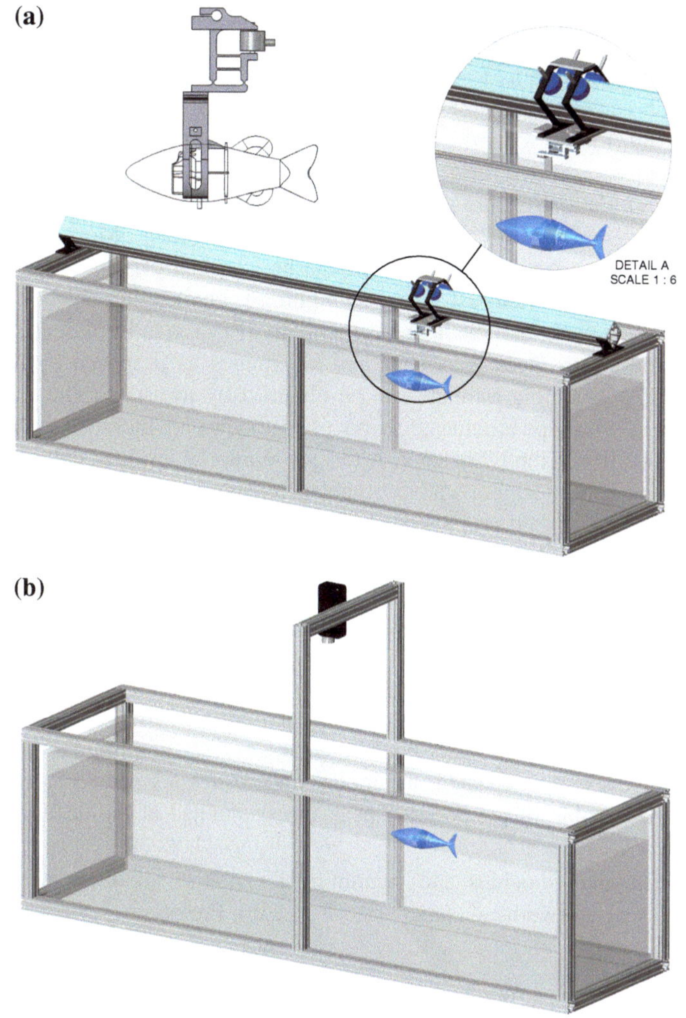

Fig. 16 Experiment setup: 2m × 0.5m × 0.5m acrylic tank surrounded by aluminum frame. **a** A beam support for an air-bearing carriage is mounted along the length of the tank. The carriage is fitted with a tension–compression load cell for propulsive force measurements. **b** An overhead camera is used for the free swimming and body kinematics experiments

Three different experiments were used to test the models in Eqs. (12) through (16): propulsive force measurements, free-swimming speed measurements, and free-swimming kinematic analysis. Details of the experimental setup are shown in Fig. 16. The main testing apparatus consists of a 2 m long, 0.5 m wide, and 0.5 m deep water tank. The tank was built using transparent acrylic sheets to provide visibility of its contents from a wide variety of angles. An aluminum frame surrounds the tank and it is used to mount cameras, lighting, and other testing equipment. An aluminum beam can be mounted on top and along the length of the tank to support a carriage. Prototypes can be mounted to the carriage bottom which is outfitted with load cells to enable propulsive force measurements. The carriage frame is fabricated entirely of low-density plastics to minimize inertia and it is supported by four vacuum preloaded air bearings to minimize friction. To accommodate the surface finish required to use air bearings in a long test section two long float glass strips are glued to the upper beam surfaces. The glass surface finish satisfies the bearings surface roughness requirements. More details of the carriage design and bearing arrangement can be found in [21].

Measuring propulsive thrust in free swimming is challenging. Propulsive forces can be estimated using particle image velocimetry (PIV) as in Epps et al. [11], or by analyzing the vortex wake geometry and kinematics as proposed by Valdivia y Alvarado [27]. A simple estimate can be obtained by measuring the static forces generated as a prototype is actuated to move. Prototypes are mounted to the carriage and held static inside the water tank. A tension–compression load cell arrangement in the carriage bottom supports a prototype and enables measurements of the forces generated during forward propulsion (see Fig. 16a). As the prototypes are actuated to move, the inline forces generated are recorded by the load cell arrangement. All experimental results in this section are obtained using the small carangiform prototype design shown in Fig. 10. The prototype used had a length $l = 0.15$ m and an average density $\rho = 900\,\text{kg/m}^3$.

Figure 17 shows measured static thrusts versus caudal fin flapping frequencies along with the model prediction based on Eq. (12) for a small bass prototype. Static thrust measurements are an upper bound estimate of free-swimming thrust values at a given frequency. As expected, the model underestimates measured static thrust values but is capable of predicting a peak thrust attained at the design frequency $\omega_d = 17(\text{rad s}^{-1})$.

Free-swimming velocities and swimming kinematics are recorded inside the same tank using an overhead camera (see Fig. 16b). Prototypes have three markers

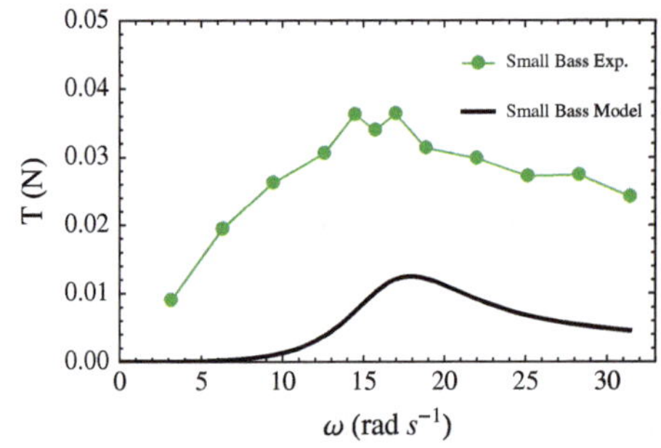

Fig. 17 Measured static thrust T versus model prediction on a small carangiform prototype. Static thrust measurements are expected to be larger than free-swimming values at a given frequency. The model underestimates, as expected, static thrust values but correctly predicts that a peak thrust is attained at the design frequency $\omega_d = 17(\text{rad s}^{-1})$

attached along their dorsal (top) edge: one marker at the frontal tip, a second marker close to the center of mass, and a third marker at the tip of the caudal fin. A MATLAB routine is used to post-process footage of the free-swimming runs and track marker locations. The largest prototype tested inside the tank had a body length $l = 0.3$ m so wall effects were negligible. Figure 18a shows a sample image of the overhead view of a bass prototype during free swimming, a series of superimposed processed images showing the trajectories of the three markers are shown in Fig. 18b, and the corresponding marker trajectories versus time are shown in Fig. 18c. Figure 19a shows measured caudal fin peak-to-peak flapping amplitude H_l for the same bass prototype, obtained from the trajectory of the marker placed at

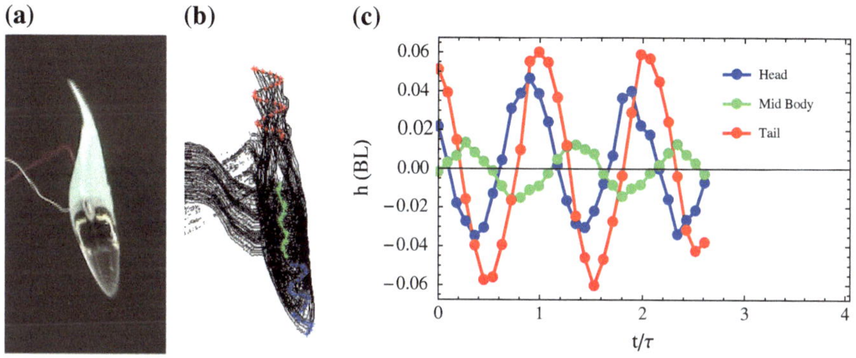

Fig. 18 **a** *Top view* of a small carangiform prototype during free swimming. **b** Superimposed contours of the prototype body during free swimming. Markers are located at the front tip (*blue*), center of mass (*green*), and caudal fin tip (*red*). **c** Trajectories of body markers, with respect to body average trajectory, plotted versus time. Both wavelength λ and wave number κ can be calculated from the phase differential between font tip and caudal fin marker trajectories. H_l is obtained from the amplitude of the caudal fin marker trajectory

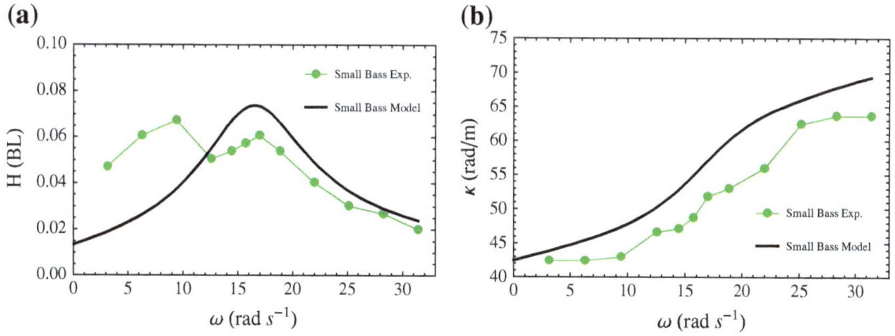

Fig. 19 **a** Caudal fin peak-to-peak flapping amplitude H_l measurements and model prediction versus fin flapping frequency. **b** Measured wave number κ and model prediction versus flapping frequency

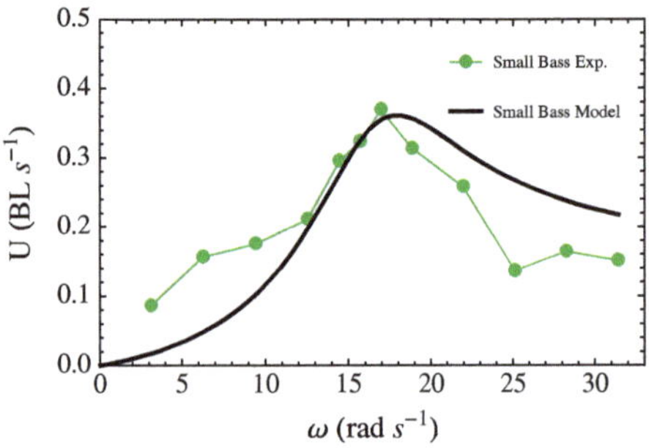

Fig. 20 Measured average swimming velocity U versus model prediction on a small carangiform prototype. The model uses a drag coefficient C_d estimated by towing the unactuated body at different speeds. The model correctly predicts a peak velocity occurs at the design frequency $\omega_d = 17(\text{rad s}^{-1})$

the caudal fin tip, versus the prediction from Eq. (15). The measured wave number κ, based on the phase between the trajectories of the front and caudal fin markers, and the prediction from Eq. (16) is shown in Fig. 19b. Average swimming speed U versus caudal fin flapping frequencies is shown in Fig. 20 along with model predictions based on Eq. (13).

To measure propulsive efficiency, the voltage v supplied and the current i consumed by all actuators were recorded. The prototypes used in the experiments are connected to an external power source. The connection cables are flexible and thin so as to minimize any interference with the prototype's natural swimming motions. The total propulsive efficiency η can be decomposed as,

$$\eta = \eta_a \times \eta_b \times \eta_h \tag{17}$$

where η_a is the actuation efficiency, η_b is the body transmission efficiency, and η_h is the hydrodynamic efficiency of swimming motions. η_a depends on the actuation technology and control approach used. η_h depends on body motions. η_b is the efficiency at which an underactuated soft body transfers actuation power to the body–liquid medium interface and it is a key performance parameter to evaluate the approach. The individual efficiencies are given by [21],

$$\eta_a = \frac{M \frac{\partial h}{\partial x \partial t}\big|_a}{vi} \tag{18}$$

$$\eta_b = \frac{Um\omega^2 H_l^2 \left(1 - \frac{U}{V}\right)}{M \sin(\kappa a) \frac{\partial h}{\partial x}\big|_a} \tag{19}$$

$$\eta_h = \frac{1}{2}\left(1 + \frac{U}{V}\right) - \frac{1}{2}\left(\frac{U}{\omega}\right)^2 \left(1 - \frac{U}{V}\right)^{-1} \left(\frac{\frac{\partial h}{\partial x}\big|_l}{H_l}\right)^2 \tag{20}$$

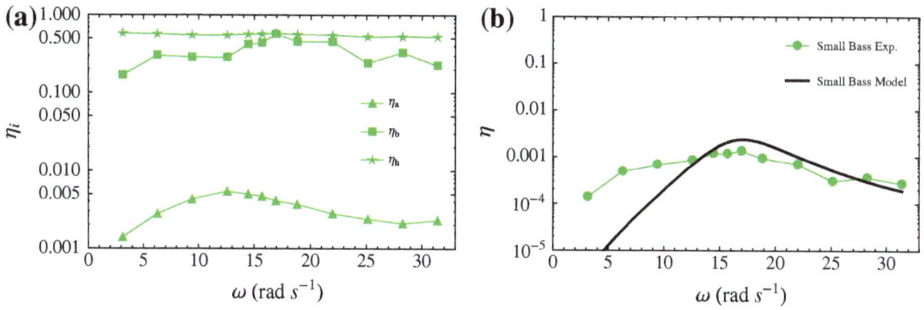

Fig. 21 **a** Predicted body efficiency η_b, predicted hydrodynamic efficiency η_h, and measured actuator efficiency η_a. **b** Measured total efficiency η against model prediction

Figure 21 shows measured and estimated system efficiencies. Predicted values for η_b and η_h, and measured values η_a versus fin flapping frequency are shown in Fig. 21a. The body and hydrodynamic efficiencies cannot be easily measured but model predictions, using measured values for H_l, U, V, and κ, provide some intuition on the mechanism performance. Both predicted body and hydrodynamic efficiencies display a maximum value at the design frequency $\omega_d = 17\,\text{rad}\,\text{s}^{-1}$. Hydrodynamic efficiencies agree with values estimated by earlier studies. Predicted body efficiencies are encouraging, at the design frequency the soft body transmission has an efficiency in the order of 50 %. A traditional discrete-stiff mechanism would require a complex transmission and likely result in a much lower body efficiency. Actuation efficiency is low, as servos are driven to oscillate the constant change in acceleration draws current whose corresponding power is lost as heat in the actuator coils. Figure 21b shows model predictions and measurements for the total system efficiency η. Measured total efficiency agrees with model predictions and its low values are explained by the low actuation efficiency.

Figure 22 shows snapshots of the wake generated by the prototype captured using fluorescent dye flow visualizations. Details of the setup and image processing can be found in [11]. Flow visualization and PIV studies confirmed the existence of

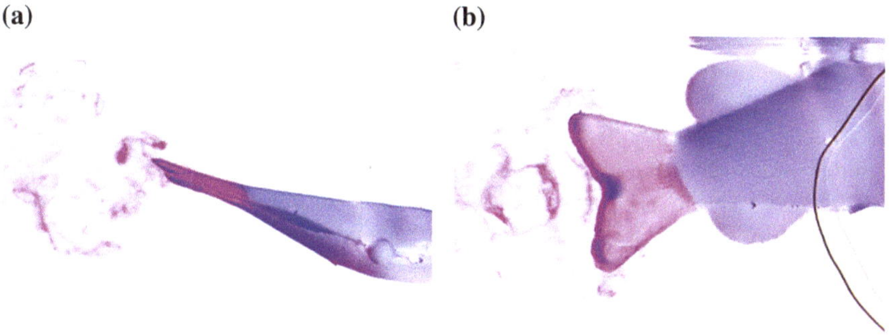

Fig. 22 Dye flow visualization of carangiform prototype: **a** *Top view*, a single vortex is just shed into the wake. **b** *Side view*, a chain of vortices akin to an inverted von karman street can be seen streaming from the caudal fin

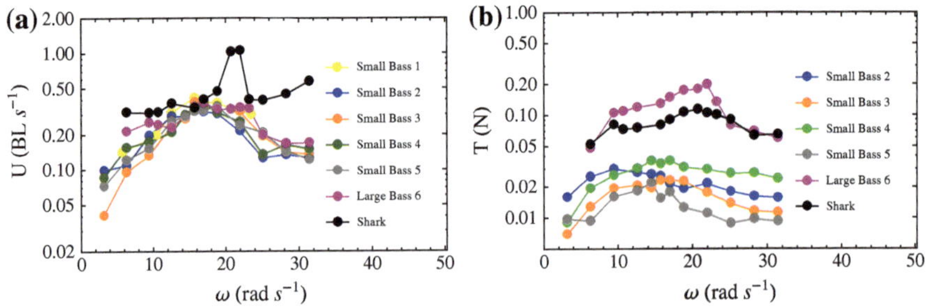

Fig. 23 **a** Measured average swimming velocities U for different prototypes. **b** Measured static thrusts T for different prototypes

a thrust generating wake streaming from the prototype's caudal fin [11]. Figure 22a shows a top view of the prototype swimming and a vortex just being released from the edge of the caudal fin. Figure 22b shows a side view of the prototype swimming. Two vortices can be seen forming a chain emanating from the caudal fin edge. Due to errors in body kinematics, the wake generated was not exactly an inverted von karman wake or 2S wake, but rather a series of vortex pairs forming a wide wake similar to a 2P wake.

Many more prototypes, both carangiform and thunniform, were tested to validate the design methodology. The results, regardless of scale, confirmed a peak in performance at the design frequencies. Figure 23 shows average swimming speeds and static thrusts for several other prototypes. The maximum swimming speed achieved at a design frequency is $U_{\max} \sim 1\,BL\,\mathrm{s}^{-1}$. Among the prototypes built for the original study on underactuated soft robot fish, the first failure due to operational wear occurred after 3 years of regular monthly testing. The failure mode thus far has only been related to the servomotors used for actuation.

6 Discussion

The design methodology introduced in Sect. 2 was tested using several carangiform and thunniform prototypes. Sect. 3 presented examples of robots and actuation unit designs and Sect. 4 described the manufacturing techniques used to fabricate underactuated soft robots. The performance results for a small carangiform design summarized in Sect. 5 confirm the approach basic claim: the natural dynamics of a flexible viscoelastic body can be excited by a simple mechanism to replicate complex bio-inspired locomotion kinematics and performance. In addition, the efficiencies displayed by the soft body transmissions are high and failure modes are related to actuation. Body transmissions maintain their mechanical robustness for extended time periods.

The implementation of the proposed approach on fish-like forward locomotion kinematics (Eq. 1) highlights several advantages. A continuous flexible body with a characteristic heterogeneous material distribution is simple, robust, and protects all encapsulated delicate components. Mechanism robustness is tied to material properties and polymer materials present several desirable attributes for operation within liquid environments. The platinum cure silicones used in this study display low permeability, ultraviolet and high temperature resistance, as well as corrosion and biofouling resistance. Furthermore, a continuous body has by definition an infinite number of DOFs which enable more natural motions.

The experiment results in Sect. 5 confirm that a flexible body can be manufactured to display natural dynamics that mimic body motions during fish locomotion. The caudal fin oscillation amplitude H_l, wave number κ, average swimming speed U, and propulsive forces T agree qualitatively with model predictions. A coarse discretization of the required material distributions leads to errors in body motions which in turn diminish the prototype performance with respect to real fish.

While these results are encouraging, a soft continuous structure is inherently more challenging to control because of the large number of DOFs involved. Furthermore, the choice of underactuation as a means to simplify the mechanism limits the controllable state space. The proposed approach exploits the natural body dynamics to achieve low-level control tasks. Actuation does not force the mechanism to track a particular trajectory, the natural dynamics of the mechanism should match the desired trajectories. As a result, if different behaviors are needed then new structural dynamics are also required. A solution requires dynamic control of local material properties. Different approaches can be taken toward achieving material property control (e.g., using temperature phase changes, granular packing, etc.) but low power [23] or passive approaches should be favored for mobile robot applications.

Power sources and actuation still require stiff and rigid components. At present rechargeable batteries are still the most practical option to power robots. However, standard battery energy densities are not sufficient to satisfy the requirements for long range and long-term deployments. Work on different forms of energy transduction and opportunistic use of environmental resources is essential to advance this area.

Actuation is an area that is closer to having full or partial soft solutions. Work on various artificial muscle actuators, a term broadly used for actuation mechanisms that are similar in function to animal muscle, has matured and practical implementations are closer to being a reality [3].

7 Conclusions

The methodology presented in Sect. 2 was first applied to mobile robots mimicking carangiform and thunniform fish where target body kinematics (traveling waves) are restricted to a single plane [21, 28–30]. However, the concept can easily be applied to more complex kinematics. Valdivia y Alvarado et al., applied the

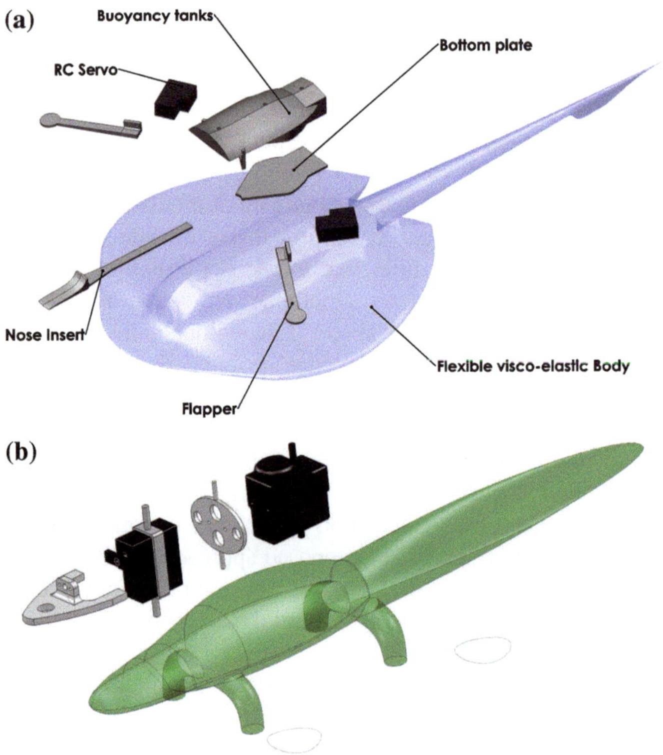

Fig. 24 **a** Batoid design: traveling waves with azimuthal and radial components for thrust vectoring. **b** Salamander design: torsional mode coupled with body mode to achieve walking and swimming

methodology to design batoid-like robots (see Fig. 24a) [8, 9, 22, 24, 26] and salamander-like robots (see Fig. 24b) [25] where required body motions involve vibrations with components along different planes. In the case of batoids, locomotion requires three-dimensional traveling waves along large pectoral fins. In the case of salamanders, locomotion involves a combination of traveling waves (swimming) and standing waves coupled with body torsion (walking). The mechanisms required to excite body modes are as simple as the resulting robots.

Several groups have also adopted similar approaches to improve mechanical robustness and test locomotion performance and control ideas. Kruusmaa et al. (FILOSE project) designed and built robotic fish inspired by the present approach [1, 10]; Strefling et al. [19] developed a jet-driven fluttering fluid-conveying flexible tail propulsion mechanism which exploited tail flexibility; Marchese et al. [16] developed soft fish-like robots to test pneumatic and hydraulic actuation and control three-dimensional motion; Kopman et al. [13] designed a robotic fish propelled by a compliant tail. These works along with many others collectively demonstrate the advantages found by exploiting body flexibility.

Due to the nature of the approach its application is better suited for periodic motions where the environment favors inertia forces. Bio-inspired locomotion in fluids when high Reynolds numbers define the dynamics is an ideal example, but several applications in manipulation and grasping fit those requirements just as well.

Modeling plays a key role in finding proper material distributions to accomplish desired body motions. However, a balance must be struck between the model complexity and practical manufacturing limitations. At present, continuous material distributions cannot be easily manufactured. A discrete approximation is needed so that casting or other deposition-based manufacture can be used. Complex models provide more accurate distributions but simpler models can often be solved analytically and provide more physical intuition.

The required material distributions are unique to a particular target motion and actuation setup. Different motions are still possible with a given material distribution but only the original target kinematics are matched accurately. In order to better adapt to different or changing motion requirements a dynamic control of material properties at the local level is needed. Using controlled temperature phase changes as in Cheng et al. [6, 7], jamming as in Kime et al. [12], or small pre-strain changes with embedded mechanisms as in Valdivia y Alvarado and Bhat [23] can enable such changes and should be further explored.

The simplicity and robustness of underactuated soft robots is an ideal attribute and enabling feature for mobile robots targeting long-term deployments in harsh environments. The examples presented herein use inorganic silicone polymers but organic and biodegradable compounds can also be used to diminish environmental concerns when a fleet of soft robots explores an ecosystem. Challenges remaining include power, actuation, and electronics which still use standard hardware components.

Acknowledgments The authors would like to acknowledge the help and contributions of past and present collaborators in this project: A. Mazumdar, S. Chin, B. Epps, B. Blackburn, F. Yuen, and A. Cloitre.

References

1. Akanyeti et al (21 authors) (2014) FILOSE: a svenning* robot. IEEE Robot Autom Mag
2. Antman SS (2005) Nonlinear problems of elasticity, 2nd edn. Springer, New York
3. Bar-Cohen Y (2001) Electroactive polymer (EAP) actuators as artificial muscles: reality, potential, and challenges. SPIE Press, Bellingham
4. Barrett DS, Triantafyllou MS, Yue DKP, Grosenbaugh MA, Wolfgang MJ (1999) Drag reduction in fish-like locomotion. J Fluid Mech 392:183–212
5. Breder CM (1926) The locomotion of fishes. Zoologica (NY) 4(5):159–297 (Society for Industrial and Applied Mathematics, Philadelphia)
6. Cheng N, Ishigami G, Hawthorne S, Chen H, Hansen M, Telleria M, Playter R, Iagnemma K (2010) Design and analysis of a soft mobile robot composed of multiple thermally activated joints driven by a single actuator. In: Proceedings of the IEEE international conference on robotics and automation

7. Cheng N, Gopinath A, Wang L, Iagnemma K, Hosoi AE (2014) Thermally tunable, self-healing composites for soft robotic applications. Macromol Mater Eng. doi:10.1002/mame.201400017
8. Cloitre A, Subramaniam V, Patrikalakis N, Valdivia y Alvarado P (2012) Design and control of a field deployable Batoid Rotor. In: Proceedings of the IEEE international conference on biomedical robotics and biomechatronics (BioRob), Rome, Italy, pp 707–712
9. Cloitre A, Arensen B, Patrikalakis N, Youcef-Toumi K, Valdivia y Alvarado P (2014) Propulsive performance of an underwater soft biomimetic batoid robot. In: Proceedings the 24th international ocean and polar engineering conferences (ISOPE), Busan, Korea
10. El Daou H, Salumae T, Chambers LD, Megill WM, Kruusmaa M (2014) Modelling of a biologically inspired robotic fish driven by compliant parts. Bioinspr Biomim 9(1):016010
11. Epps B, Valdivia y Alvarado P, Youcef-Toumi K, Techet A (2009) Swimming performance of a biomimetic compliant fish-like robot. Exp Fluids 47(6):927–939
12. Kim Y, Cheng S, Kim S, Iagnemma K (2013) A novel layer jamming mechanism with tunable stiffness capability for minimally invasive surgery. IEEE Trans Robot 29(1):1031–1042
13. Kopman V, Laut J, Acquaviva F, Rizzo A, Porfiri M (2014) Dynamic modeling of a Robotic Fish propelled by a compliant tail. IEEE J Ocean Eng 99:1–13
14. Kubow TM, Full RJ (1999) The role of the mechanical system in control: a hypothesis of self-stabilization in hexapedal runners. Phil Trans R Soc Lond B 354:849–862
15. Lighthill MJ (1975) Mathematical biofluiddynamics. Society for Industrial and Applied Mathematics, Philadelphia
16. Marchese AD, Onal CD, Rus D (2014) Autonomous soft robotic fish capable of escape maneuvers using fluidic elastomer actuators. Soft Robot 1(1):75–87
17. McGeer T (1990) Passive dynamic walking. Int J Robot Res 9(2):62–82
18. Merz R, Prinz FB, Ramaswami K, Terk M, Weiss L (1994) Shape deposition manufacturing. In: Proceedings of the solid freeform fabrication symposium, University of Texas, Austin. Accessed 8–10 Aug 1994
19. Strefling PC, Helium AM, Mukherjee R (2012) Modeling, simulation, and performance of a synergistically propelled ichthyoid. IEEE/ASME Trans Mech 17(1):36–45
20. Timoshenko S, Young DH, Weaver W Jr (1974) Vibration problems in engineering. Wiley, New York
21. Valdivia y Alvarado P (2007) Design of biomimetic compliant devices for locomotion in liquid environments, doctoral dissertation. MIT, Cambridge
22. Valdivia y Alvarado P (2011) hydrodynamic performance of a soft body under-actuated batoid robot In: Proceedings IEEE international conference on robotics and biomimetics (ROBIO), Phuket Island, Thailand, pp 1712–1717
23. Valdivia y Alvarado P, Bhat S (2014) Whisker-like sensors with tunable follicle sinus complex for underwater applications In: Proceedings SPIE bioinspiration, biomimetics, and bioreplication IV conference, San Diego, CA, 9–12 March 2014
24. Valdivia y Alvarado P (In preparation) Design of soft batoid-like underwater robots
25. Valdivia y Alvarado P (In preparation) Design of soft amphibian robots
26. Valdivia y Alvarado P, Chin S, Larson W, Mazumdar A, Youcef-Toumi K (2010) A soft body under-actuated approach to multi degree of freedom biomimetic robots: a stingray example In: Proceedings IEEE RAS/EMBS international conference on biomedical robotics and biomechatronics (BIOROB), Tokyo, Japan, pp 473–478
27. Valdivia y Alvarado P, Sekar KS (Under Review) Modeling wake topology and thrust production in batoid-inspired oscillating fins
28. Valdivia y Alvarado P, Youcef-Toumi K (2003) Modeling and design methodology for an efficient underwater propulsion system. In: Proceedings IASTED international conference on robotics and applications, Salzburg, Austria, pp 161–166

29. Valdivia y Alvarado P, Youcef-Toumi K (2005) Performance of machines with flexible bodies designed for biomimetic locomotion in liquid environments. In: Proceedings IEEE international conference on robotics and automation (ICRA), Barcelona, Spain, pp 3324–3329
30. Valdivia y Alvarado P, Youcef-Toumi K (2006) Design of machines with compliant bodies for biomimetic locomotion in liquid environments. ASME J Dyn Sys Meas Cont 128:3–13

iSplash: Realizing Fast Carangiform Swimming to Outperform a Real Fish

Richard James Clapham and Huosheng Hu

Abstract This paper focuses on the linear swimming motion of Carangiform fish and presents two novel prototypes: *iSplash*-I and *iSplash*-II, which have overcome some of the previously known challenges, in particular the straight line swimming speed of robotic fish. The first generation *iSplash*-I improved the kinematic pattern by deploying a full-body length swimming motion to coordinate anterior, mid-body, and posterior displacements. The second generation *iSplash*-II achieved consistent untethered stabilized swimming speeds of 11.6 BL/s (i.e., 3.7 m/s), with a frequency of 20 Hz during the field trials, outperforming real carangiform fish in terms of average maximum velocity (measured in body lengths/second) and endurance, the duration that top speed is maintained.

Keywords Robotic fish · Carangiform swimming · Full-Body length · Swimming speed · Maximum velocity

1 Introduction

To navigate through a marine environment, a robotic vehicle requires mobility to effectively contend with the physical forces exerted by the surrounding fluid. Live fish can coordinate their body motions in harmony with the surrounding fluid generating large transient forces efficiently, as opposed to rigid hull underwater vehicles (UV) powered by rotary propellers [1–2]. For a man-made vehicle

R.J. Clapham (✉) · H. Hu
School of Computer Science and Electronic Engineering, University of Essex, Colchester CO4 3SQ UK
e-mail: r.j.c@ieee.org; richard.j.clapham@gmail.com
URL: http://www.isplash-robotics.com

H. Hu
e-mail: hhu@essex.ac.uk

© Springer-Verlag Berlin Heidelberg 2015

R. Du et al. (eds.), *Robot Fish*, Springer Tracts in Mechanical Engineering,
DOI 10.1007/978-3-662-46870-8_7

to achieve greater locomotive capability there is potential to engineer a structure that can accurately replicate the wave form of swimming fish.

Bainbridge's intensive observational studies measured live fish to attain an average maximum velocity of 10 body lengths/second (BL/s) [3]. A single high performance of a Cyprinus carpio was noted, achieving the swimming speed of 12.6 BL/s (1.7 m/s) with a distance traveled per tail beat or Swimming number (Sw) of 0.7. Endurance at the highest velocities is limited, burst speeds can only be maintained for short durations of approximately one second. Velocities were measured to decrease to 7 BL/s in 2.5 s of swimming, to 5 BL/s in 10 s, and to 4 BL/s in 20 s.

Although engineers have focused on hydrodynamic mechanisms, currently published robotic fish are unable to gain the locomotive efficiencies of live fish, proving a complex challenge. There are two limitations in particular: (i) They cannot achieve accurate replication of the linear swimming motion as free swimming robotic fish generate kinematic parameter errors and therefore reduced propulsion; (ii) They have low force transfer due to the complexity of developing the power train, limited by mass, volume, force, frequencies, and internal mechanical losses. Some examples of novel design approaches and their maximum velocities are Barrett's hyper-redundant Robotuna, achieving a maximum velocity of 0.65 body lengths/second (BL/s) (0.7 m/s) [4], Anderson's VCUUV with 0.5 BL/s (1.2 m/s) [5], Yu's discrete structure with 0.8 BL/s (0.32 m/s) [6], Essex's G9 with 1.02 BL/s (0.5 m/s) [7], Wen's carangiform with 0.98 BL/s (0.58 m/s) [8], and Valdivia y Alvarado's compliant method with 1.1 BL/s (0.32 m/s) [9]. The straight line speed of current robotic fish, peaking at 1 BL/s, is typically unpractical for marine-based environments.

In particular the kinematic errors are due to the lateral (F_L) and thrust (F_T) forces not being optimized and as a consequence excessive anterior destabilization in the yaw plane due to the concentration of posterior thrust creates reaction forces around the center of mass [7, 9]. In turn the anterior creates posterior displacement errors. As a result the body wave motion along the full length of body has large matching errors in comparison to the swimming patterns of live fish leading to reduced performance and high cost of transport.

Research Objectives This research project considered the factors contributing to the linear swimming speed of current robotic fish and initially proposed four main objectives: (i) to introduce a new swimming pattern to reduce the kinematic parameter errors by coordinating transverse displacements along the body length; (ii) to allow for efficient energy transfer by engineering a mechanism that takes into account hardware and material constraints so that propulsion is not restricted; (iii) to develop a prototype to improve stability in the vertical and specifically the horizontal plane, by optimizing the lateral and thrust forces around the center of mass; and (iv) to validate the proposed swimming motion by realizing a mechanism capable of consistent free swimming operation, measuring its achievement in terms of speed, thrust, and energy consumption over a range of frequencies.

The remainder of the chapter is organized as follows: Sect. 2 presents the design, construction, and experimental results of the first generation prototype, namely *iSplash*-I, and introduces a new approach to coordinated full-body length swimming motion. Section 3 describes the new mechanical drive system and fabrication technique and experimental results of the second generation, *iSplash*-II. Finally, a brief conclusion and future work are given in Sect. 4.

2 *iSplash*-I: A New Robotic Fish with Full-Body Coordination

2.1 Design Methodology

Traditional Approach Modeling from body and/or caudal fin (BCF) swimmers, the selected carangiform swimming mode can be identified by the wave length and amplitude envelope. The Cyprinus carpio (common carp) has been chosen specifically for its high locomotive performance [3, 10].

We only consider modeling within the confinements of the horizontal plane where the kinematics of propulsion is commonly reduced to the form of a traveling wave, concentrated to the posterior, varying in amplitude along the length, smoothly increasing toward the tail [2]. Present robotic swimmers adopted this method which limits undulatory motions, typically to <1/2 the body length toward the posterior and the wave form motion consists of one positive phase and one negative phase. The commonly adopted model proposed in [4], is in the form of:

$$y_{\text{body}}(x,\ t) = \left(c_1 x + c_2 x^2\right)\sin(kx + \omega t) \tag{1}$$

where y_{body} is the transverse displacement of the body; x is the displacement along the main axis starting from the nose of the robotic fish; $k = 2\pi/\lambda$ is the wave number; λ is the body wave length; $\omega = 2\pi f$ is the body wave frequency; c_1 is the linear wave amplitude envelope, and c_2 is the quadratic wave. The parameters $P = \{c_1, c_2, k, \omega\}$ can be adjusted to achieve the desired posterior swimming pattern.

Proposed Full-Body Swimming Motion Propulsion of carangiform swimming is associated with the method of added mass [11]. Each propulsive segment of the traveling wave creates a force against the surrounding water generating momentum. This causes a reaction force (F_R) from water onto the propulsive segment. F_R normal to the propulsive segment is decomposed into the lateral F_L component which can lead to energy loss and anterior destabilization and the thrust F_T component providing propulsion increases in magnitude toward the tail. The overall magnitude of added mass passing downstream is approximately measured as the water mass accelerated and its acceleration (Fig. 1).

Therefore, it is proposed that initiating the starting moment of added mass upstream and optimizing the F_L and F_T forces around the center of mass would increase the overall magnitude of thrust contributing to increased forward velocity. In consideration of this, we designed a novel robotic fish which can operate in two

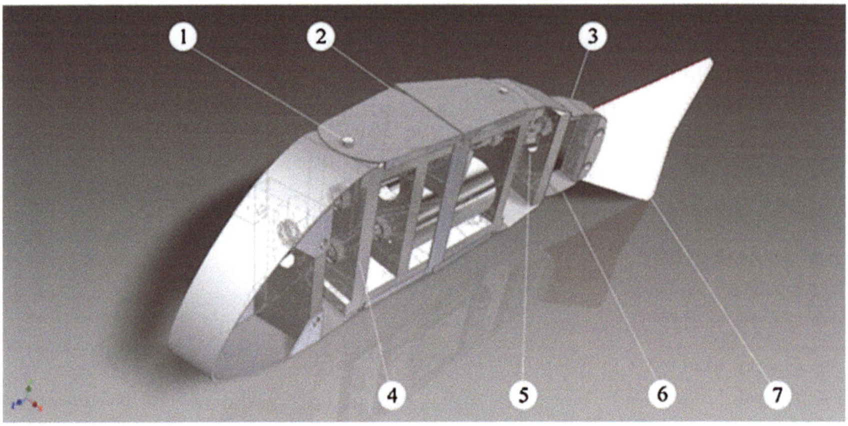

Fig. 1 *iSplash*-I: *1* Anterior joint; *2* Mid-body links; *3* Posterior with thick peduncle; *4* Transmission system; *5* Driven tail plate; *6* Tendons; *7* Compliant fin

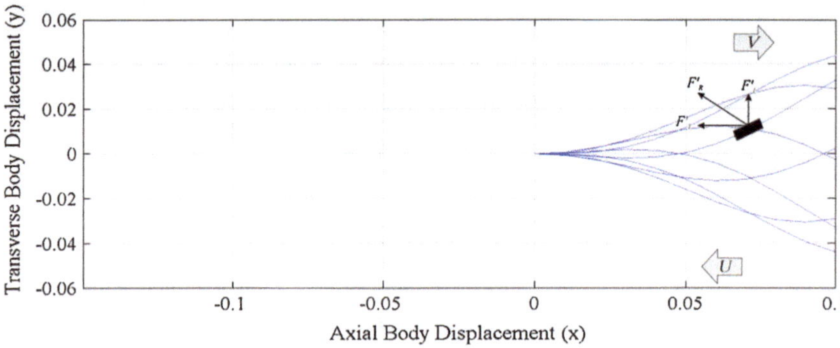

Fig. 2 Mode 1: Wave form is confined to the posterior 2/5. Parameters have been determined from experimental tests. Showing the thrust generation by the method of added mass (Adapted from Webb [11])

swimming patterns: (i) Applying a traditional rigid mid-body and anterior. Concentrating the undulations and degrees of freedom (DOF) to the posterior end of the body length which will be described as Mode 1, illustrated in Fig. 2; (ii) Based on intensive observation and fluid flow assumptions a new full-body carangiform swimming pattern is introduced. The coordination of anterior, mid-body, and posterior body motions are proposed in an attempt to reduce kinematic parameter errors, this will be described as Mode 2, illustrated in Fig. 3.

The models midline and body motion parameters were first established based on observation and published data from literature providing an initial engineering reference. The wave form motion first developed for a discrete rigid anterior

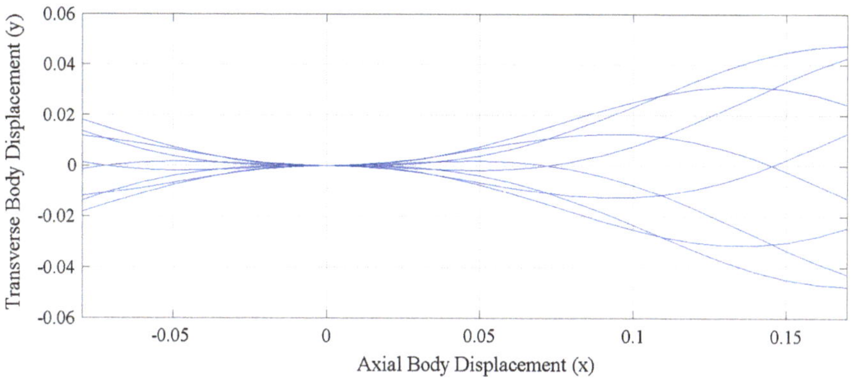

Fig. 3 Mode 2: Full-body coordination. The kinematic parameters have been determined from experimental tests

prototype in [7] can be extended to represent the full-body motions of Mode 2 in the form:

$$y_{\text{body}}(x,t) = \left(c_1 x + c_2 x^2\right) \sin(\text{k}x + \omega t) - c_1 x \sin(\omega t) \tag{2}$$

The relationships between the defined parameters $P = \{0.44, 0, 21.6, 8\}$ shown in Fig. 3 can first be found by evaluating the x location pivot at 0. In the kinematic pattern of Mode 2, the fraction of body length displaced is equal to the anguilliform swimming mode but reflects changes in the wave form. The anguilliform swim pattern is defined by large amplitude undulations propagating from nose to tail. The newly introduced Mode 2 applies an oscillatory motion to head and mid-body and pivots the entire body around a single point associated with the carangiform fish swimming motion [2].

Although this is the first account of applying full-body actuation to a research prototype fish, mechanisms such as "vortex peg" and "undulating pump" and flow visualization techniques have been proposed [12, 13] from published biological studies, indicating a possible fluid body interaction that contributes to propulsive thrust is generated upstream to the posterior section. The muscle activity in the anterior has been measured to be low, suggesting that accurate modeling of the kinematics could be more significant than anterior force in improving energy transfer.

As previously mentioned, anterior destabilization has been difficult to control [6, 7, 9], as passive rigid anterior mechanisms recoil around the center of mass. Free swimming robotic fish have excessive head swing, similar in magnitude to the posterior which increases drag. The proposed Mode 2 drives the anterior into the unwanted yaw direction, in an attempt to reduce amplitude errors by optimizing the F_R around the center of mass. It has also been noted in [2], that the morphological adaptations of reduced depth at the peduncle, increased depth of body toward the anterior and vertical compression minimize recoil forces.

2.2 Construction Method

Mechanical Design Mechanical structure limitations set a great challenge when modeling the displacements within the traveling wave. Current methods typically adopt either a discrete assembly [6, 7] or compliant structure [9] but both are seen to have limitations. A construction method using structural compliance combined with a rigid discrete assembly is proposed. The arrangement distributes three actuated joints and one coupled joint along the axial length shown in Fig. 4.

Mode 1 disregards transverse displacements of links I, II, III whereas Mode 2 actuates all links along the axial length to provide anterior and mid-body transversal displacements. The development allowed for both Modes of operation to be applied to the same prototype by adjusting the configuration. Uniform material properties were chosen for links I–III and stiffness distribution begins at joint 3 and continues to the tail tip. To provide the undulatory motion a compliant caudal fin is attached to the link V and is actuated by tendons anchored to the main housing rear bulkhead. The developed mechanism allows for the expansion of the tendons and material stiffness of the caudal fin to be adjusted experimentally to provide the targeted curves during free swimming at various frequencies.

The approximation of a traveling wave using links I–V and turning angles of joints 1–4 are shown in Fig. 4. Details of the fully discretized body wave fitting method are given in [6, 7]. The location of joints in the series can be determined by parameterized fitment to a spatial and time dependent body wave. The discrete

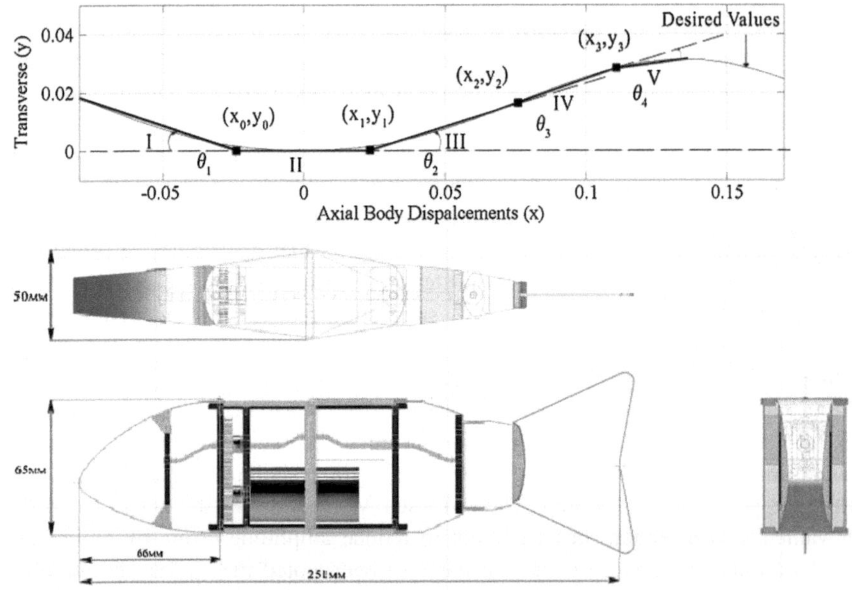

Fig. 4 Link approximation (*top*); *1* Plan; *2* Profile; *3* Front (*bottom*)

construction method can be defined as a series of links or N links. N being the number of links after joint 1 typically <6 due to structural limitations, more links reduce curve alignment errors. The aim of the design is to improve complexity of motion without an increase of structural parts. The link end points are shown in Fig. 4, it can be seen that the arrangement of links distributed along the length of the body provides an accurate curve alignment reducing large errors and excrescences in the outer profile. In addition we have observed that the aerofoil section NACA (12)520 based on camber, chord, and thickness can be utilized to illustrate the outer structure profile of Mode 2, which we propose contributes to the fluid flow interaction.

Power Transmission System The developed transmission system providing rotary power to linear oscillations is illustrated in Figs. 5 and 6. All actuated links are directly driven by the five-bearing crankshaft providing an equal power distribution. The developed mechanical design required high-precision engineering of the chassis and crankshaft to avoid deadlock and reduce friction. The driven link amplitudes are determined by the offset cranks, L_3 represents one of the discrete links of the structure. The maximum amplitude of the link length L_3 at point P_2 is determined by the predetermined maximum crank offset P_1. The coordinates of P_1 (P_{1x}, P_{1y}) and P_2 (P_{2x}, P_{2y}) can be derived by:

$$\begin{cases} P_{1x} = A + B\cos\theta_1 + C\cos\theta_2 \\ P_{1y} = B\sin\theta_1 + C\sin\theta_2 \end{cases} \begin{cases} P_{2x} = P_{1x} + D\sin\theta_2 \\ P_{2y} = P_{1y} + D\sin\theta_2 \end{cases} \tag{3}$$

The length of L_3 can be derived by $L_3^2 = P_{2x}^2 + P_{2y}^2$. Assume that ω_1 is the angular velocity of the link L_3, and the velocity vector V_{P2} is perpendicular to L_3. We have:

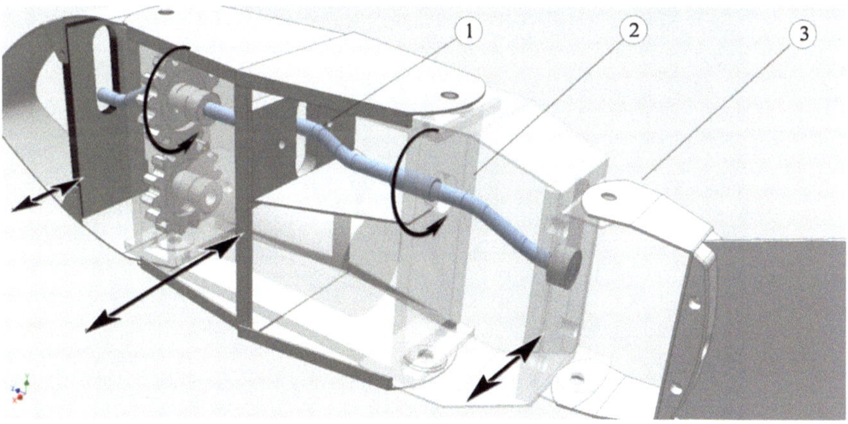

Fig. 5 Power transmission system: *1* Transision plate; *2* Crankshaft; *3* Free end of link and connecting pivot

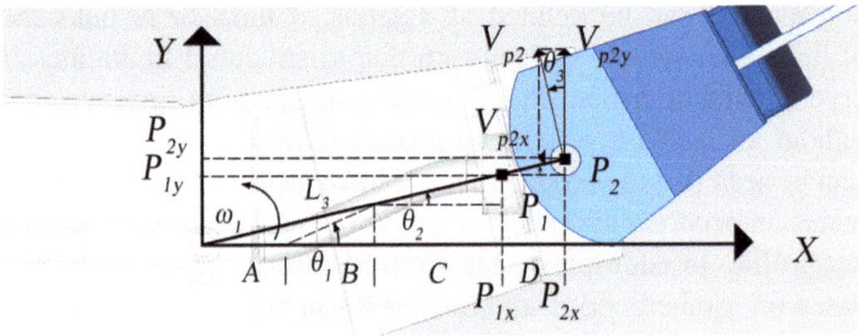

Fig. 6 Schematic drawing of the tail offset drive crank and linkage

$$\begin{cases} V_{p2x} = -\omega_1 L_3 \sin\theta_3 \\ V_{p2y} = \omega_1 L_3 \cos\theta_3 \end{cases} \tag{4}$$

where V_{p2x} and V_{p2y} are the decomposed vectors of the velocity vector $V_{P2} = \omega_1 L_1$.

Fabrication *iSplash*-I shown in Fig. 7 was engineered as a morphological approximation of the common carp. The physical specifications are given in Table 1. We devised a structurally robust prototype allowing for consistency of operation at high frequencies, as force has to be applied to the water and reactively, the opposing force is applied to the vehicle. All structural parts were precision engineered, hand fitted, and assembled. A consideration of the development took hardware and material constraints into account, so that geometric and kinematic parameters are not affected. The hydrostatic streamlined profile was optimized

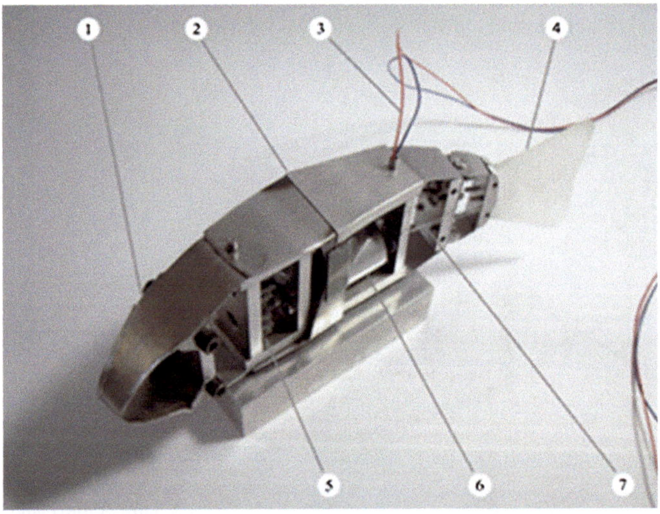

Fig. 7 Inner Structure of *iSplash*: *1* Steel space frame; *2* Mid-body driven plate; *3* External source cables; *4* Polypropylene caudal fin; *5* Aluminum main bulkhead; *6* Electric motor; *7* Offset crankshaft

Table 1 Physical parameters of *iSplash*-I

Parameters	Specific value
Body size: m (L × W × H)	0.25 × 0.05 × 0.062
Body mass (Kg)	0.367
Maximum velocity BL/s (m/s)	3.4 (0.88)
No-load maximum frequency (Hz)	8
Actuator	Single electric motor
Power supply	12 V Pb external battery supply
Fabrication	Low tolerance engineering
Materials	Aluminum, mild steel, stainless
Swimming mode	Linear locomotion
Tail material	Polypropylene
Outer structure and skin material	Polystyrene and polypropylene
Caudal fin aspect ratio (AR)	1.73

by favorably positioning the maximum thickness of the cross section, reducing pressure drag. In [10], the cross section has been measured to be optimal at 0.2 of the body length. These aspects relate to amount of resistance during forward motion and were taken into consideration within the design.

Increasing endurance is a desirable feature of a UV. Current robotic fish are still limited to short operational times as energy losses can be produced in many stages of the mechanical transfer. Recent designs have found that it is advantageous to utilize a single electrical motor for actuation [9]. The classical actuator is still the most effective way of providing power at high frequencies and reduces energy consumption over multilink discrete assemblies. Mass and volume distribution are key principles of stability in the horizontal and vertical planes. A single actuator power transmission system can be positioned in the optimum location. In contrast multilink servo assemblies are limited as mass and volume are confined to the posterior (Table 2).

Table 2 Comparison of test results between Modes 1 and 2

Parameters	Mode 1	Mode 2
Reynolds number Re (10^5)	1.4	1.7
Strouhal number (St)	0.48	0.41
Maximum thrust (N)	0.63	1.17
Consistent maximum velocity BL/s (m/s)	2.2 (0.55)	2.8 (0.70)
Frequency (Hz)	6.1	6.6
Max power consumption air (W)	3.48	3.76
Max power consumption water (W)	5.76	7.68
Swimming number (Sw)	0.36	0.42
Head swing amplitude (m)	0.044	0.018
Tail swing amplitude (m)	0.044	0.044
Test run distance (m)	0.5	0.5

The body has open-loop stability if the relative position of buoyancy is higher than the center of mass as the surrounding fluid counterbalances the gravitational weight [14]. Therefore, the hydrostatic buoyancy level and stability were solved by adjusting material properties and configuration. Stability was found to be particularly difficult to maintain during free swimming at high frequencies.

Lastly, the inner structure of the prototype is negatively buoyant. A significant development of the prototype was a watertight skin that allowed unrestricted flexing of the external surface and provided the volume needed to maintain neutral buoyancy.

2.3 Experimental Procedure and Results

Field Trials A series of experiments were undertaken in order to verify the proposed swimming pattern by assessing the locomotive performance of Modes 1 and 2 in terms of speed, thrust, and energy consumption at frequencies in the range of 2–8 Hz. Experiments were conducted within a 1 m long × 0.5 m wide × 0.25 m deep test tank. Stabilized free swimming over a distance of 0.5 m was used to measure speed with a 0.5 m acceleration distance. The prototype had sufficient space to move without disturbances from side boundaries and the free surface. Measurements were averaged over many cycles once consistency of operation was achieved and steady state swimming was obtained.

Although the prototype was measured to have a higher mechanical efficiency with an oil-filled structure, the developed skin proved inconsistent. Therefore, all runs were completed actuating the prototype with a water filled structure. This method attained consistency of operation providing stabilized swimming and maintaining the required buoyancy within the depth of the testing tank, whilst gently skimming the bottom surface. Velocity greatly reduced during runs when the skin detached, the build became negatively buoyant, destabilized, or the cross-sectional area was increased.

Swimming Pattern Observation Figure 8 shows snapshots of Mode 2 in eight instances with time intervals of 0.02 s throughout one body cycle. The midline was tracked at 50 frames per second to provide the amplitude envelopes of the anterior and posterior for comparison. Good agreement with fish kinematic data is a difficult task and current free swimming robotic fish have shown excessive head and tail amplitude errors. When comparing Modes 1 and 2, Mode 2 was found to reduce the head amplitude by over half from 0.17 (0.044 m) of the body length in Mode 1 to 0.07 (0.018 m). The tail amplitude of the common carp is 0.1 [3, 10], larger values were found to increase performance. Both Modes 1 and 2 were able to attain amplitudes of 0.17 (0.044 m). The location of the midline pivot point should be in the range of 0.15–0.25 of the body length [10]. Mode 2 has a reduced error location of 0.33 in comparison to 0.5 in Mode 1. Indicating Mode 2 greatly reduces the kinematic matching errors.

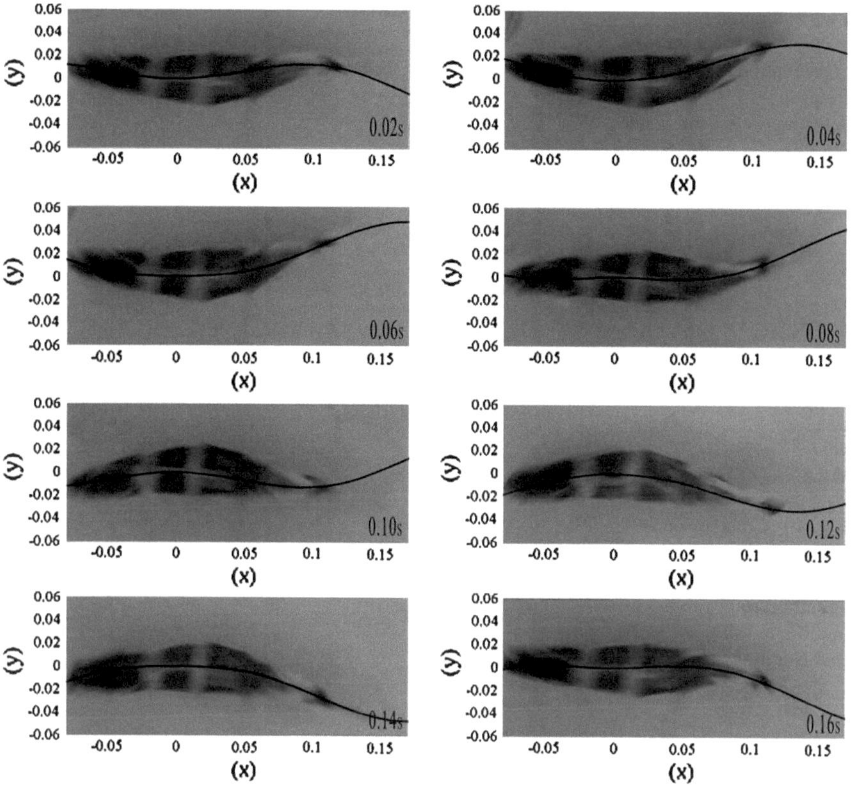

Fig. 8 Mode 2 during one body cycle, eight instances every 0.02 s

In addition it was observed that the posterior 2/5 of the body length deforms due to stiffness distribution providing a smooth transition phase between body and tail. Although high aspect ratio (AR) caudal fins have been found to produce greater efficiency [15], in initial testing a low aspect ratio tail provided higher speeds. AR is calculated using: AR = b^2/Sc where *b* squared is the fin span and Sc is the projected fin area. AR in this case was 1.73.

Experimental Results Figure 9 shows the average energy consumption in relationship to driven frequency, comparing both Modes in air and water. This comparison measured the value of the increased resistance during locomotion due to the surrounding liquid. Measuring of energy consumption and thrust took many cycles to average, as the swimming motion produces fluctuating readings within a single body motion cycle. Both Modes actuating in water resulted in an increase in energy consumption, i.e., Mode 2 increasing from 3.76 to 7.68 W and Mode 1 increasing from 3.48 to 5.76 W.

As the configurations of robotic fish show various hardware and morphological properties, the main value of comparison has become speed divided by body length

Fig. 9 Comparison of average electrical power consumption over driven frequency of both Modes, actuating in air and water

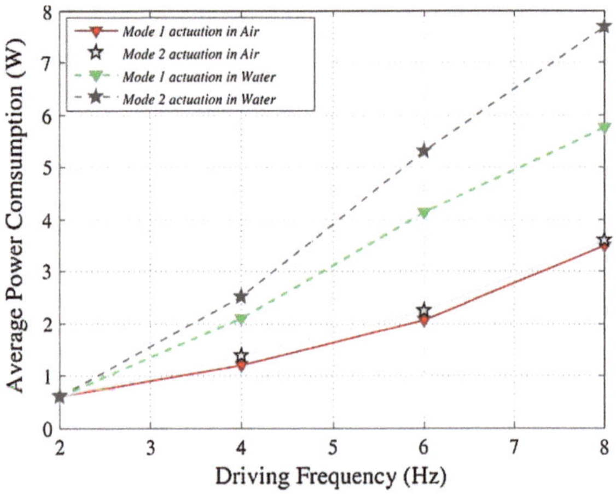

Fig. 10 Comparison of average velocities achieved by both modes, contrasted against current robotic fish and the cruising speed of a real common carp

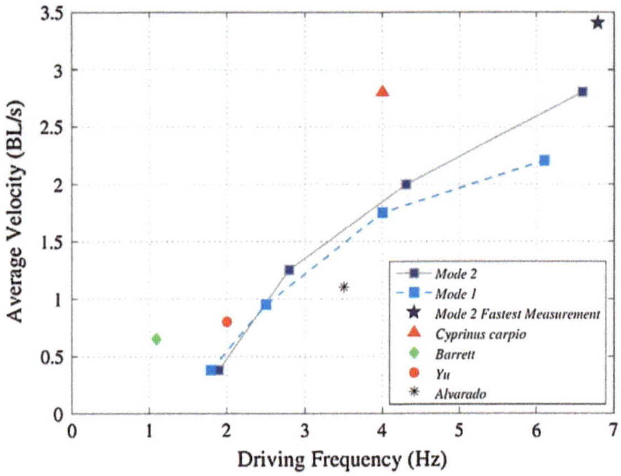

(BL/s). In this case the body length is measured from nose tip to the most posterior extremity of the tail.

The relationship between velocity and driven frequency is shown in Fig. 10. The corresponding values of Modes 1 and 2 during consistent swimming were measured and compared to current robotic fish. Mode 1 achieved maximum velocity of 2.2 Bl/s (0.55 m/s), at 6.1 Hz. Mode 2 increased maximum velocity to 2.8 BL/s (0.70 m/s) at 6.6 Hz. Mode 2 has an increased performance in comparison with current robotic fish which typically peak around 1 BL/s. An initial value of 3.4 BL/s (0.87 m/s) at 6.8 Hz was recorded by Mode 2 with an oil-filled structure. Sealing the developed skin when in contact with oil could not be maintained and skin detachment consistently affected stability and buoyancy, greatly reducing performance.

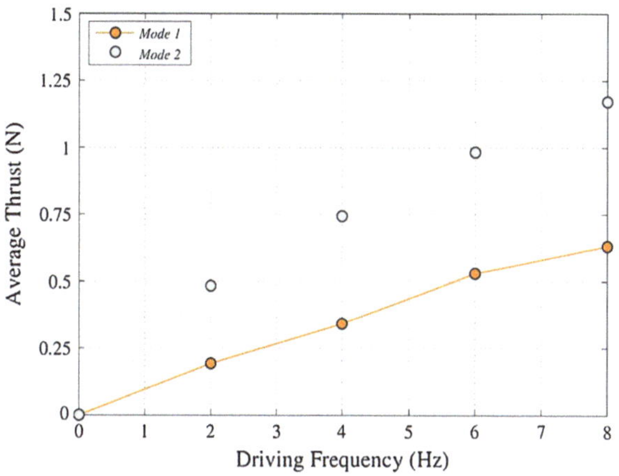

Fig. 11 Comparison of average thrust in relationship to driven frequency achieved by both Modes

We can notice that Mode 2 had an 85 % increase of thrust (Fig. 11) and a 27 % increase in velocity over Mode 1 whilst consuming only 7.68 W of power at 2.8 BL/s. As the power supply contributes to a significant portion of the total mass, high energy efficiency is important. The measured low energy consumption indicates that the next generation could carry its own power supply within a comparable geometric frame with good endurance.

A prominent parameter for analyzing BCF locomotive performance is the Strouhal number (St), defined as $St = fA/U$, where f denotes the frequency, A denotes the tail amplitude, and U is the average forward velocity. St is considered optimal within the range of $0.25 < St < 0.40$. Mode 1 has a peak St of 0.48 under the condition of $Re = 1.4 \times 10^5$ and Mode 2 consistently measured a $St = 0.41$ and peaked at a $St = 0.34$ under a condition of $Re = 2.2 \times 10^5$. A comparable live fish was measured in [16], to have a $St = 0.34$, $Re = 2 \sim 8 \times 10^5$. Applying Mode 2 shows a high-performance increase within the St optimal range and achieves the higher cruising speeds of swimming fish.

A significant relationship between velocity and driven frequency was found. As higher frequencies were applied velocity increased in both Modes, matching the reported findings of live fish [3]. From this it can be assumed that a further increase of frequency applied to this prototype may continue to increase its performance.

3 *iSplash*-II: Realizing Fast Carangiform Swimming

As previously noted, throughout the field trials *iSplash*-I was able to replicate the key swimming properties of real fish. As frequencies were raised the prototype continued to increase velocity. This matches Bainbridge's study of swimming fish, measuring no noticeable change in kinematics after tail oscillations are raised beyond 5 Hz, indicating that only an altered frequency is required to increase

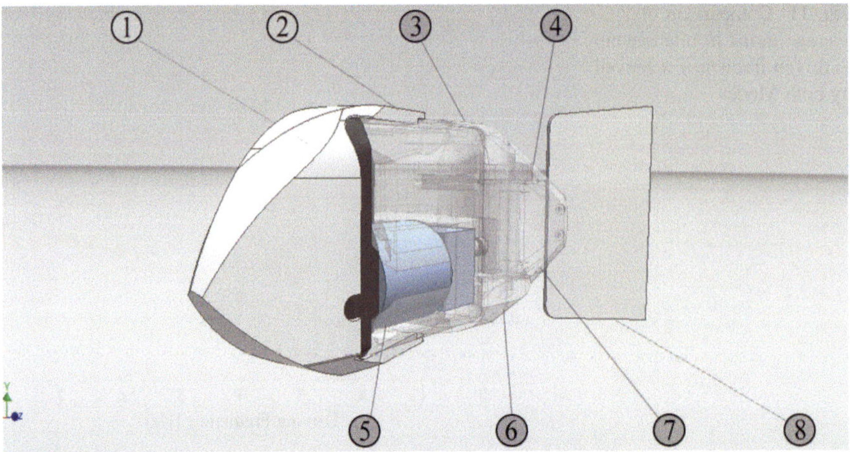

Fig. 12 *iSplash*-II: *1* Anterior link; *2* Mid-body transition links; *3* First posterior link; *4* Final posterior pivot; *5* Primary actuator; *6* Direct drive offset crank; *7* Tendon driven peduncle; *8* Compliant caudal fin

swimming speed. Hence it was expected, that combining the critical aspects of the *iSplash*-I mechanical drive system with frequencies higher than 6.6 Hz may significantly increase maximum velocity, *iSplash*-II was developed.

Figure 12 presents *iSplash*-II, a prototype 32 cm in length which has achieved consistent untethered stabilized swimming speeds of 11.6 BL/s (i.e., 3.7 m/s), with a frequency of 20 Hz. The prototype was developed with a new fabrication technique and mechanical drive system, effectively transmitting large forces at high frequencies to obtain high-speed propulsion. The F_L and F_T forces were optimized around the center of mass, generating accurate kinematic displacements and increasing the magnitude of added mass.

Research Objectives This prototype was developed with the aim of achieving the fastest swimming speeds of real fish with seven main objectives:

(i) to devise a prototype which operates in the two swimming patterns, for further investigation of the carangiform swimming motion to be conducted;
(ii) to significantly increase force transfer by achieving a high power density ratio in combination with an efficient mechanical energy transfer;
(iii) to achieve unrestricted high force swimming by realizing a prototype capable of carrying a high powered energy supply;
(iv) to develop a structurally robust mechanical drive system based on the critical properties proposed in [17], capable of intensively high frequencies of 20 Hz;
(v) to greatly reduce forward resistance by engineering a streamlined body considering individual parts' geometries and alignment throughout the kinematic cycle;

(vi) to stabilize the free swimming prototype's unsteady oscillatory motion during intensively high frequencies to achieve a more efficient force transfer;

(vii) to conduct a series of experiments measuring the prototype's achievements in terms of kinematic data, speed, thrust, and energy consumption in relation to driven frequency.

3.1 Construction Method

Mechanical Design In order to increase the swimming speed a new mechanical drive system was required, able to effectively transmit large forces at intensively high tail oscillation frequencies. In consideration of this, a feasible design structure to fit the linear swimming patterns of both modes was developed. Although a power train utilizing a single motor with continuous rotation is more complex to develop without large internal mechanical loss [4], it is still advantageous in comparison to multilink servos or smart materials, which are limited by force, frequency, volume, and mass distribution [6–8]. Therefore a single actuator was deployed, as shown in Fig. 13.

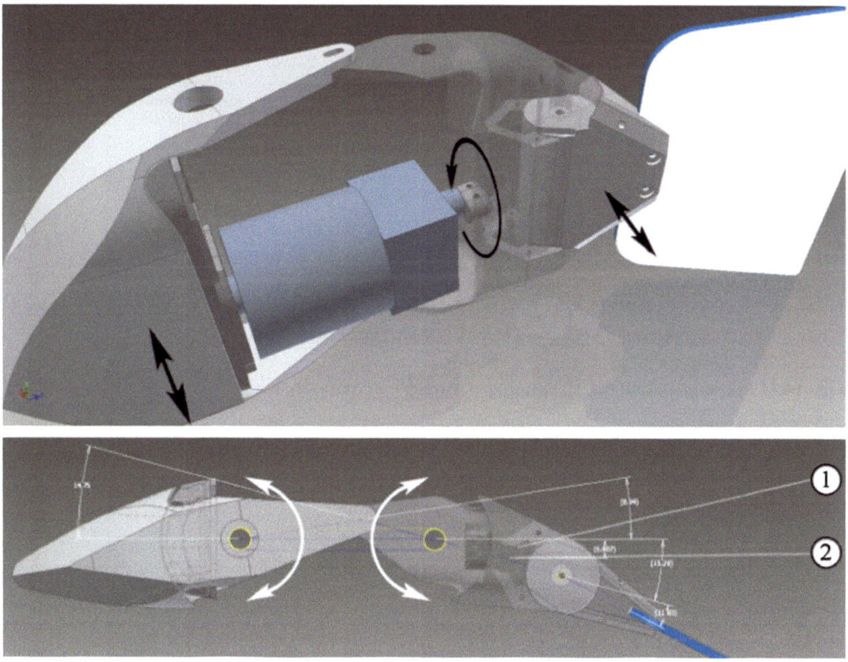

Fig. 13 Power transmission system: *1* Offset crank; *2* Posterior link

As the build required a high power density ratio, the structural arrangement was governed by the dimensions of the large electrical motor 83 mm long × 50 mm diameter. This required a slight increase in body length from 250 mm to 320 mm and a significant adjustment to the link structure to take the mass of the actuator into consideration, removing the coupled mid-body joint and the associated discrete linkages. The discrete construction method, defined as a series of links or N links, aims to achieve accurate midline kinematic parameters whilst minimizing complexity of the mechanical drive and linkages. The sequence of links can generate the required swimming motion by locating the joints to the spatial and time-dependent body wave. The fully discretized body wave fitting method is given in [6, 7].

The assembly of *iSplash*-II is illustrated in Fig. 14, showing the four joints distributed along the axial length. Three rigid links are coupled to a compliant fourth link and caudal fin with stiffness distribution, devised to generate a smooth body to tail transition phase of the posterior undulations. The developed modular build allowed for both Modes of operation to be applied to the same prototype by adjusting the configuration. Links III and IV are actuated to generate the posterior kinematics of operational Mode 1, Mode 2 actuates all links along the axial length to provide anterior, mid, and posterior body displacements.

It was proposed in [17], that the outer profile of the coordinated full-body swimming pattern, represented by the aerofoil section NACA (12)520 aids the fluid flow interaction, producing greater locomotive speeds. In consideration of the simplified link assembly and estimated center of mass, the head and tail amplitudes were increased. As can be seen in Fig. 15, the approximation of a traveling wave

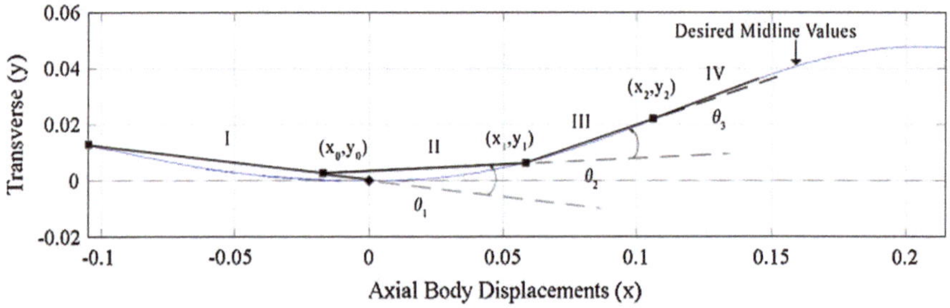

Fig. 14 Link approximation, illustrating accurate kinematic matching

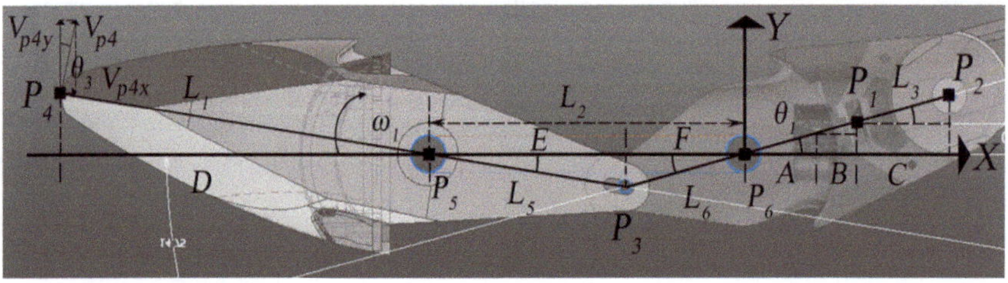

Fig. 15 Schematic drawing of the offset drive crank and linkages

using link end points I–IV and turning angles of joints 1–3 of the reduced link arrangement provides an accurate curve alignment agreeable with the form of (2), therefore reducing errors and excrescences in the outer profile and achieving accuracy with the required aerofoil section.

Power Transmission System The leading tail discrete link III is directly driven by the single bearing crankshaft attached to the output shaft of the primary actuator, increasing power distribution to the posterior. As link III is actuated, link IV is passively displaced. This final posterior linkage IV, coupled to the compliant caudal fin is anchored by four expandable tendons attached to the main chassis rear bulkhead, crossing through linkage III. The anterior link I is transmitted motion by paired linkages fixed at points P_5 and P_6, located at the top and bottom of the main chassis. The developed mechanical design required precision fitment of the chassis, crankshaft, cantilevers, and linkages to reduce internal mechanical losses, avoid deadlock, and reduce friction.

Illustrated in Figs. 13, 14, and 15 is the developed power train transmitting rotary power to linear oscillating links. All driven link amplitudes are determined by the single offset crank. L_3 represents the leading tail discrete link of the structure.

The maximum amplitude of the link length L_3 and L_1 at point P_2 and P_4, respectively, are determined by the predetermined maximum crank offset P_1. The coordinates of P_1 (P_{1x}, P_{1y}), P_2 (P_{2x}, P_{2y}), P_3 (P_{3x}, P_{3y}), and P_4 (P_{4x}, P_{4y}) can be derived by:

$$\begin{cases} P_{1x} = A + B \\ P_{1y} = (A + B)\tan\theta_1 \end{cases} \begin{cases} P_{2x} = P_{1x} + C\cos\theta_1 \\ P_{2y} = P_{1y} + C\sin\theta_1 \end{cases} \tag{5}$$

$$\begin{cases} P_{3x} = -F\cos\theta_1 \\ P_{3y} = -F\sin\theta_1 \end{cases} \qquad \begin{cases} P_{4x} = -(L_2 + D) \\ P_{4y} = -P_{3y}D/E \end{cases} \tag{6}$$

The length of L_1 can be derived by $L_1^2 = P_{4x}^2 + P_{4y}^2$. Assume that ω_1 is the angular velocity of the link L_1, and the velocity vector V_{P4} is perpendicular to L_1. We have:

$$\begin{cases} V_{p4x} = -\omega L_1 \sin\theta_3 \\ V_{p4y} = \omega L_1 \cos\theta_3 \end{cases} \tag{7}$$

where V_{p4x} and V_{p4y} are the decomposed vectors of the velocity vector $V_{P4} = \omega_1 L_1$.

Fabrication The prototype *iSplash*-II is shown in Fig. 16 with the physical specifications given in Table 3. The entire body was digital modeled and formed using 3D printing techniques, at layers of 0.09 mm in PLA filament. This method produced precise 3D structural geometries of the individual segments and predetermining alignment tolerances throughout the complete kinematic cycle. It was a key challenge to develop a high power density build, small in size with high structural strength. The individual printed parts were optimized for robustness through physical strength tests and computational stress analysis, highlighting

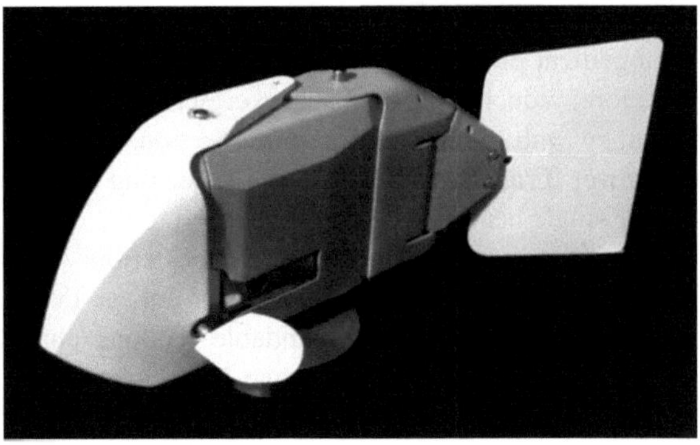

Fig. 16 *iSplash*-II

Table 3 Physical parameters of *iSplash*-II

Parameters	Specific value
Body size: m (L × W × H)	0.32 × 0.048 × 0.112
Body mass (Kg)	0.835
Actuator	Single electric motor
Actuator mass (Kg)	0.63
Power supply	11.1 V onboard LiPo battery
Manufacturing technique	3D Printing
Materials	PLA filament, acetal, stainless
Primary swimming mode	Linear locomotion
Additional maneuverability	Vertical plane
Additional control surfaces	Pectoral fins
Caudal fin material	Polypropylene
Thickness of caudal fin (mm)	2.3
Caudal fin aspect ratio (AR)	1.6

initial areas of weakness. These parts were reprinted many times in order to realize high frequency actuation. As PLA filament has a low melting point softening at approximately 60 °C, material wear at the pivots and actuated surfaces was reduced by acetal bushes and inserts, at the cost of additional weight.

It was necessary for the body size to be compact, as increasing the build geometric magnitude will increase the resistance during forward motion and therefore the power consumption required [4–17]. An accurate approximation of the streamlined body shape of the common carp was achieved within the horizontal plane illustrated in Fig. 13. The maximum thickness of the cross section is measured optimal at 0.2 of the body length [10] and was favorably positioned therefore reducing pressure drag.

The static stability in the horizontal and vertical planes is affected by material density distribution. For linear locomotive research open-loop stability is beneficial, this was achieved by the relative position of buoyancy being higher than the center of mass, as the surrounding fluid counterbalances the gravitational weight [14]. The short body length greatly increased the difficulty in achieving open-loop stability as the finest weight change in structure of individual pieces distributed across the assembly dramatically affected stability and buoyancy. This was solved by collaborating the individual parts of the modular build by adjusting the geometries and the inner structure's weight to strength configuration.

The prototype was designed with increased stability in roll and pitch as the large mass of the electric motor, 0.6 kg, 75 % of the total mass, was positioned low within the structure. To achieve a short body length, contain the embedded system and 11.1 V LiPo power supply and counteract the large mass of the primary actuator the build volume was increased vertically. This aided stability, as the lightweight PLA material and increased height positioned the center of buoyancy at the top of the prototype.

Mobility within the vertical plane was achieved to maintain a stable mid-tank trajectory during free swimming. Two rigid morphological approximations of pectoral fins were developed and positioned at the leading bulk head of the main chassis, actuated by a single servo motor. A cross beam anchored on both sides of the centralized motor was formed to link, support, and actuate the control surfaces. The addition of pectoral fins required a compact mechanism to be devised due to the very restrictive space available.

3.2 Experimental Procedures and Results

Field Trials A series of experiments, identical to the first generation were conducted in order to verify the prototype by evaluating the locomotive performance of Modes 1 and 2 in terms of kinematic parameters, speed, force, and energy consumption at frequencies within the range of 5–20 Hz. In comparison to the tests of the first prototype the number of cycles was required to be raised to increase the accuracy of data. Measurements were taken once consistency of operation was achieved and stabilized free swimming was obtained. The test results are summarized in Table 4. Experiments were conducted within a test tank, 5 m long × 2 m wide × 1.5 m deep. Free swimming between two fixed points at a distance of 4 m was used to evaluate maximum speed. The prototype had sufficient space to move without disturbances from side boundaries and the free surface, capable of consistent untethered swimming at mid-height of the tank aided by adjusting the angle of pectoral fins during swimming.

Locomotion at high speeds was unachievable without extensive stability optimization. Once achieved, an accurate straight line trajectory was possible. A thorough description of the improvements undertaken on the mechanical structure and the extent of the intensive destabilization are beyond the scope of this chapter. In addition, the devised mechanical drive system was found to be very robust, showing no signs of structural failure throughout the field trials whilst

Table 4 Comparison of test results between Modes 1 and 2

Parameters	Mode 1	Mode 2
Maximum velocity BL/s (m/s)	11.6 (3.7)	11.6 (3.7)
Acceleration time to maximum velocity (s)	0.6	0.6
Frequency (Hz)	20	20
Reynolds number Re (10^6)	1.2	1.2
Strouhal number (St)	0.34	0.34
Maximum thrust (N)	9	9
Max power consumption air (W)	120	120
Max power consumption water (W)	120	120
Swimming number (Sw)	0.58	0.58
Head swing amplitude (m)	0.003	0.013
Tail swing amplitude (m)	0.063	0.063
Body length displaced (%)	51	76

actuating at intensively high frequencies over long periods and accidentally hitting the walls of the test tank.

Swimming Pattern Observation The frame sequence of Mode 2 in eight instances, at time intervals of 0.006 s throughout one complete body cycle at 19 Hz is illustrated in Fig. 17. The obtained midline was tracked at 50 frames per second and is plotted against the desired amplitude envelopes of the anterior and posterior for comparison. When observing the midline of Mode 2, it can be seen that the desired full-body coordination presented in [17] was not achieved. As previously described in Section III–A the build required a simplified link structure due to power density constraints. Although the estimated midline curve alignment tested during stationary actuation was accurate, the excessive mass of the primary actuator held the main chassis (the entire length of link II) fixed in line with the forward heading and no single pivot point was obtained. Consequently, the swimming motion during locomotion was found to produce matching errors over the full-body in comparison to the desired swimming pattern of *iSplash*-I.

Comparing both Modes, taking into consideration that the mid-body was held rigid, the anterior amplitude of Mode 2 was measured to be 0.04 (0.013 m) of the body length, equivalent to the common carp, whereas Mode 1 was found to generate <0.01 (0.003 m) head amplitude. In addition, the large centralized mass arrangement and increased depth of body effectively minimized recoil forces and aided the stability of the posterior, allowing for accurate posterior amplitude and large thrust forces to be generated.

Figure 18 shows that the developed posterior structure can accurately mimic the undulatory parameters of real fish, as the components of link IV can be adjusted experimentally to provide the targeted midline during free swimming at various frequencies. Both Modes were able to generate accurate amplitudes of 0.1 of the body length and attain large tail amplitudes of 0.2 (0.063 m) which was found to significantly increase performance. This value is twice the size of the observed

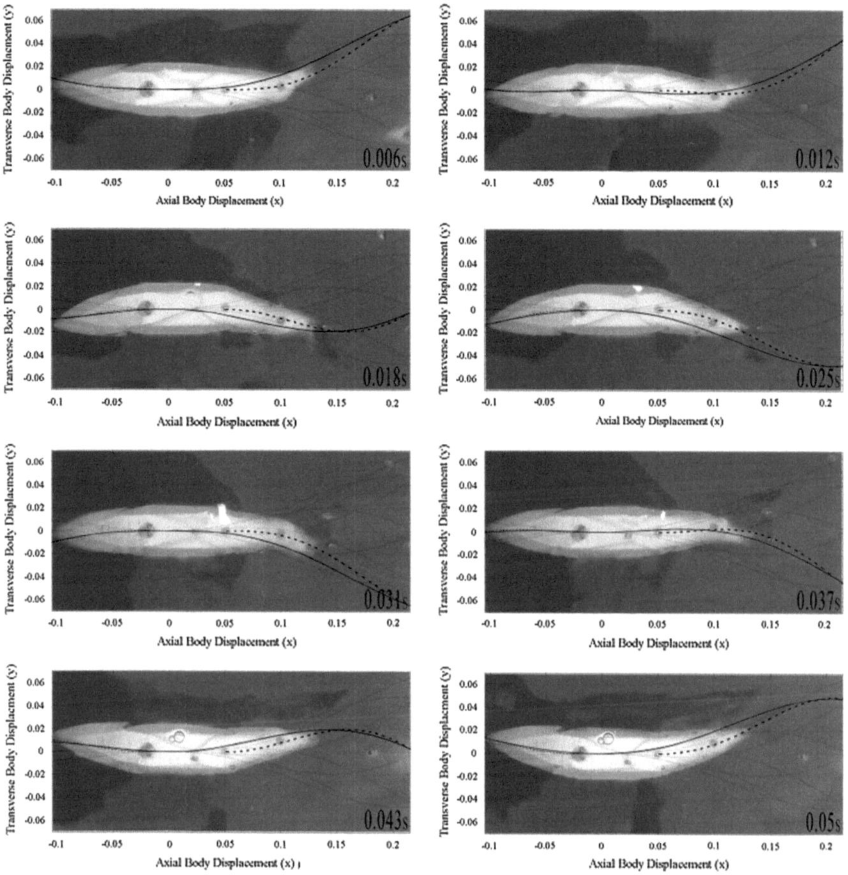

Fig. 17 Frame sequence of Mode 2 during one full-body cycle, eight instances every 0.006 s. The desired midline (—) (illustrated in Fig. 3) and the generated kinematics (illustrated in Fig. 2) from locomotion at 20 Hz (- -) are shown for comparison. We can see that the prototype achieved an anterior amplitude value of 0.04 L (0.013 m) and a tail amplitude value of 0.20 L (0.063 m)

value of the common carp at 0.1 and is increased over the first generation at 0.17. This generated amplitude is greater than the highly efficient swimming motion of a dolphin measured at 0.175 [10].

Replicating the form of the first generation the caudal fin was constructed with a low aspect ratio (AR). Although not yet thoroughly investigated, this tail was measured to achieve the highest maximum velocity and acceleration during the initial field trails. AR is defined as: AR = b^2/Sc where b squared is the fin span and Sc is the projected fin area. In this case the AR was 1.6.

Experimental Results In Fig. 18 the average energy economy in relationship to driven frequency is shown, comparing both operational Modes in air and water. It can be seen that both Modes actuating in water consumed a maximum 120 W at 20 Hz. This measurement was obtained by a connecting tethered power supply and

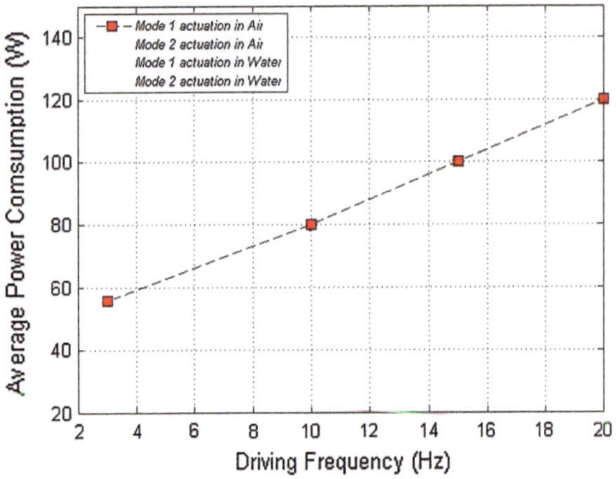

Fig. 18 Comparison of average electrical power consumption over driven frequency of both modes, actuating in air and water. No noticeable change was measured for both modes during actuation in air and in water

no noticeable increase in energy consumption was measured due to a resistance of the surrounding liquid. The result of high energy consumption can be significantly improved as the tests indicated large mechanical gains when actuating the mechanical drive system without link IV, reducing from a 120 to 70 W consumption. This was a result of pressure increase at higher velocities, as link IV was actuated; the tendons were required to be tighter to provide the desired posterior kinematics, putting increased strain on the mechanism. Despite the high energy consumption, the prototype can maintain an operational time of approximately 10 min at maximum velocity (estimated by video recording multiple runs), far surpassing the endurance of live fish, as equivalent burst speeds can only be maintained for short times of around one second. We can assume that engineering a greater mechanically efficient drive of link IV in the next generation may greatly improve endurance, relating to an estimated reduced energy consumption of approximately 50 %.

As illustrated in Fig. 19, the developed high power density ratio build can generate an incredible amount of force of up to 9 N. This can be effectively transferred in the water, accelerating both Modes to maximum velocity in approximately 0.6 s. The relationship between velocity (speed divided by body length) and driven frequency is shown in Fig. 20. The corresponding values of Modes 1 and 2 during consistent swimming are shown and compared to current robotic fish. Both operational Modes can achieve an average maximum velocity of 11.6 Bl/s, (i.e., 3.7 m/s) at 20 Hz, increasing performance in comparison with *iSplash*-I and current published robotic fish which typically peak around 1 BL/s. This result also outperforms the average maximum velocity of real fish measured at 10 BL/s. The values illustrated in Fig. 20 shows that applying the operational Mode 2 swimming pattern had no effect on performance due to kinematic alignment errors, discussed in Section IV–B, therefore it is predicted that the magnitude of added mass in both modes is equal. Hence, we can estimate that accurately applying

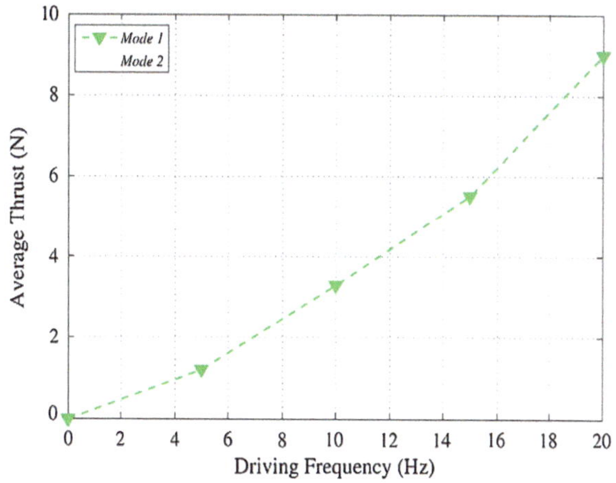

Fig. 19 Comparison of average thrust in relationship to driven frequency. Modes 1 and 2 have equivalent measurements

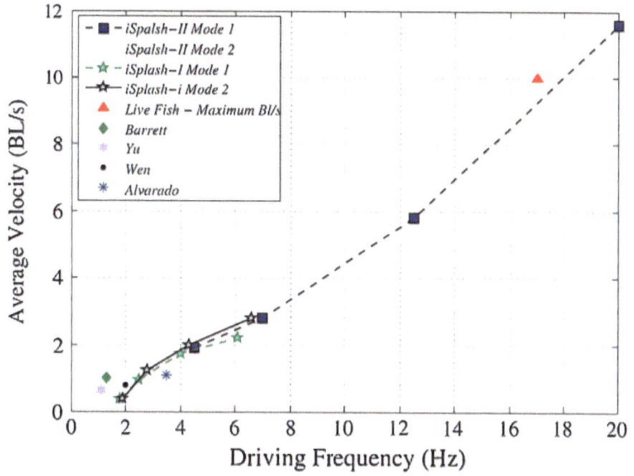

Fig. 20 Comparison of average velocities achieved by both modes, against robotic and live fish. Modes 1 and 2 measured equal velocities

the coordinated full-body swimming pattern of *iSplash*-I may increase speed by a further 27 %.

A prominent parameter for analyzing BCF locomotive performance is the Strouhal number (St), defined as St = fA/U, where f denotes the frequency, A denotes the tail amplitude, and U is the average forward velocity. St is considered optimal within the range of 0.25 < St < 0.40 [13]. The measured St = 0.34 under the condition of Re = 1.2×10^6, in both Modes is within the desired range. The prototypes Swimming number (Sw) is highly efficient, measuring a Sw of 0.58 in comparison to the previous build with a Sw of 0.42 and close to the particular efficient common carp with a Sw of 0.70 [3, 10].

For this research we undertook experiments to gain knowledge if raising driven frequencies greater than the previous build of 6.6 Hz would continue to increase speed without peak or decline. This was achieved measuring a continued increase in velocity up to intensively high frequencies of 20 Hz. Mimicking the swimming properties of real fish, frequency has become the key variable to enhance the linear locomotive performance of the *iSplash* platforms.

4 Conclusion and Future Work

This paper describes the development and experimental analysis of *iSplash*-I and *iSplash*-II. The experimental results of the first generation show conclusively that by coordinating the full-body length of the carangiform swimming motion a 27 % increase in performance, in terms of linear swimming speed, is gained over the traditional posterior confined wave form. The introduced full-body swimming motion can coordinate anterior, mid-body, and posterior displacements, and is able to reduce the kinematic errors seen in existing free swimming robotic fish. *iSplash*-I achieved a maximum velocity of 3.4 BL/s and consistently achieved a velocity of 2.8 BL/s at 6.6 Hz with a low energy consumption of 7.68 W.

A few points can be made to account for the increase in speed: The magnitude of propulsive force was increased by initiating the starting moment of added mass upstream; the developed structural arrangement allowed for smooth transition of flow along the length of body; anterior and/or mid-body vortices were formed, coordinated, and propagated downstream; lateral and thrust forces were optimized around the center of mass; a reduction in drag resistance due to reduced anterior amplitude errors. These points can be clarified in future experimental analysis utilizing visualization techniques which accurately measure the surrounding fluid flow. During the field trials the first generation prototype was found to increase its velocity as increased frequencies were applied in either swimming pattern, which indicated, that higher swimming speeds may be gained by raising frequency further.

The second generation was developed to realize the fastest speeds of live fish. A high-performance prototype was developed, robust, compact, naturally buoyant, carrying its own power supply, with a high power density and able to effectively transmit large forces at intensively high tail oscillation frequencies for untethered high-speed propulsion. Although the desired kinematics over the full-body could not be attained due to the power density requirements (with the primary actuator 75 % of the total mass), the devised assembly was able to reduce the recoil around the center of mass, therefore generating an effective propulsive mechanism. As a result, large posterior forces and tail amplitudes 0.2 of the body length (with smooth generated undulations from mid-body to tail tip) were attained. The prototype was able to accelerate to steady state swimming in an approximate time of 0.6 s, maintain an endurance at maximum speed for approximately 10 min (greater than the measurement of real fish of approximately one second), realize a highly efficient stride rate (Sw), and attain high tail oscillatory frequencies without early peak, decline, or mechanical failure.

iSplash-II, a 32 cm untethered carangiform swimmer, 0.835 kg, formed in PLA filament, consistently achieved a maximum velocity of 11.6 BL/s (i.e., 3.7 m/s) at 20 Hz with a stride rate of 0.58 and a force production of 9 N. The prototype is capable of outperforming the recorded average maximum velocity of real fish measured in BL/s, attaining speeds adequate for real world environments.

Future Work Our future research will focus on the following aspects: (i) continue to raise driven frequency to achieve greater speeds over the fastest real fish. As the build showed no signs of failure an initial aim of 40 Hz can be made; (ii) to accurately emulate the kinematic parameters of the full-body swimming motion of the first generation, indicating that maximum velocity may increase a further 27 %; (iii) to replace the drive mechanism of link IV, to significantly improve the energy consumption; (iv) to optimize the tail amplitude, shape, 3D deformation, and magnitude; (v) to apply the behavioral technique of burst and coast, as live fish generating 10 BL/s at the burst stage reduce the cost of transport by approximately 50 % [18]; (vi) to develop mobility within the horizontal plane with estimated turning diameter of <1 L

Acknowledgments Our thanks go to Richard Clapham senior for his technical assistance and financial contribution toward the project. This research was financially supported by a University of Essex Scholarship, as well as a research grant FI03005.

References

1. Bandyopadhyay PR (2002) Manoeuvring hydrodynamics of fish and small underwater vehicles. J Integr Comp Biol 42(1):102–117
2. Lighthill J (1975) Mathematical biofluiddynamics. Society for Industrial and Applied Mathematics, Philadelphia
3. Bainbridge R (1957) The speed of swimming of fish as related to size and to the frequency and amplitude of the tail beat. J Exp Biol 35:109–133
4. Barrett DS, Triantafyllou MS, Yue DKP, Grosenbaugh MA, Wolfgang MJ (1999) Drag reduction in fish-like locomotion. J Fluid Mech 392:183–212
5. Anderson JM, Chhabra NK (2002) Maneuvering and stability performance of a robotic tuna. J Integr Comp Biol 42(5):1026–1031
6. Yu J, Tan M, Wang S, Chen E (2004) Development of a biomimetic robotic fish and its control algorithm. IEEE Trans Syst Man Cybern Part B Cybern 34(4):1798–1810
7. Liu J, Hu H (2010) Biological inspiration: from carangiform fish to multi-joint robotic fish. J Bionic Eng 7:35–48
8. Wen L, Wu GH, Liang JH, Li JL (2010) Hydrodynamic experimental investigation on efficient swimming of robotic fish using self-propelled method. Int J Offshore Polar Eng 20:167–174
9. Valdivia P, Alvarado Y, Youcef-Toumi K (2003) Modeling and design methodology for an efficient underwater propulsion system. In: Proceedings IASTED international conference on robotics and applications, Salzburg
10. Nagai M (1999) Thinking fluid dynamics with dolphins. Ohmsha, LTD, Japan
11. Webb PW (1984) Form and function in fish swimming. Sci Am 251:58–68
12. Rosen MW (1959) Water flow about a swimming fish, China Lake, US Naval Ordnance Test Station TP 2298, p 96

13. Wolfgang MJ, Anderson JM, Grosenbaugh MA, Yue DK, Triantafyllou MS (1999) Near-body flow dynamics in swimming fish. J Exp Biol 202:2303–2327
14. Lauder GV, Drucker EG (2004) Morphology and experimental hydrodynamics of fish control surfaces. IEEE J Ocean Eng 29:556–571
15. Triantafyllou GS, Triantafyllou MS, Grosenbaugh MA (1993) Optimal thrust development in oscillating foils with application to fish propulsion. J Fluids Struct 7:205–224
16. Videler JJ (1993) Fish swimming. Chapman and Hall, London
17. Clapham RJ, Hu H, (2014) *iSplash*-I: high performance swimming motion of a carangiform robotic fish with full-body coordination. In: Proceedings of IEEE international conference on robotics and automation, Hong Kong
18. Videler JJ, Weihs D (1982) Energetic advantages of burst-and-coast swimming of fish at high speeds. J Exp Biol 97:169–178

IPMC-Actuated Robotic Fish

Zheng Chen, Hilary Bart-Smith and Xiaobo Tan

Abstract Excellent swimmers, such as tuna, rays, and goldfish, take advantage of their flexible fins, compliant bodies, and swimming bladders to achieve fast, highly maneuverable, and energy-efficient locomotion. Ionic polymer-metal composites (IPMCs) present attractive opportunities for implementation in flexible underwater propulsion systems due to their intrinsic compliancy and underwater actuation capability. IPMCs can also perform as lightweight and compact catalysts for water electrolysis, which can be used to generate gas for buoyancy control. In this chapter, the potential of IPMCs in underwater propulsion is explored, including caudal fin propulsion, pectoral fin propulsion, and buoyancy control. Enabling technologies, including fabrication methods, modeling and control strategies, and design approaches, are developed for creating bio-inspired robots using IPMC as artificial muscle and buoyancy engine. Three types of underwater robots have been developed to evaluate their performance. First, a robotic fish propelled by an IPMC caudal fin is developed to evaluate its caudal fin propulsion. Second, a bio-inspired robotic cownose ray propelled by two IPMC actuated pectoral fins is demonstrated to evaluate its pectoral fin propulsion. Third, a buoyancy control device enabled by IPMC-enhanced electrolysis is developed to explore its buoyancy control performance.

Z. Chen (✉)
Wichita State University, 1845 Fairmount ST, Wichita, KS 67260-0083, USA
e-mail: Zheng.Chen@wichita.edu

H. Bart-Smith
University of Virginia, 122 Engineer's Way, Charlottesville, VA 22904-4746, USA
e-mail: hb8h@virginia.edu

X. Tan
Michigan State University, 428 S. Shaw Lane, Rm. 2120 Engineering Building, East Lansing, MI 48824, USA
e-mail: xbtan@egr.msu.edu

© Springer-Verlag Berlin Heidelberg 2015

R. Du et al. (eds.), *Robot Fish*, Springer Tracts in Mechanical Engineering,
DOI 10.1007/978-3-662-46870-8_8

1 Introduction

Nature has produced a plethora of aquatic animals that have developed unique swimming techniques to locomote in oceans, lakes, and rivers. Recent research suggests that the high performance observed in tuna, batoid rays, and goldfish, takes advantage of their flexible fins, compliant body, and swimming bladders, to achieve fast, maneuverable, and energy-efficient locomotion. Recent years have witnessed significant efforts in the development of bio-inspired underwater robots that mimic aquatic animals, such as robotic fish [4, 20, 22, 23, 26, 36, 49], robotic jelly fish [37, 52, 54], and robotic rays [15, 19, 53]. In most of these examples, traditional electric motors were used to generate the flapping and heaving motions observed in nature. However, electric motor-driven propulsion has the following disadvantages: (1) they are too bulky for small-scale robots; (2) they generate unfavorable acoustic noise, which makes the robots detectable and unfriendly to marine life; (3) last but most important, electric motors cannot generate compliant actuation without force feedback control, which limits its propulsion efficiency. If force feedback control is included, extra sensors and complicated control strategies are needed, greatly increasing the size and complexity of the robots.

Compliant underwater propulsion [44] requires a compliant actuator coupled with a compliant structure [3, 18, 21, 50] to mimic the musculoskeletal structures in biology [27]. To achieve such propulsion, studies have been focused on electro-active polymers (EAPs) to achieve the large deformations demonstrated in nature [6]. Due to their similarities to biological muscles, EAPs are often called artificial muscles [6]. EAPs can be divided into two categories: ionic EAPs and dielectric EAPs. Dielectric EAPs can generate large deformation with large force [11, 39, 48]. However, they require high actuation voltage (typically higher than 1 kV), which limits their applications in underwater propulsion. Ionic polymer-metal composites (IPMCs) are an important category of ionic EAPs due to their compliancy and underwater actuation capability [46]. An IPMC consists of an ion exchange membrane coated with noble metal electrodes, such as gold or platinum. Application of a small voltage (<2 V) to the IPMC leads to cation transport to the cathode side, which introduces swelling effect on the cathode side and shrinking effect on the anode side. Eventually, the IPMC bends to the anode side and thus realize the actuation effect. Figure 1 shows the actuation mechanism of IPMC. An IPMC strip (0.5 cm wide, 2 cm long, 0.2 mm thick) can generate more than 10 % of strain, up to 10–20 MPa stress, with about 20 mW power consumption [45]. Bennett and Leo, who worked with Nafion–platinum IPMC, compared water and ionic liquid as electrolyte and reported the cycle life being 10^3 for the former and 3×10^5 for the latter [7]. IPMCs offer excellent opportunities for underwater applications due to their actuation capabilities, compliance, and low weight. Moreover, when the applied voltage is higher than 1.48 V, due to water electrolysis, hydrogen and oxygen gases are generated at the cathode and anode sides, respectively [33], as shown in Fig. 1. This has implications in buoyancy control applications. Compared to other existing buoyancy control approaches, the IPMC-enabled

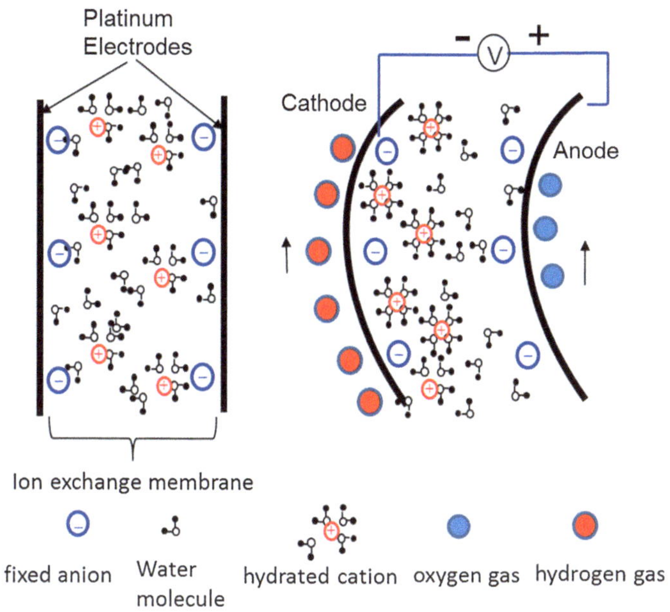

Fig. 1 Actuation mechanism of IPMC (cross-sectional view)

buoyancy control enables us to design a compact, lightweight, stealthy, and energy-efficient buoyancy control device [51].

In this chapter, we will present three key technologies for creating IPMC-enabled bio-inspired robots, including (1) a physics-based and control-oriented modeling approach that can capture the intrinsic actuation dynamics of IPMC and the hydrodynamics of robotic fish; (2) a fabrication technology for creating IPMC actuating membranes capable of generating 3D kinematic motions; and (3) an IPMC-enabled buoyancy control scheme that can achieve depth control. Based on the modeling approach, Chen et al. [13] developed a model of robotic fish propelled by an IPMC caudal fin. Based on the fabrication technology, Chen et al. [15] developed a small-size and free swimming robotic Cownose ray propelled by two pectoral fins. Based on the buoyancy control, Um et al. [51] developed a compact and energy-efficient depth control device for bio-inspired robots. These results are briefly summarized below.

- **Modeling of Robotic fish propelled by an IPMC caudal fin**

Significant effort has been devoted toward IPMC-powered underwater robots. A robotic fish propelled by an IPMC caudal fin was developed in Tan's group at Michigan State University, shown in Fig. 2 [13]. Inspired by biological fish fins, where passive, collagenous membranes are driven by muscle-controlled fin rays [26], a passive, plastic fin was attached to the tip of IPMC to enhance propulsion.

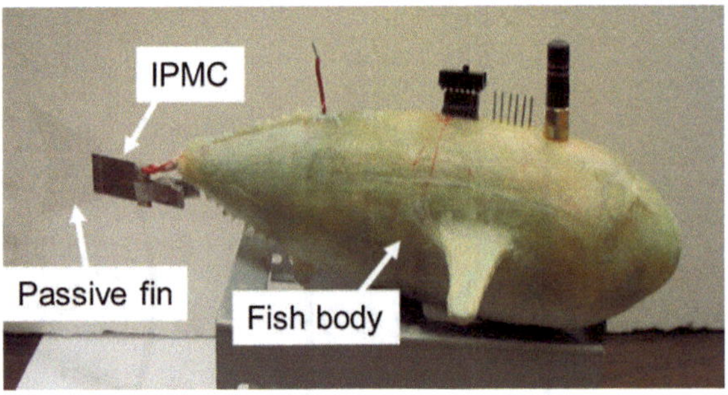

Fig. 2 Robotic fish propelled by an IPMC caudal fin. © 2009 IEEE. Reprinted, with permission, from [13]

The fish was designed to be fully autonomous and serve as a mobile, aquatic sensing platform. It consists of a rigid body and an IPMC caudal fin.

An accurate model is desirable for both optimal design [56] and control simulations [57] of an underwater robot. For instance, Boyer et al. [9] presented a dynamic model for three-dimensional (3D) eel-like robot. Morgansen et al. [36] investigated geometric methods for modeling and control of free-swimming and rigid fin-actuated underwater vehicles. There have been limited studies related to modeling of flexible fins. A finite-element method was adopted by Yim et al. [55] to model the motion of an IPMC actuator underwater, where an empirical RC circuit was used to predict the bending moment of IPMC under actuation. The added-mass effect due to acceleration of surrounding water was ignored in modeling hydrodynamic interactions, and the authors presented only simulation results. Modeling of IPMC actuators in underwater operations was also studied by Brunetto et al. [10]. However, the actuation dynamics of IPMC was represented by a frequency-dependent coupling term, which was essentially the empirical frequency response and did not capture the fundamental physics of IPMC. Furthermore, the experimental results presented by Brunetto et al. [10] were limited to a clamped IPMC beam in water, and no attempt was made to validate the model on a free-swimming robot. Porfiri and co-workers investigated the hydrodynamics of an IPMC beam using numerical computation [1] and digital particle image velocimetry (DPIV) measurements [40]. Although their work was very interesting, it was limited to a clamped IPMC beam only and the actuation dynamics of IPMC was not considered. Further studies carried out by Porfiri and co-workers considered free-swimming of IPMC actuated robotic fish [5]. However, the actuation dynamics of IPMC was simplified as an RC circuit, which cannot fully capture the ion transportation dynamics and distributed surface resistance effect in IPMC.

In Sect. 2, we will present a model for IPMC-propelled robotic fish that captures the intrinsic actuation physics of IPMC, and the complex hydrodynamic interactions between IPMC and fluid, and is validated in experiments on a free-swimming robotic fish prototype. The proposed model incorporates the interactions of the passive fin with both the IPMC actuator and the fluid, allowing us to simultaneously capture the passive fin's role in boosting propulsion and its loading effect on the

Fig. 3 Atlantic cownose ray (courtesy of www.elasmodiver.com) [15]. © 2011 by ASME. Reproduced by permission of ASME. All rights reserved

IPMC beam. The model is used to predict the cruising speed of robotic fish, given a periodic actuation voltage to the IPMC fin.

- **Robotic Cownose ray propelled by two IPMC pectoral fins**

The cownose ray (*Rhinoptera bonasus*, shown in Fig. 3) demonstrates excellent swimming capabilities, generating highly efficient thrust via flapping of dorsally flattened pectoral fins [42]. Many efforts have been directed towards building a bio-inspired pectoral fin structure to mimic the swimming behavior of the ray, such as electric motor-actuated rigid plates [34] and tensegrity structures [35]. Due to size limitations of the electric motors and tensegrity structures, they are not suitable for small-scale robots (of the order of 5–10 cm) [20, 41, 49]. A bio-inspired actuating material, which is lightweight, compliant, resilient, and capable of generating 3D deformation with low power consumption, is highly desirable in constructing a free swimming and small-scale robotic cownose ray.

Traditional IPMC can only generate bending motion. In the propulsion mechanism of rays, 3D kinematic motions on the pectoral fin play an important role in generating highly efficient propulsion and maneuvering [42]. In order to obtain an actuating membrane capable of generating 3D deformation, lithography-based and surface machining-based approaches have been employed to pattern the electrodes of the IPMC to create active and passive areas in a Nafion membrane [14, 38]. By individually controlling the bending of each active area, 3D deformations of the membrane have been demonstrated. However, the stiffness of the Nafion in the passive area limited its capability of generating twisting motions. Punning et al. [41] developed pectoral fins for ray-like underwater robot by assembling separated IPMC beams with a latex foil. But their robot did not achieve free swimming capability because of high power consumption and low propulsion efficiency.

In Sect. 3, a bio-inspired and free swimming robotic cownose ray propelled by artificial pectoral fins will be presented. The pectoral fin consists of one IPMC as artificial muscle in the leading edge and a passive PDMS membrane in the trailing edge. The body matches the morphology of a cownose ray (through computer axial tomography scans). The fin has been characterized by some key factors as related to

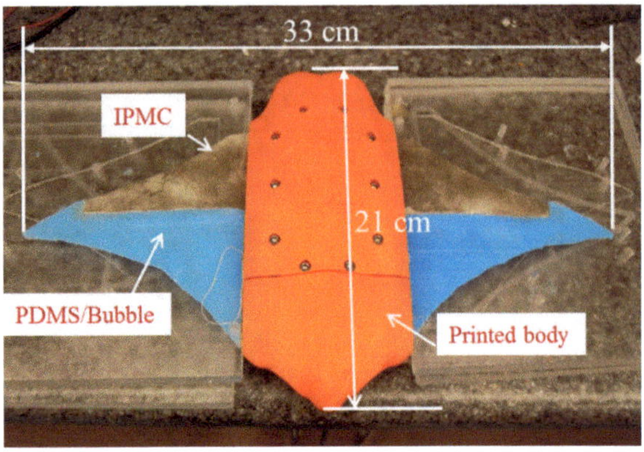

Fig. 4 Robotic cownose ray propelled by two IPMC pectoral fins [15]. © 2011 by ASME. Reproduced by permission of ASME. All rights reserved

the function of the robot: tip deflection, twist angle, and power consumption. The overall shape of the fin mimics that of the biology. Experimental results have shown that the robot (shown in Fig. 4) is capable of free swimming [15].

- **Buoyancy control device with IPMC enhanced water electrolysis**

3D maneuvering, such as diving and ascending, are more difficult to achieve and control in underwater devices. A simple and energy-efficient depth control method is critical in the design of 3D maneuverable underwater robots. The most common method utilized in mechanical systems is a servomotor to control the volume of a piston-cylinder system containing compressed air [19]. This usually yields reliable results with relatively fast response time. However, due to the limitations in scaling the servo motor, this method is not feasible for implementation in small devices. In order to build more compact and efficient depth control devices, researchers have turned to biology for inspiration.

Nature has evolved many novel and effective depth control mechanisms suitable for a variety of environments. For example, sperm whales (Fig. 5a) achieve

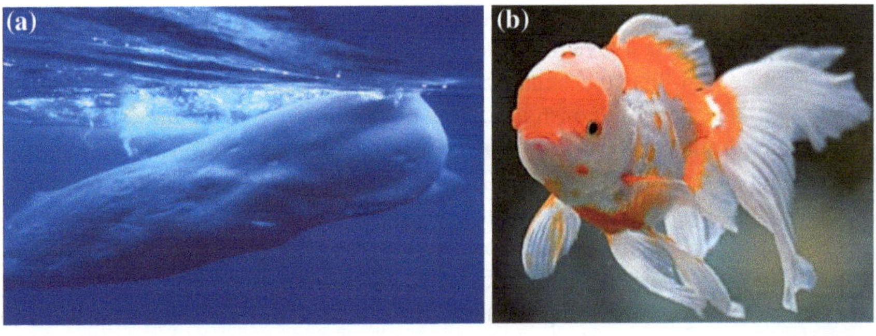

Fig. 5 Depth control in biology (courtesy of www.elasmodiver.com). © 2011 IEEE. Reprinted, with permission, from [51]. **a** Sperm whale (*Physeter macrocephalus*), **b** fancy goldfish (*Carassius auratus*)

buoyancy control by using their spermaceti oil. An adult sperm whale contains about 4 tons of spermaceti oils in their spermaceti organ, which represents approximately 8 % of its total mass [47]. The spermaceti oil has a low melting point and its density depends largely on the temperature of the oil. By manipulating the arterial blood flow through the spermaceti organ, the sperm whales can regulate the temperature of the oil and are thus able to control their buoyancy. There have been recent demonstrations of buoyancy control concepts that manipulate the temperature to change the density of oil [47] or wax [32]. However, the response times are slow (of the order of 10 min), and it is inefficient for small devices because a constant power must be supplied to maintain the temperature of the oil while cruising at a certain depth.

Akle et al. [2] developed a bio-inspired and compact buoyancy engine by displacing sodium and cathium ions out of seawater inside, which changes the water density. However, it has been demonstrated that the volume change rate was very low. Rayfinned fish, such as that depicted in Fig. 5b, change the buoyancy of their body using a swim bladder [8]. Expansion of the bladder results in increased volume, thus making the body more positively buoyant and vice versa. Inspiration for the buoyancy control device presented in this chapter comes from these ray-finned fish.

In Sect. 4, a buoyancy control device using IPMC-enhanced water electrolysis will be presented (shown in Fig. 6). Electrolysis is used to generate gases in order to displace water and increase buoyancy. A novel depth control device using electrolysis has been designed and built. The device design incorporates an artificial bladder where the volume of gas generated by electrolysis is controlled via a solenoid valve, thus changing the device's buoyancy. A set of gold electrodes, separated by an IPMC film, is used as a lightweight and compact electrolysis generator. Tests demonstrate the feasibility of this approach for depth control.

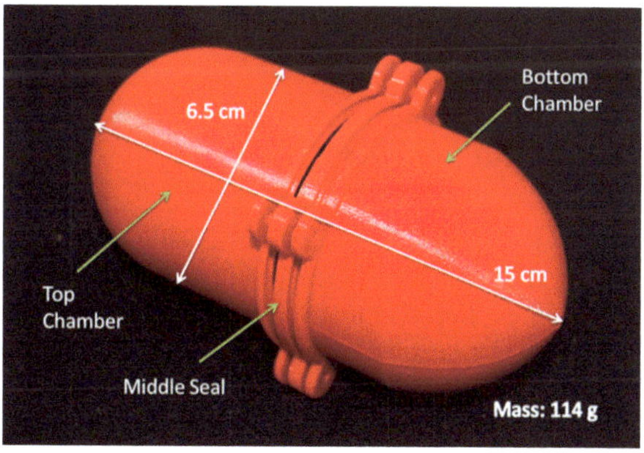

Fig. 6 Buoyancy control device enabled by IPMC enhanced water electrolysis

2 Robotic Fish Propelled by an IPMC Caudal Fin and Its Modeling

Based on the elongated-body propulsion theory of Lighthill [29, 30] the steady-state velocity of a swimming fish is related to the bending displacement and the slope of the tail end. The key of the modeling work is then to derive the motion of the hybrid tail under IPMC actuation and hydrodynamic interactions. Modeling of the hybrid tail starts with a fourth-order partial differential equation (PDE) that can capture the beam dynamics of IPMC in fluid. It incorporates the hydrodynamic force acting on the IPMC beam and the driving force introduced by actuation of IPMC. A mode-summation method is employed to obtain the solution to the PDE. The distributed bending moment generated by an actuation input is obtained using a physical, yet compact, model that captures the internal ion dynamics of IPMC [12]. In order to evaluate the actuation-induced generalized forces for the mode equations, we decompose the distributed moment into a distributed force along the length and a concentrated bending moment at the beam end, which would generate the same bending moment along the length. The model is then extended to capture the interactions between the IPMC and the passive fin. The hydrodynamic force acting on the passive fin is replaced by a concentrated moment and the force acting at the tip of IPMC actuator, which can be incorporated into the beam dynamics to obtain an analytical expression for the motion of hybrid tail. Finally, a speed model can be derived based on the Lighthill theory and actuation model of the IPMC hybrid tail.

In this section, we first review Lighthill's theory on elongated body propulsion. IPMC beam dynamics in fluid is discussed next, considering general force and moment inputs. This is followed by detailed consideration of actuation-induced bending moment in the model, as well as the load contribution to the IPMC beam from the passive fin. Then, the model for computing the speed of IPMC-propelled robotic fish is obtained by merging Lighthill's theory and the hybrid tail dynamics. Finally, the speed model was validated through experiment. Most of the modeling work in this section was published in Chen et al. [13].

2.1 Lighthill Theory

A body is considered elongated if its cross-sectional area changes slowly along its length. The robotic fish described in Sect. 1 is thus elongated and Lighthill's theory [29] applies. Suppose that the tail is bending periodically with the bending displacement at z denoted by $w(z, t)$. See Fig. 7 for notation [13]. At the steady state, the fish will achieve a periodic, forward motion with some mean speed U. In the discussion here, the word "mean" refers to the average over one period. The mean thrust T produced by the tail can be calculated as

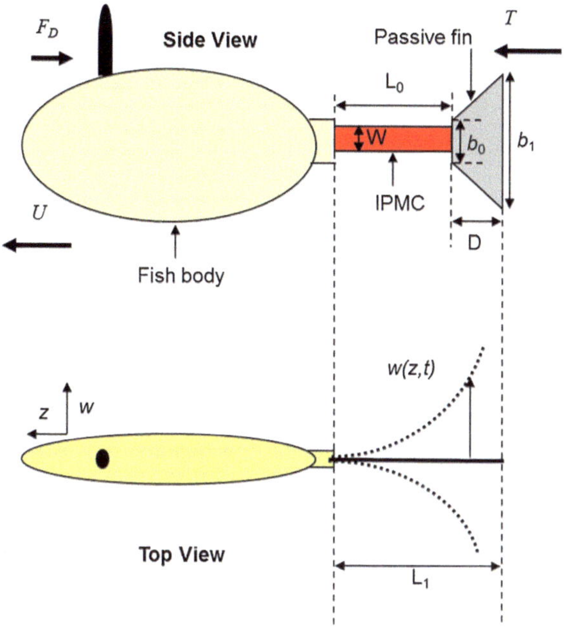

Fig. 7 Definition of variables in the robotic fish [13]. © 2009 IEEE. Reprinted, with permission, from [13]

$$\bar{T} = \left[\frac{m}{2}\left(\overline{\left(\frac{\partial w(z,t)}{\partial t}\right)^2} - U^2\overline{\left(\frac{\partial w(z,t)}{\partial z}\right)^2}\right)\right]_{z=L} \tag{1}$$

where $z = L_1$ denotes the end of tail, $\overline{(\cdot)}$ denotes the mean value, and m is the virtual mass density at $z = L_1$, expressed as

$$m = \frac{1}{4}\pi S_c^2 \rho_w \beta \tag{2}$$

where S_c is the width of the tail at the end $z = L_1$, ρ_w is the fluid density, and β is a nondimensional parameter close to 1. Equation (1) indicates that the mean thrust depends only on the lateral velocity ($\partial w/\partial t$) and the slope ($\partial w/\partial z$) at the tail end. A cruising fish, under inviscid flow conditions, will experience a drag force F_D as

$$F_D = \frac{C_D \rho_w U^2 S}{2} \tag{3}$$

where S is the wetted surface area and C_D is the drag coefficient. At the steady state, the mean thrust T is balanced by the drag F_D, from which one can solve the cruising speed U as

$$\overline{U} = \sqrt{\frac{m \cdot \overline{\left(\frac{\partial w(L,t)}{\partial t}\right)^2}}{C_D \rho_w S + m \cdot \overline{\left(\frac{\partial w(L,t)}{\partial z}\right)^2}}} \tag{4}$$

Since the speed of the fish is related to the lateral velocity and the slope of the trailing edge, one needs to fully understand the actuation dynamics of the tail.

2.2 IPMC Beam Dynamics in Fluid

In order to obtain the full actuation model of IPMC, we start with a fourth-order PDE for the dynamic deflection function $w(z, t)$ [17] as

$$YI\frac{\partial^4 w(z,t)}{\partial z^4} + C\frac{\partial w(z,t)}{\partial t} + \rho_m A\frac{\partial^2 w(z,t)}{\partial t^2} = f(z,t) \tag{5}$$

where Y, I, C, ρ_m, and A denote the effective Young's modulus, the area moment of inertia, the internal damping ratio, the density, and the cross-sectional area of the IPMC beam, respectively, and $f(z, t)$ is the distributed force density acting on the beam.

Converting (5) into the Laplace domain, we obtain

$$YI\frac{\partial^4 w(z,s)}{\partial z^4} + Csw(z,s) + \rho_m As^2 w(z,s) = F(z,s) \tag{6}$$

The force on the beam consists of two components, the hydrodynamic force F_{hydro} from water and the driving force F_{drive} due to the actuation of IPMC

$$F(z,s) = F_{\text{hydro}}(z,s) + F_{\text{drive}}(z,s). \tag{7}$$

The hydrodynamic force acting on the IPMC beam can be expressed as [43]

$$F_{\text{hydro}}(z,s) = -\rho_w \frac{\pi}{4} W^2 s^2 \Gamma_1(\omega) w(z,s), \quad 0 \le z \le L \tag{8}$$

where W is the width of the IPMC beam, $\Gamma_1(\omega)$ is the hydrodynamic function for the IPMC beam subject to an oscillation with radial frequency ω, and ρ_w is the density of fluid. The hydrodynamic function for a rectangular beam can be represented as [43].

$$\Gamma_1(\omega) = \Omega(R_e)\left[1 + \frac{4iK_1(-i\sqrt{iR_e})}{\sqrt{iR_e}K_0(-i\sqrt{iR_e})}\right] \tag{9}$$

where the Reynolds number R_e of a vibrated beam in water is given as

$$R_e = \frac{\rho_w W^2 \omega}{4\eta}.$$

K_0 and K_1 are modified Bessel functions of the third type, $\Omega(R_e)$ is the correction function associated with the rectangular beam cross section [43], and η is the viscosity of fluid. With (7) and (8), the beam dynamics Eq. (6) can be written as

$$YI\frac{\partial^4 w(z,s)}{\partial z^4} + Csw(z,s) + (\mu_m + m_d\Gamma_1)s^2 w(z,s) = F_{\text{drive}}(z,s), \tag{10}$$

where $m_d = \rho_w (\pi/4)W^2$ is the added mass and $\mu_m = \rho_m A$ is the mass of IPMC per unit length. Under harmonic oscillation with frequency ω, we can denote

$$\mu_v = \mu_m + m_d \,\text{Re}(\Gamma_1) \tag{11}$$

$$C_v = C - m_d\omega \text{Im}\,(\Gamma_1) \tag{12}$$

where μ_v is the equivalent mass of IPMC per unit length in water and C_v is the equivalent damping coefficient of IPMC in water. Re(·) and Im(·) are the functions that get the real part and the imaginary part from a complex value, respectively. Equation (12) means that the damping of IPMC vibration in water includes both the internal damping in IPMC and the frequency-dependent external damping caused by fluid. With (11) and (12), (10) can be written as [10]

$$YI\frac{\partial^4 w(z,s)}{\partial z^4} + Csw(z,s) + \mu_v s^2 w(z,s) = F_{\text{drive}}(z,s) \tag{13}$$

According to the mode analysis method, we can express the solution to (13) as the sum of different modes [31] as

$$w(z,s) = \sum_{i=1}^{\infty} \varphi_i(z) q_i(s) \tag{14}$$

where $\varphi_i(z)$ is the beam shape for the ith mode and $q_i(s)$ is the corresponding generalized coordinate. The mode shape $\varphi_i(z)$ takes the form

$$\varphi_i(z) = \cosh(\lambda_i z) - \cos(\lambda_i z) - \beta_i(\sinh(\lambda_i z) - \sin(\lambda_i z)) \tag{15}$$

where λ_i can be obtained by solving

$$1 + \cos(\lambda_i L)\cosh(\lambda_i L) = 0 \tag{16}$$

and

$$\beta_i = \frac{\sinh(\lambda_i L) - \sin(\lambda_i L)}{\cosh(\lambda_i L) + \cos(\lambda_i L)} \tag{17}$$

The generalized coordinate $q_i(s)$ can be represented as

$$q_i(s) = f_i(s) Q_i(s) \tag{18}$$

where $f_i(s)$ is the generalized force

$$Q_i(s) = \frac{1}{s^2 + 2\xi_i \omega_i s + \omega_i^2} \tag{19}$$

and the natural frequency ω_i and the damping ratio ξ_i for the ith mode are

$$\omega_i = \frac{C_i^2}{L^2} \sqrt{\frac{YI}{\mu_v(\omega_i)}} \tag{20}$$

$$\xi_i = \frac{C_v(\omega_i)}{2\mu_v(\omega_i)\omega_i} \tag{21}$$

and $C_i = \lambda_i L$. Noting that $\Gamma_1(\omega)$ is almost a constant value in the frequency region around ω_i, one can consider $\mu_v(\omega_i)$ as a constant in (20). Therefore, ω_i can be obtained approximately. Then, with ω_i, ξ_i can be obtained from (21). The generalized force $f_i(s)$ is obtained from F_{drive} as

$$f_i(s) = \frac{1}{M_i} \int_0^L F_{\text{drive}}(z, s) \phi_i(z) \mathrm{d}z \tag{22}$$

where M_i is the generalized mass

$$M_i(s) = \int_0^L \mu_v \varphi_i^2(z) \mathrm{d}z = \mu_v L \tag{23}$$

The next step is to derive the generalized force $f_i(s)$ from the moment generated by IPMC actuation and from the hydrodynamic force acting on the passive fin but transmitted to the IPMC beam.

2.3 Actuation Model of the Tail

Chen and Tan [12] investigated the electrical dynamics of IPMC to obtain the moment generated within IPMC, but there, the beam dynamics in water was not considered. In the following, we will incorporate both electrical dynamics and hydrodynamic interactions into a full-actuation model for IPMC hybrid tail in water.

With distributed surface resistance, we can relate the actuation-induced bending moment $M_{\mathrm{IPMC}}(z, s)$ at point z to the actuation voltage $V(s)$ by an infinite-dimensional transfer function [12] as

$$M_{\mathrm{IPMC}}(z,s) = \frac{\alpha_0 W K k_e(\gamma(s) - \tanh(\gamma(s)))V(s)}{(s\gamma(s) + K\tanh(\gamma(s)))} \cdot \frac{\left(\sinh\left(\sqrt{B(s)}z\right)\tanh\left(\sqrt{B(s)}L\right) - \cosh\left(\sqrt{B(s)}z\right)\right)}{1 + r_2\theta(s)} \tag{24}$$

with

$$\theta(s) \triangleq \frac{sW\kappa_e\gamma(s)(s+K)}{h(s\gamma(s) + K\tanh(\gamma(s)))}, \quad B(s) \triangleq \sqrt{r_1\left(\frac{\theta(s)}{1 + r_2\theta(s)/W} + \frac{2}{R_p}\right)},$$

$$\gamma(s) \triangleq h\sqrt{\frac{s+K}{d}}, \quad K \triangleq \frac{F^2 dC^-}{k_e RT}(1 - C^-\Delta V),$$

where α_0 is an electromechanical coupling constant, d is the ionic diffusivity, R is the gas constant, F is Faraday's constant, T is the absolute temperature, C^- is the anion concentration, ΔV is the volumetric change, x is the coordinate defined in the thickness direction, κ_e is the effective dielectric constant of the polymer, r_1 is the electrode resistance per unit length in the length direction, r_2 is the electrode resistance per unit length in the thickness direction, and R_p is the through-polymer resistance per unit length. W, L, and h are the width, length, and half thickness of the IPMC beam, respectively. We replace the moment $M_{\mathrm{IPMC}}(z, s)$ induced by actuation by three components: a distributed force density $F_d(z, s)$ acting along the length, a concentrated force $F_c(L, s)$, and a moment $M(L, s)$ acting at the IPMC tip $z = L$, where

$$F_c(L,s) = \frac{\partial M_{\mathrm{IPMC}}(z,s)}{\partial z}\Big|_{z=L}, \tag{25}$$

$$F_d(z,s) = \frac{\partial^2 M_{\mathrm{IPMC}}(z,s)}{\partial z^2}, \tag{26}$$

$$M(L,s) = M_{\mathrm{IPMC}}(L,s). \tag{27}$$

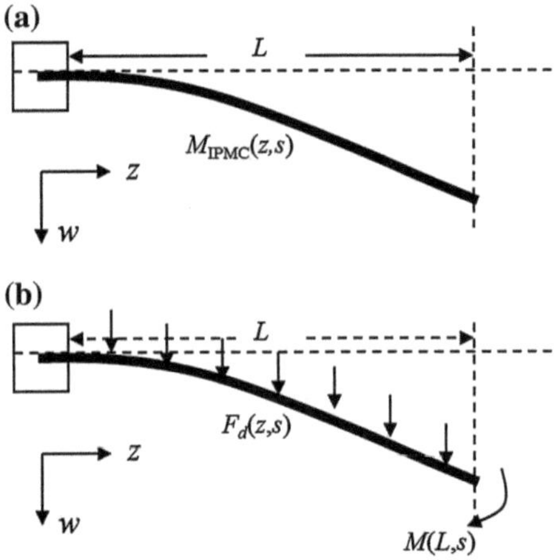

Fig. 8 Original moment in (**a**) is replaced by a distributed force density and a concentrated bending moment in (**b**). © 2009 IEEE. Reprinted, with permission, from [13]

The rationale of this replacement is that these components can generate the same bending moment as $M_{\text{IPMC}}(z, s)$. See Appendix A for the details of this justification. With (24), it can be verified that $F_c(L, s) \equiv 0$. Then, the generalized force can be obtained as [43]

$$f_{1i}(s) = \frac{1}{M_i}\left(\int_0^L F_d(z,s)\phi_i(z)\mathrm{d}z + M(L,s)\phi_i'(L)\right). \tag{28}$$

- **With IPMC only**

Figure 8 shows that the original moment is replaced by a distributed force density and a concentrated moment [13]. Note that for the IPMC only case, L_I defined in Fig. 7 is the same as L defined in Fig. 8. With the aforementioned replacement, we can derive the models for the IPMC tail-only case.

With (22) and (23), the generalized force (24) can be written as

$$f_{1i}(s) = H_{fi}(s)V(s), \tag{29}$$

where

$$\begin{aligned} H_{fi}(s) = &\frac{a-b}{2M_i}(a_L + b_L - c_L - d_L - \beta_i(a_L - b_L + jc_L - jd_L)) \\ &+ \frac{\alpha_0 WKk_e(\gamma(s) - \tanh(\gamma(s)))}{M_i(s\gamma(s) + K\tanh(\gamma(s)))}\frac{\varphi_i'(L)}{(1 + r_2\theta(s))\cosh(cL)}, \end{aligned}$$

and

$$a = \frac{\alpha_0 WKk_e(\gamma(s) - \tanh(\gamma(s)))}{(s\gamma(s) + K\tanh(\gamma(s)))} \frac{B(s)}{1 + r_2\theta(s)},$$

$$b = a\tanh\left(\sqrt{B(s)L}\right), \quad c = \sqrt{B(s)},$$

$$a_L = \frac{\sinh((c + \lambda_i)L)}{c + \lambda_i}, \quad b_L = \frac{\sinh((c - \lambda_i)L)}{c - \lambda_i},$$

$$c_L = \frac{\sinh((c + j\lambda_i)L)}{c + j\lambda_i}, \quad d_L = \frac{\sinh((c - j\lambda_i)L)}{c - j\lambda_i}.$$

See Chen et al. [13] for detailed derivation.

From (14), one can then get the transfer function $H_1(L, s)$ relating $w(L, s)$ to $V(s)$ as

$$H_1(L,s) = \frac{w(L,s)}{V(s)} = \sum_{i=1}^{\infty} \varphi_i(L)H_{fi}(s)Q_i(s). \tag{30}$$

We can also derive the transfer function $H_{1d}(L, s)$ relating the slope of the beam $\partial w(z, s)/\partial z$ at $z = L$ to the input voltage $V(s)$ as

$$H_{1d}(L,s) = \frac{(\partial w(z,s)/\partial z|_{z=L})}{V(s)} = \sum_{i=1}^{\infty} \varphi_i'(L)H_{fi}(s)Q_i(s). \tag{31}$$

- **With a hybrid tail**

From (1) and (2), the tail width S_c at the end has a significant impact on the speed U. One could increase S_c by simply using a wider IPMC beam. Due to the IPMC actuation mechanism, however, a too wide beam (i.e., plate) will produce curling instead of bending motion, and is thus not desirable. Therefore, it has been chosen to increase the edge width by attaching a passive plastic piece, as illustrated in Fig. 9. While such a hybrid tail is expected to increase the thrust, one has to also consider that the extra hydrodynamic force on the passive fin adds to the load of IPMC, and may reduce the bending amplitude. Therefore, it is necessary to model these interactions carefully.

The hydrodynamic force acting on the passive fin can be written as [43]

$$f_{\text{tail}}(z,s) = -\frac{\pi}{4}\rho_w s^2 b(z)^2 \Gamma_2(\omega) w(z,s), \quad L_0 \le z \le L_1, \tag{32}$$

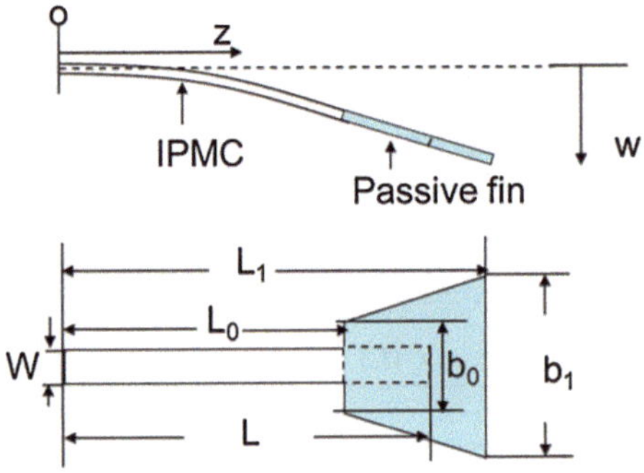

Fig. 9 Illustration of an IPMC beam with a passive fin. The lower schematic shows the definitions of dimensions. © 2009 IEEE. Reprinted, with permission, from [13]

where $\Gamma_2(\omega)$ is the hydrodynamic function of the passive fin. Note that the hydrodynamic force acting on the active IPMC beam has been incorporated in (10), and therefore, only the hydrodynamic force on the passive fin needs to be considered here. Since the passive fin used is very light, its inertial mass is negligible compared to the propelled virtual fluid mass, and is thus ignored in the analysis here. Considering that the passive fin is rigid compared to IPMC, its width $b(z)$ and deflection $w(z, s)$ can be expressed as

$$b(z) = \frac{b_1 - b_0}{L_1 - L_0}(z - L_0) + b_0, \tag{33}$$

$$w(z, s) = w(L_0, s) + w'(L_0, s)(z - L_0), \tag{34}$$

where b_0, b_1, L, L_0, and L_1 are as defined in Fig. 9. Then, one can calculate the moment introduced by the passive fin: for $L_0 \leq z \leq L_1$,

$$\begin{aligned} M_{\text{fin}}(z, s) &= \int_{L_0}^{L_1} f_{\text{tail}}(\tau, s)(\tau - z)\mathrm{d}\tau \\ &= \int_{L_0}^{L_1} f_{\text{tail}}(\tau, s)(\tau - L_0)\mathrm{d}\tau + (L_0 - z)\int_{L_0}^{L_1} f_{\text{tail}}(\tau, s)\mathrm{d}\tau, \end{aligned} \tag{35}$$

If we define

$$M_{\text{tail}}(s) = \int_{L_0}^{L_1} f_{\text{tail}}(\tau, s)(\tau - L_0)\mathrm{d}\tau, \tag{36}$$

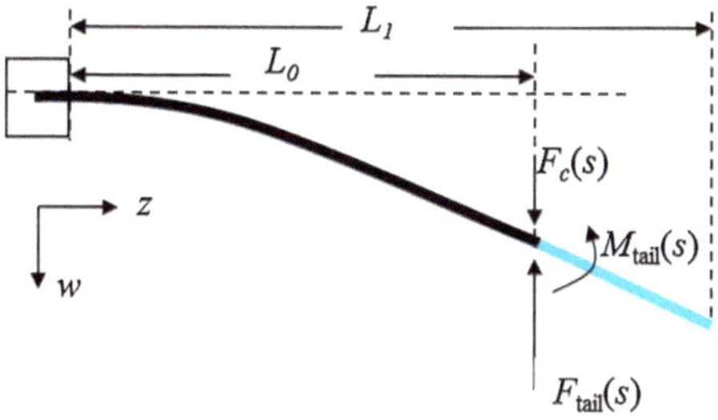

Fig. 10 Forces and moments acting on the hybrid tail. © 2009 IEEE. Reprinted, with permission, from [13]

$$F_{\text{tail}}(s) = \int_{L_0}^{L_1} f_{\text{tail}}(\tau, s)\mathrm{d}\tau. \tag{37}$$

Then (35) an be written as

$$M_{\text{fin}}(z, s) = M_{\text{tail}}(s) + F_{\text{tail}}(L_0 - z). \tag{38}$$

Figure 10 shows the forces and moments acting on the hybrid tail.

$$\begin{aligned} f_{2i}(s) = &\frac{1}{M_i}\left(\int_0^{L_0} F_d(z, s)\varphi_i(z)\mathrm{d}z + \varphi_i(L_0)F_{\text{tail}}(s)\right) \\ &+ \frac{(M_{\text{tail}}(s) + M(L_0, s))\varphi_i'(L_0)}{M_i}, \end{aligned} \tag{39}$$

where M_{tail} and F_{tail} are defined in (36) and (37), respectively, $F_d(z, s)$ and $M(L_0, s)$ are defined in (26) and (27), respectively.

Then, the transfer functions relating $w(L_0, s)$ to $V(s)$ and that relating to $w'(L_0, s)(s) \triangleq \partial w(z, s)/\partial z|_{z=L_0}$ to $V(s)$ can be found as

$$H_2(L_0, s) = \frac{(1 + F_s)A_s - B_sE_s}{(1 + C_s)(1 + F_s) - B_sJ_s}, \tag{40}$$

$$H_{2d}(L_0, s) = \frac{(1 + C_s)E_s - A_sJ_s}{(1 + C_s)(1 + F_s) - B_sJ_s}, \tag{41}$$

where

$$A_s = H_1(L_0, s),\ E_s = H_{1d}(L_0, s),\ D = L_1 - L_0,\ k = \frac{b_1 - b_0}{D},$$
$$M_s = \frac{\pi}{4} s^2 \Gamma_2(\omega) \rho_w,$$
$$B_s = \sum_{i=1}^{\infty} \frac{\varphi_i(L_0) Q_i(s)}{M_i} M_s \left[\varphi_i'(L_0) k_a + \varphi_i(L_0) k_b\right],$$
$$C_s = \sum_{i=1}^{\infty} \frac{\varphi_i(L_0) Q_i(s)}{M_i} M_s \left[\varphi_i'(L_0) k_b + \varphi_i(L_0) k_c\right],$$
$$F_s = \sum_{i=1}^{\infty} \frac{\varphi_i'(L_0) Q_i(s)}{M_i} M_s \left[\varphi_i'(L_0) k_a + \varphi_i(L_0) k_b\right],$$
$$J_s = \sum_{i=1}^{\infty} \frac{\varphi_i'(L_0) Q_i(s)}{M_i} M_s \left[\varphi_i'(L_0) k_b + \varphi_i(L_0) k_c\right],$$

and

$$k_a = \frac{k^2 D^5}{5} + \frac{2 k b_0 D^4}{4} + \frac{b_0^2 D^3}{3},$$
$$k_b = \frac{k^2 D^4}{4} + \frac{2 k b_0 D^3}{3} + \frac{b_0^2 D^3}{2},$$
$$k_c = \frac{k^2 D^3}{3} + k b_0 D^2 + b_0^2 D.$$

See Chen et al. [13] for the detailed derivation. From (34), (40), and (41), one can obtain the transfer functions relating the bending displacement and the slope at $z = L_1$ to the voltage input $V(s)$ as follows:

$$H_3(L_1, s) \triangleq \frac{w(L_1, s)}{V(s)} = H_2(L_0, s) + H_{2d}(L_0, s) D, \tag{42}$$

$$H_{3d}(L_1, s) \triangleq \frac{w'(L_0, s)}{V(s)} = H_{2d}(L_0, s). \tag{43}$$

2.4 Speed Model of Robotic Fish

Given a voltage input $V(t) = A_m \sin(\omega t)$ to the IPMC actuator, the bending displacement and the slope of the tail at the tip $z = L_1$ can be written as

$$w(L_1,t) = A_m|H(j\omega)|\sin(\omega t + \angle H(j\omega)), \tag{44}$$

$$\frac{\partial w(z,t)}{\partial z}|_{z=L_1} = A_m|H_d(j\omega)|\sin(\omega t + \angle H_d(j\omega)), \tag{45}$$

where $\angle(\cdot)$ denotes the phase angle, and $H(s)$ and $H_d(s)$ represent $H_3(L_1, s)$ and $H_{3d}(L_1, s)$, respectively. From (4), one can then obtain the steady-state speed U of the robotic fish under the actuation voltage $V(t) = A_m \sin(\omega t)$ as

$$\bar{U} = \sqrt{\frac{mA_m^2\omega^2|H(j\omega)|^2}{2C_D\rho_w S + mA_m^2|H_d(j\omega)|^2}}. \tag{46}$$

One can easily extend (46) to periodic signals of other forms. For instance, the prototype in Fig. 2 uses square-wave voltage signals for ease of implementation. To derive the speed U, we can write out the Fourier series of a square wave. Then the velocity of the fish actuated under a square-wave voltage with amplitude A_m can be obtained as

$$\bar{U} = \sqrt{\frac{m\left(8\omega^2A_m^2/\pi^2\right)\sum_{n=1,3,5\cdots}^{\infty}|H(jn\omega)|^2}{C_D\rho_w S + m\left(8A_m^2/\pi^2\right)\sum_{n=1,3,5\cdots}^{\infty}\left(|H_d(jn\omega)|^2/n^2\right)}}. \tag{47}$$

2.5 Speed Model Validation

In the speed model (47), all the parameters are either dimensions or physical parameters, which can be either measured or identified. See [13] for the details in parameter identification. To validate the speed model of the robotic fish, the velocities of the fish propelled by the IPMC under square-wave voltage inputs with amplitude 3.3 V and different frequencies were measured. In this experiment, the robotic fish swam freely in a tank marked with start and finish lines, and a timer recorded the time it took for the fish to travel the designated range after it reached the steady state. The capability of the model in predicting cruising speed was verified for different operating frequencies, for different tail dimensions. The speed model for a square-wave input [see (47)] was applied to the robotic fish, as described in Sect. 2. In the simulation of (47), we took the first three terms in each infinite series, which provided a good approximation to the sum of infinite series.

Four different hybrid tails were investigated, shown in Table 1. The predicted speeds match the experimental data well, as shown in Fig. 11.

Intuitively, within the actuation bandwidth of IPMC, the achieved speed increases with the actuation frequency. As the frequency gets relatively high, the bending amplitude of IPMC decreases. Thus, for each tail, there is an optimal

Table 1 Dimensions of four tails (see Fig. 9 for the definitions of dimension variables)

	L_0 (mm)	L (mm)	W (mm)	b_0 (mm)	b_1 (mm)	D (mm)
Tail 1	18	23	15	20	40	40
Tail 2	18	23	15	20	50	30
Tail 3	18	23	20	20	65	25
Tail 4	18	23	20	20	50	30

© 2009 IEEE. Reprinted, with permission, from [13]

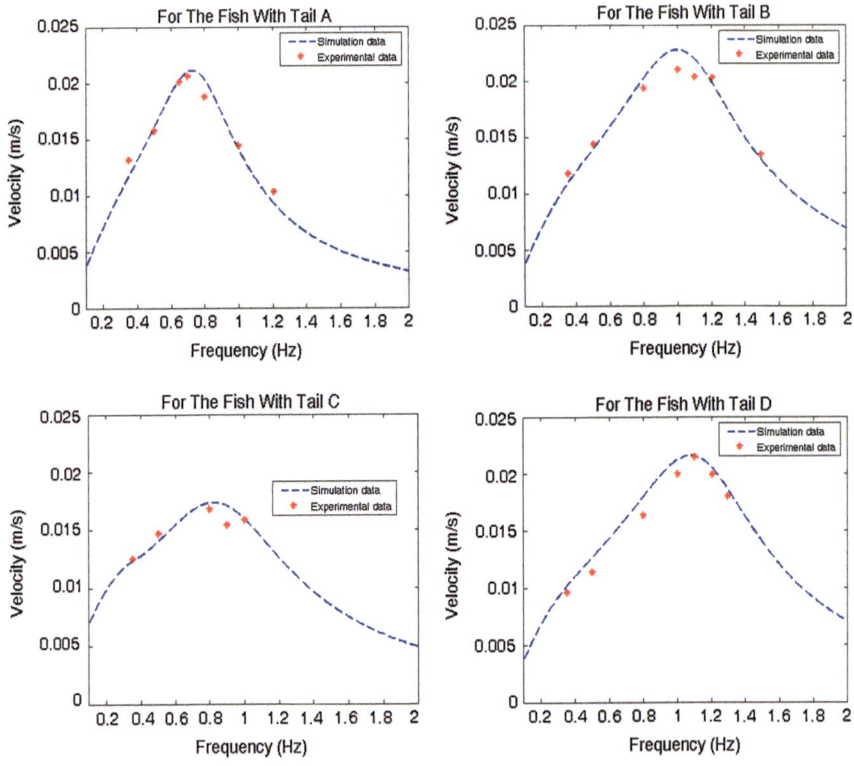

Fig. 11 Verification of the speed model for the fish with four different tails. © 2009 IEEE. Reprinted, with permission, from [13]

frequency under which the fish reaches the highest speed. Both the optimal frequency and the corresponding highest speed depend on the dimensions of both IPMC and passive fin, which can be predicted by the speed model.

3 Bio-inspired Robotic Cownose Ray Propelled by IPMC Pectoral Fin

In this section, a PDMS assembly-based fabrication process will be presented first. The pectoral fin consists of one IPMC in the leading edge and a passive PDMS membrane in the trailing edge. By applying voltage potential to the IPMC, the passive PDMS membrane follows the bending of IPMC with a phase delay, which leads to thrust generation on the fin. The overall shape of the fin mimics that of the biology. Then a bio-inspired and free swimming robotic cownose ray propelled by artificial pectoral fins will be presented. The robotic body was designed using biological data from a cownose ray and printed using a 3D printer. A light and compact on-board control unit with a lithium ion polymer battery was used for generating control signal to the IPMC in the pectoral fin. Experimental results demonstrate that the pectoral fin is capable of generating thrust and the robot is capable of free swimming. Details of this work were published in Chen et al. [15].

3.1 Fabrication of Artificial Pectoral Fin

The artificial pectoral fin must be able to generate an oscillatory motion with a twist angle observed in nature, while under hydrodynamic loads. The fin was fabricated by combining IPMC actuator with a PDMS elastomer in a mold to create a pre-defined planform shape. The planform design of the pectoral fin is shown in Fig. 12. The outline shape of the fin mimics that of the biological cownose ray. An IPMC beam is placed in the leading edge of the fin. The rest of the fin is composed by a PDMS passive membrane.

The first step in creating the artificial pectoral fin is to fabricate the IPMC actuator. Many groups have developed different IPMC fabrication processes for various purposes [16, 24, 25, 46]. In our fabrication method, we primarily follow the procedure outlined by Kim and Shahinpoor [24] but one important difference in

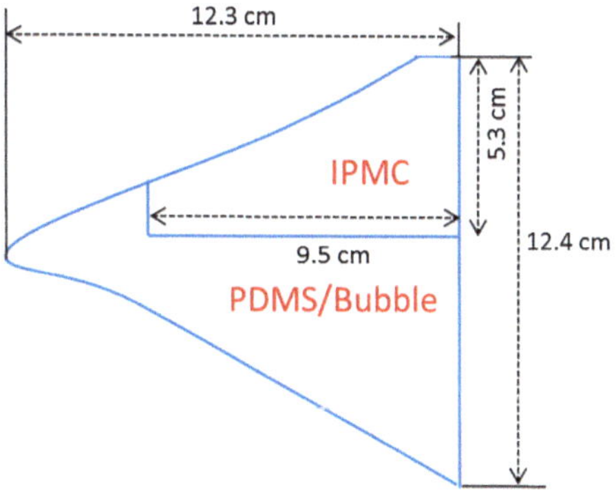

Fig. 12 Artificial pectoral fin design [15]. © 2011 by ASME. Reproduced by permission of ASME. All rights reserved

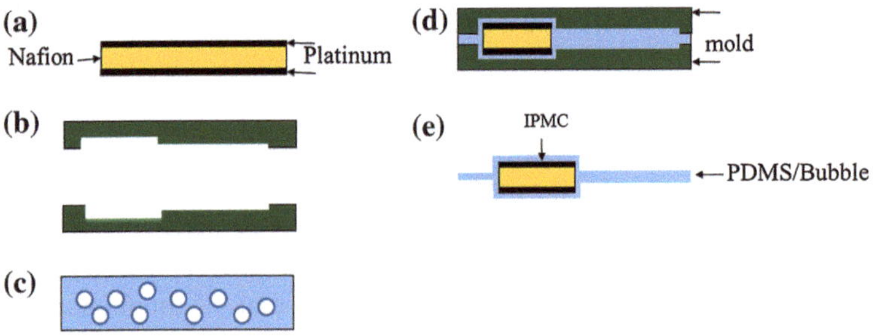

Fig. 13 Fabrication process (cross-section view) [15]. © 2011 by ASME. Reproduced by permission of ASME. All rights reserved. **a** Fabricate IPMC, **b** make a mold, **c** mix PDMS gel with glass bubbles, **d** clamp the mold containing the IPMCs and uncured PDMS, and then cure the PDMS at room temperature, **e** remove the mold

our method includes multiple platinum plating processes to reduce the surface resistance of the electrodes [28]. A Nafion (NafionTM N1110, DuPont) membrane was selected as the ion-exchange membrane in IPMC. After the electroless plating process, about 6 μm thick platinum electrodes were deposited on the Nafion surfaces with good polymer-metal adhesion. The sample was then submerged in a sodium solution (1 N) for 1 day to exchange H^+ with Na^+ to enhance the actuation performance of the IPMC.

The next step was to bond the IPMC with a PDMS elastomer membrane. The PDMS bonding process is shown in Fig. 13: (a) two Delrin polymer (McMaster) molds were made using a CNC rapid milling machine (MDX-650, Roland). Each mold had two concaved areas to house the PDMS passive membrane and the IPMC actuator in the molding process. The thickness of PDMS membrane was 1 mm and thickness of IPMC was 280 μm. (b) the IPMC was cut into the shape shown in Fig. 12. (c) 5 % glass bubbles (Glass bubble K37, 3 M Inc) were added into PDMS gel (Ecoflex 0030, Smooth-on Inc.) to gain a neutrally buoyant pectoral fin. To make it visible underwater, blue dye was added into the PDMS gel. (d) the IPMC and the PDMS gel were then clamped with the molds and the PDMS was cured at room temperature for 3 h. (e) the IPMC/PDMS artificial pectoral fin (Fig. 14) was removed from the molds.

The pectoral fin was tested in a water tank. The results suggest that the fin can generate up to 40 % tip deflection, 10° twisting angle with less than 1 W of power consumption [15]. Propulsion model for the IPMC-enabled pectoral fin is still under investigation.

3.2 Design and Test of Robotic Cownose Ray

- **On-board Control Circuit**

The on-board circuit provided a square wave voltage signal to the IPMC actuator in the pectoral fin. Figure 15a shows the schematic of the circuit. A 555-timer was

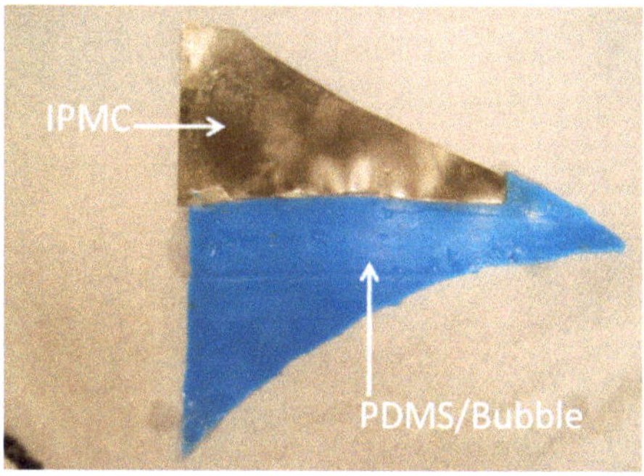

Fig. 14 IPMC/PDMS artificial pectoral fin [15]. © 2011 by ASME. Reproduced by permission of ASME. All rights reserved

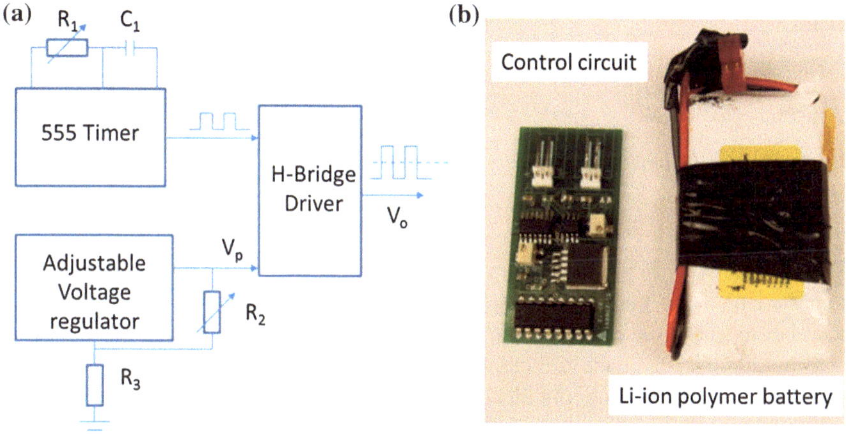

Fig. 15 On-board control unit [15]. © 2011 by ASME. Reproduced by permission of ASME. All rights reserved. **a** Schematic of circuit board, **b** picture of PCB board and battery

used to generate a frequency tunable square wave. The amplitude of the voltage signal, V_p, was controlled by an adjustable voltage regulator. An H-bridge driver was used to draw up to 2 A output peak current. A rechargeable 7.3 V Lithium Ion Polymer battery (1700 mAh, AA Portable Power Corp) was selected as the power source for the robot. Figure 15b shows the picture of PCB board and battery.

- **Bio-inspired Body Design**

The robotic body was designed using Inventor (Autodesk). A set of CT Scan data from a biological cownose ray was used in the body design. The CT Scan data provided cross sectional images of the cownose ray's body. We took 10 cross sectional images of the center body and lofted them to create a 3-dimensional robotic body. Since a small scale robot was our interest, the robotic body was scaled down compared to that of the biological cownose ray. The overall the body

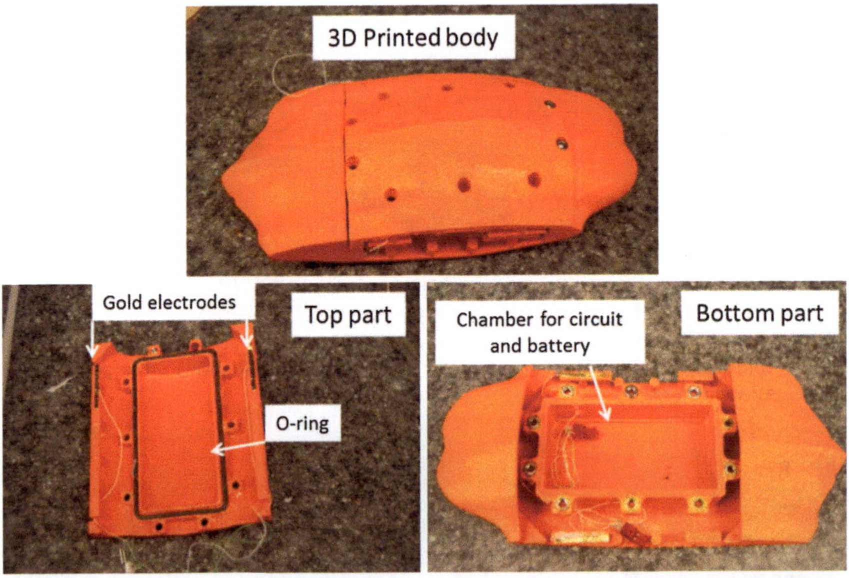

Fig. 16 Bio-inspired robotic body [15]. © 2011 by ASME. Reproduced by permission of ASME. All rights reserved

dimensions were 21 cm long, 5 cm wide, and 5 cm thick (Fig. 16). The body consisted of a top part and a bottom part. In the bottom part, there was a cubic chamber to house the circuitry. In the top part, there was an O-ring that was used to seal the chamber to make the circuitry waterproof. Both parts had two gold electrodes that were used to apply electric signals to the IPMCs. The body was printed using a Fused Deposition Modeling (FDM) machine (*u*Print Plus, Dimension). The fully assembled robot was 21 cm long, 33 cm wide, 5 cm thick, and weighted 119 g. The free-swimming robot is shown in Fig. 4.

- **Free Swimming Test**

The robot was tested in a water tank (1.5 m wide, 4.7 m long, and 0.9 m deep), which is shown in Fig. 17. The robot was neutrally buoyant underwater. In the free swimming test, the operating frequency of the square wave actuation voltage was tuned at 0.157 Hz and the amplitude was set at 3.3 V.

A digital video camera (VIXIA HG21, Canon) was used to capture the videos of the swimming robot. Figure 18 shows six snap shots of the swimming robot from top view. Each snap shot was taken every 5 s. The speed of the robot was extracted from the movie using the Edge Detection program in the Labview. The swimming speed shown in Fig. 18 was 0.7 cm/s. Since the body length was 21 cm, the speed in body length per second was 0.034 BL/s with less than 2 W power consumption.

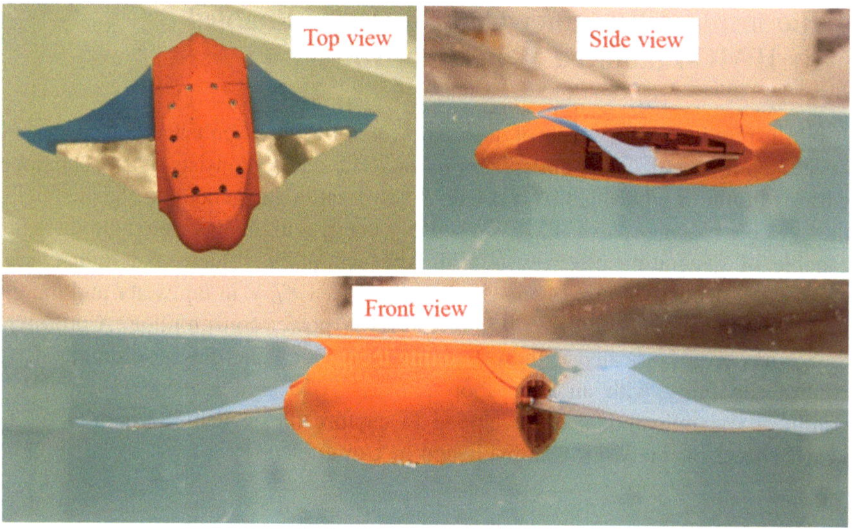

Fig. 17 Pictures of the robot in underwater test [15]. © 2011 by ASME. Reproduced by permission of ASME. All rights reserved

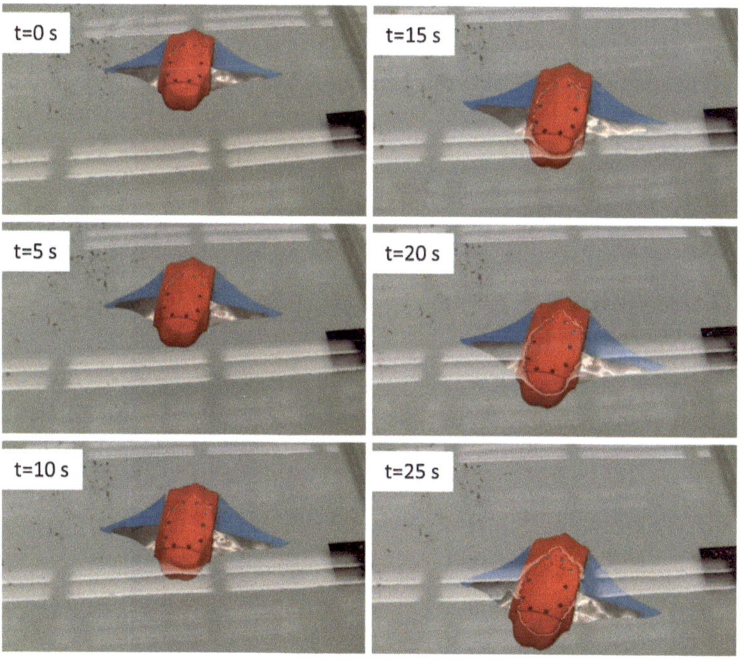

Fig. 18 Snap shots of the free swimming robot [15]. © 2011 by ASME. Reproduced by permission of ASME. All rights reserved

4 Bio-inspired Buoyancy Control Device Enabled by IPMC-Enhanced Water Electrolysis

A bio-inspired buoyancy control device powered by IPMC enhanced electrolysis is developed for the depth of underwater robots. The device consists of a gas chamber whose volume is controlled by a solenoid valve and a gas generator, thus changing the device's buoyancy. An IPMC is used as a catalyst in the gas generator since it is capable of generating electrolysis with low activation voltage and high energy efficiency. The activation voltage can be as low as 1.87 V in tap water and the gas generate rate can reach at 0.135 mL/s with 3.8 W power consumption. A prototype device has been fabricated with 3D printing technology and demonstrated its depth control capability with an open loop control. To design a feedback control, a nonlinear dynamic model is developed to capture the unstable dynamics of the device. Most of this work was published in Um et al. [51].

4.1 Description of Buoyancy Control Device

The buoyancy control device consists of a gas chamber whose volume is controlled by electrolysis gas generator and a solenoid valve, thus changing the device's buoyancy. A set of gold electrodes, separated by an IPMC film, is used as a lightweight and compact electrolysis generator. The overall schematic of the device is shown in Fig. 19. The gas generation mechanism and gas releasing mechanism will be described as follows.

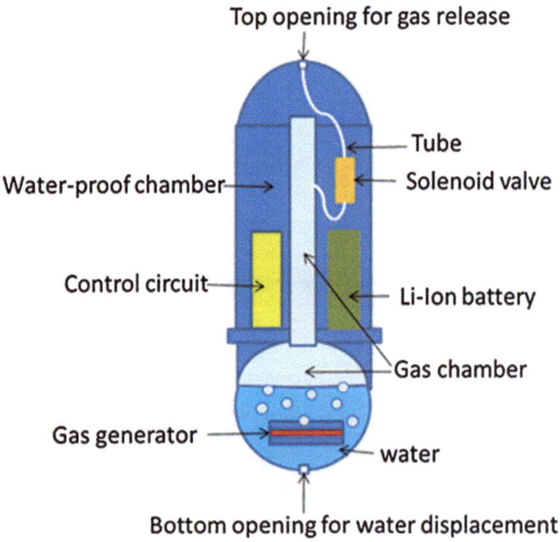

Fig. 19 Schematic Illustration of the novel buoyancy control device. © 2011 IEEE. Reprinted, with permission, from [51]

- **Gas Generation Mechanism and Control**

When an electric current is applied between positively and negatively charged electrodes in ionized water, a chemical reaction occurs, in which pure oxygen and hydrogen gases are produced at the anode and cathode electrodes, respectively. Most electrolysis experiments are performed in ionic solutions, which are usually prepared by adding salt, acid, or a base. However, adding electrolytes is not feasible in this case because the device must be able to generate gas in regular tap water. Electrolysis in tap water is much slower because of the limited number of ions present. In order to enhance the electrolysis process, an IPMC is placed directly in between the electrodes [33].

The mechanism to control the release of gas, and thereby depth, uses a two-way solenoid valve. When the valve is actuated, the gas formed during electrolysis escapes from the device and water supersedes. As a result, the density of the device increases, causing it to become more negatively buoyant. When the valve is closed, water cannot enter through the bottom opening because the pressure inside the device is equal to that of outside. Thus the device is able to maintain the same depth by having a constant buoyant force. In order to travel up or downwards, gas is either produced or released in a controlled manner.

4.1.1 Fabrication of the Device

The parts were drawn using Autodesk Inventor 2010 and printed using a FDM machine (uPrint Plus by Dimension). Figure 20 shows the design of the device. The device can be divided into three chambers. Assembled device is shown in Fig. 6. The device was approximately 15 cm tall, 6.5 cm in diameter, and had a mass of 114 g.

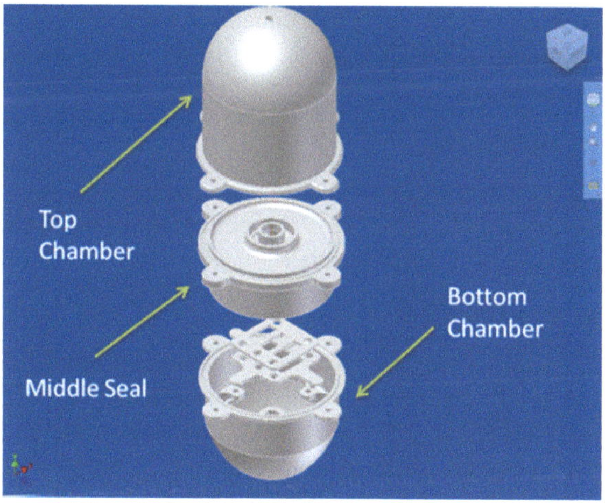

Fig. 20 Design of the device using Autodesk. © 2011 IEEE. Reprinted, with permission, from [51]

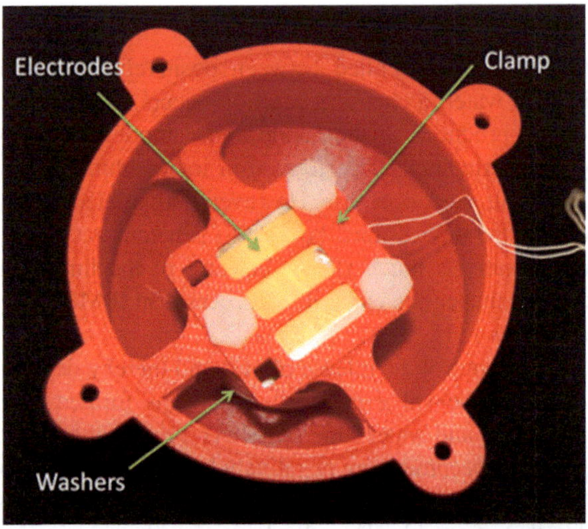

Fig. 21 Bottom chamber. © 2011 IEEE. Reprinted, with permission, from [51]

- **Bottom Chamber**

The bottom chamber encased the gas generator and hold water (Fig. 21). The water level varied depending on the depth location of the device. When the device was below the surface, the gas generator was submerged under water to ensure that the electrolysis can take place to allow the device to float up. The bottom chamber also had room to secure up to eight metal washers such that the device was initially about 80–90 % submerged with addition of washers as dead weights.

On the bottom of the chamber, there was a small opening to allow water to enter or escape the device. This opening ensured that the inside and outside pressures were kept equal. An IPMC enhanced water electrolysis generator was placed in the bottom chamber. The size of the IPMC was 2 cm by 3 cm with about 150 μm in thickness (Fig. 22).

- **Middle Seal**

The purpose of the middle seal was to provide waterproofing of the electronics in the top chamber (Fig. 23). The bottom inner surface was concaved to direct the gas products to the gas chamber.

- **Top Chamber**

The top chamber was partitioned into a waterproof chamber and a gas chamber. The waterproof chamber contained all the electronics, 7.4 V Li-ion battery and Solenoid valve (Fig. 24). The gas chamber contained a mixture of air and gases produced at the electrodes in the bottom chamber. A tube connected the gas chamber to the device surroundings via solenoid valve.

Fig. 22 IPMC enhanced electrolysis generator. © 2011 IEEE. Reprinted, with permission, from [51]

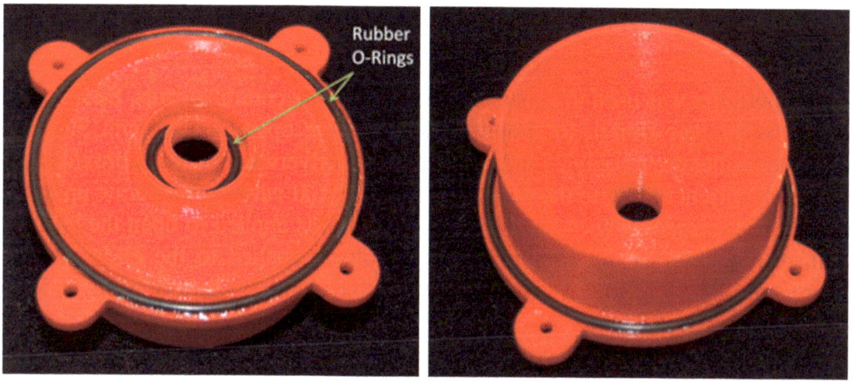

Fig. 23 Middle seal. © 2011 IEEE. Reprinted, with permission, from [51]

Fig. 24 Top Chamber with electronic components and gas chamber. © 2011 IEEE. Reprinted, with permission, from [51]

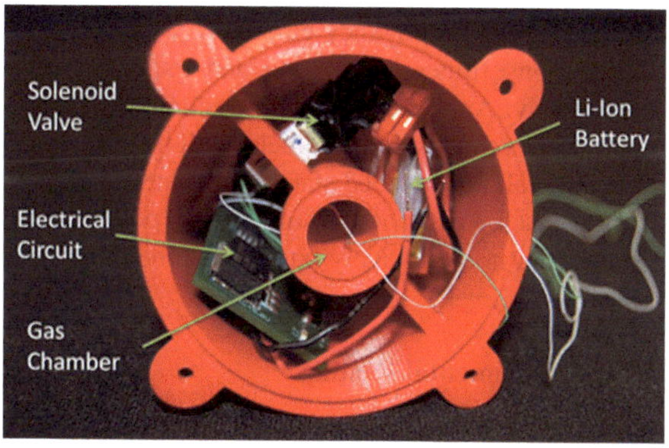

- **On-board Circuit**

The on-board circuit provided actuation voltage signals to the electrodes and solenoid valve (Fig. 25). A rechargeable 7.4 V, 400 mAh AA Portal Power Corp Lithium Ion Polymer battery was used as a power source and PIC12F508

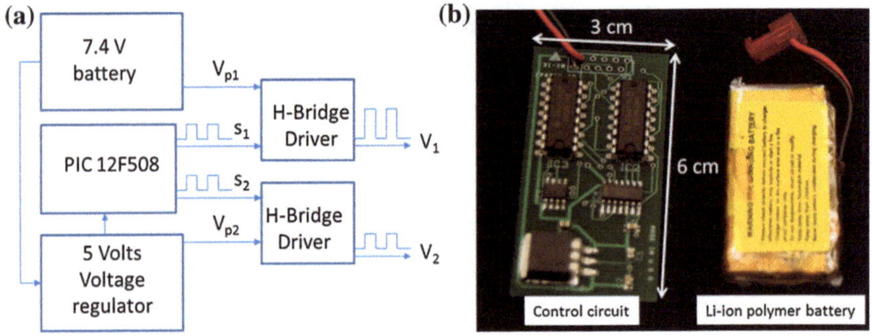

Fig. 25 Circuit schematics. © 2011 IEEE. Reprinted, with permission, from [51]. **a** Schematic of the circuit, **b** picture of the printed circuit board and battery

microcontroller was used to generate two square wave control signals, S_1 and S_2. A square wave was chosen to simplify testing of the open-loop system. Since the microcontroller draws only 25 mA, two H-bridge drivers were used to provide up to 2 A peak current output to the electrodes and solenoid valve, which draw up to 500 and 80 mA, respectively. A 7.4 V voltage (V_{p1}), was applied to open the valve. A voltage regulator set the amplitude of the voltage applied to the electrodes and microcontroller, V_{p2}, to 5 V. Mass of the circuitry and the battery were 11.5 and 19.1 g, respectively.

The body of the depth control device is made from acrylonitrile butadiene styrene (ABS) using a FDM machine. The device contains an on-board electrical circuit with a microcontroller, rechargeable battery, and artificial bladder (shown in Fig. 6) [51]. The device has a mass of 114 g and a peak power consumption of 1.2 W. Experimental results using open-loop control show that the device is capable of controlling its buoyancy efficiently with no noise and low power consumption. The response times for sinking and rising are approximately 4.67 s and 180 s, respectively. Applications for this technology include integration into bio-inspired underwater vehicles. Such vehicles will be able to explore freshwater lakes or rivers in its natural, unperturbed state. It also has advantages in: easy transport, exploring hard-to-reach areas, and consuming less power. Furthermore, multiple vehicles can collect data in a massively parallel manner due to their small size and efficiency.

4.1.2 Experimental Results

- **Gas Generation Rate Test**

An experiment was set up to measure the gas generation rate at different voltage levels. Voltages ranging from 2.0 to 6.0 V at 0.5 V intervals were applied to the electrodes using an Agilent DC Power Supply (Model #E3646A). The hydrogen

Fig. 26 Gas flow rate versus power consumption. © 2011 IEEE. Reprinted, with permission, from [51]

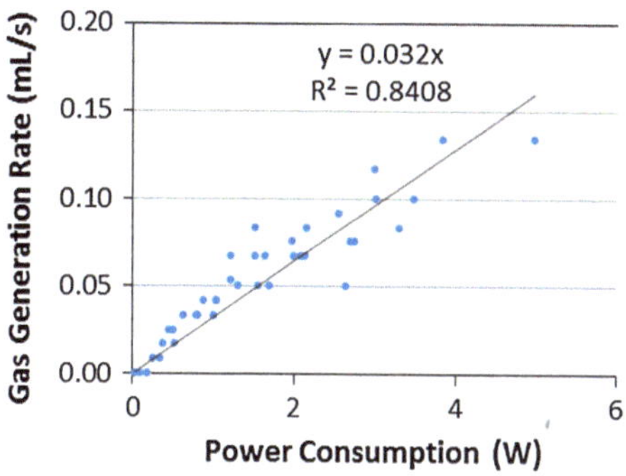

and oxygen gases generated were collected using water displacement technique with 50 mL graduated cylinder.

The results (Fig. 26) indicated a fairly strong linear relationship between gases generated and power consumption. The proportionality constant was approximately 0.032 mL/J. The on-board circuit was set up such that the output voltage was 5 V. However the actual voltage measured across the electrodes was 4 V due to low capacity of the battery. At this voltage, the average current and power consumption based on five trials were 0.3 A and 1.2 W, respectively. The average gas generation rate was 0.048 mL/s. A linear model predicted 0.040 mL/s at this power so the model and the result were in a fair agreement.

- **Diving Test**

The depth control device was tested in a water tank (1.5 m wide, 4.7 m long, and 0.9 m deep). The tank was filled with tap water at a room temperature. A critical mass was found when the device was about 95 % submerged under water, at which a slight decrease in buoyancy caused the device to sink. The critical mass of the device was experimentally found to be 283 g. Because the mass of the device was 114 g, 169 g of metal washers were added to the bottom chamber as dead weights. This mass was also equal to the payload for this particular device. The micro-controller was programmed such that there was an initial 3 min delay to allow time for assembly and fastening of the bolts. After the initial delay, the solenoid valve turned on for 12 s to allow gas to escape such that the device sunk to the bottom of the tank. Then the solenoid valve was turned off and 5 V was applied to the electrode plates for up to 15 min. Gases generated by the electrolysis filled the gas chamber, which displaced the water inside the bottom chamber. Thus, the device became more positively buoyant. The time it took to rise back up to the top was measured. Figure 27 shows the timing of the control signals.

It took approximately 4.7 s for the device to sink to the bottom of the tank and 180 s to rise back up to the top. The power consumption for sinking and rising were 0.5 W and 1.2 W, respectively. Figure 28 shows snapshots of a successful demonstration of the open-loop proof-of-concept depth control device.

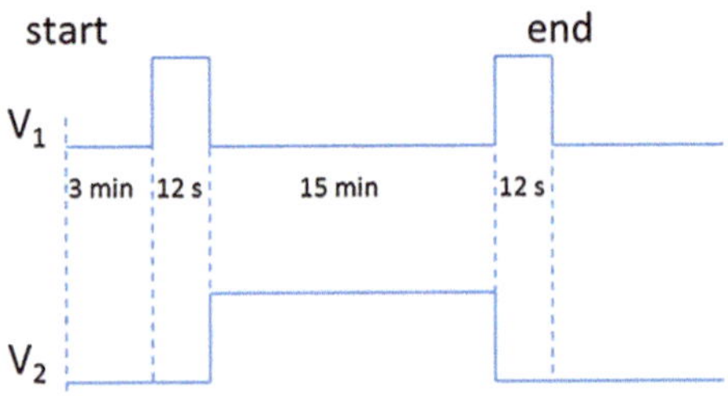

Fig. 27 Timing of the control signals. © 2011 IEEE. Reprinted, with permission, from [51]

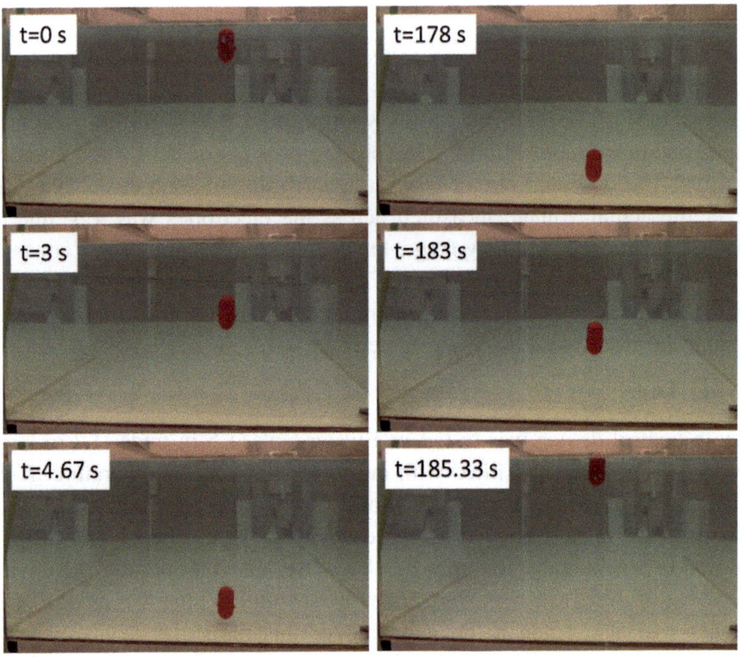

Fig. 28 Snap shots of the device in diving test. © 2011 IEEE. Reprinted, with permission, from [51]

5 Conclusion

In this Chapter, three IPMC-enabled technologies are presented for creating bio-inspired robotic fish. First, a physics-based and control-oriented modeling approach has been developed for robotic fish propelled by an IPMC caudal fin. The model incorporates the actuation dynamics of IPMC and hydrodynamics of robotic fish. The model can capture the steady-state speed of the fish, which enables real-time control of the fish. The model is geometrically scalable, which enables optimal design of the robotic fish.

Second, a fabrication technology for creating IPMC actuating membrane capable of generating 3D kinematic motion has been developed. Based on this technology, artificial pectoral fins were fabricated and a small, free-swimming robotic Cownose ray was developed. The experimental results show that the robotic Cownose was able to swim at 0.034 BL/s with less than 2 W power consumption.

Third, a buoyancy control strategy using IPMC-enhanced water electrolysis has been developed. A compact and power-efficient depth-control device was fabricated and tested to verify the buoyancy control strategy. This technology can be combined with any free swimming unmanned underwater vehicle.

Acknowledgements This research was supported in part by Office of Naval Research (ONR) Grant N000140810640, National Science Foundation (NSF) CAREER Grant ECCS 0547131, ONR under the Multidisciplinary University Research Initiative (MURI) Grant N00014-08-1-0642, and the David and Lucille Packard Foundation.

References

1. Abdelnour K, Mancia E, Peterson SD, Porfiri M (2009) Hydrodynamics of underwater propulsors based on ionic polymer-metal composites: a numerical study. Smart Mater Struct 18(8):085006 (11 pp)
2. Akle B, Habchi W, Abdelnour R, Blottman JI, Leo DJ (2012) Biologically inspired highly efficient buoyancy engine. In: Lakhtakia A (ed) Proceedings of the SPIE conference on bioinspiration, biomimetics, and bioreplication, vol 8339, 833900 (7 pp)
3. Akanyeti O, Ernits A, Fiazza C, Toming G, Kulikovskis G, Listak M, Raag R, Salumäe T, Fiorini P, Kruusmaa M (2010) Myometry-driven compliant-body design for underwater propulsion. In: Proceedings of the IEEE international conference on robotics and automation, pp 84–89
4. Alvarado PV, Youcef-Toumi K (2006) Design of machines with compliant bodies for biomimetic locomotion in liquid environments. J Dyn Syst Meas Contr 128(1):3–13
5. Aureli M, Kopman V, Porfiri M (2010) Free-locomotion of underwater vehicles actuated by ionic polymer metal composites. IEEE/ASME Trans Mechatron 15(4):603–614
6. Bar-Cohen Y (2000) Electroactive polymers as artificial muscles: capabilities, potentials and challenges. In: Robotics, pp 188–196
7. Bennett M, Leo D (2004) Ionic liquids as stable solvents for ionic polymer transducers. Sens Actuators A 115(1):79–90
8. Bond CE (1996) Swim bladder. Saunders College Publishing, Orlando
9. Boyer F, Porez M, Khalil W (2006) Macro-continuous computed torque algorithm for a three-dimensional eel-like robot. IEEE Trans Robot 22(4):763–775
10. Brunetto P, Fortuna L, Graziani S, Strazzeri S (2008) A model of ionic polymer–metal composite actuators in underwater operations. Smart Mater Struct 17(2):025029 (12 pp)
11. Carpi F, Rossi DD, Kornbluh R, Pelrine R, Sommer-Larsen P (2008) Dielectric elastomers as electromechanical transducers: fundamentals, materials, devices, models and applications of an emerging electroactive polymer technology. Elsevier, New York
12. Chen Z, Tan X (2008) A control-oriented and physics-based model for ionic polymer-metal composite actuators. IEEE/ASME Trans Mechatron 13(5):519–529
13. Chen Z, Sharata S, Tan X (2010) Modeling of biomimetic robotic fish propelled by an ionic polymer metal composite caudal fin. IEEE/ASME Trans Mechatron 15(3):448–459
14. Chen Z, Tan X (2010) Monolithic fabrication of ionic polymer-metal compo-site actuators capable of complex deformation. Sens Actuators A Phys 157(2):246–257

15. Chen Z, Um T, Zhu J, Bart-Smith H (2011) Bio-inspired robotic cownose ray propelled by electroactive polymer pectoral fin. In: Proceedings of the ASME international mechanical engineering congress and exposition, pp 64174:1–8
16. Chung C, Fung P, Hong Y, Ju M, Lin C, Wu T (2006) A novel fabrication of ionic polymer-metal composites (IPMC) actuator with silver nano-powders. Sens Actuators B Chem 117 (2):367–375
17. Clough RW, Penzien J (1993) Dynamics of structures. McGraw-Hill, NewYork
18. Daou HE, Salumäe T, Toming G, Kruusmaa M (2012) A bio-inspired compliant robotic fish: design and experiments. In: Proceedings of the IEEE international conference on robotics and automation, pp 5340–5345
19. Gao J, Bi S, Xu Y, Liu C (2007) Development and design of a robotic manta ray featuring flexible pectoral fins. In: Proceedings of the IEEE international conference on robotic and biomimetics, pp 519–523
20. Guo S, Fukuda T, Asaka K (2003) A new type of fish-like underwater micro-robot. IEEE/ ASME Trans Mechatron 8(1):136–141
21. Guo J, Yen WK (2008) Power reduction by controlling joint compliance for the propulsion of a biomimetic underwater vehicle. In: Proceedings of the 17th world congress the international federation of automatic control, vol 17, pp 15618–15623
22. Hu H, Liu J, Dukes I, Francis G (2006) Design of 3D swim patterns for autonomous robotic fish. In: Proceedings of the 2006 IEEE/RSJ international conference on intelligent robots and systems, pp 2406–2411
23. Kim B, Kim DH, Jung J, Park O (2005) A biomimetic undulatory tadpole robot using ionic polymer-metal composite actuators. Smart Mater Struct 14(6):1579–1585
24. Kim KJ, Shahinpoor M (2002) A novel method of manufacturing three-dimensional ionic polymer-metal composites (IPMCs) biomimetic sensors, actuators, and artificial muscles. Polymer 43(3):797–802
25. Kim SJ, Lee IT, Kim YH (2007) Performance enhancement of IPMC actuator by plasma surface treatment. Smart Mater Struct 16(1):N6–N11
26. Lauder GV, Anderson EJ, Tangorra J, Madden PGA (2007) Fish biorobotics: kinematics and hydrodynamics of self-propulsion. J Exp Biol 210(16):2767–2780
27. Lauder GV, Madden PGA, Tangorra JL, Anderson E, Baker TV (2011) Bioinspiration from fish for smart material design and function. Smart Mater Struct 20(9):094014
28. Lee SJ, Han MJ, JunKim S, Jho J, Lee HY, Kim YH (2006) A new fabrication method for IPMC actuators and application to artificial fingers. Smart Mater Struct 15(5):1217–1224
29. Lighthill MJ (1960) Note on the swimming of slender fish. J Fluid Mech 9(2):305–317
30. Lighthill MJ (1970) Aquatic animal propulsion of high hydromechanical efficiency. J Fluid Mech 44(2):265–301
31. Lu P, Lee K (2003) An alternative derivation of dynamic admittance matrix of piezoelectric cantilever bimorph. J Sound Vib 266(4):723–735
32. McFarland D, Gihespy I, Honary E (2003) Divebot: A diving robot with a whale-like buoyancy mechanism. Robotica 21(4):385–398
33. Millet P, Pineri M, Durand R (1989) New solid polymer electrolyte composites for water electrolysis. J Appl Electrochem 19(2):162–166
34. Moored K, Smith W, Hester J, Chang W, Bart-Smith H (2008) Investigating the thrust production of a myliobatoid-inspired oscillating wing. Adv Sci Technol 58:25–30
35. Moored K, Bart-Smith H (2009) Investigation of clustered actuation in tensegrity structures. Int J Solids Struct 46(17):3272–3281
36. Morgansen KA, Triplett BI, Klein DJ (2007) Geometric methods for modeling and control of free-swimming fin-actuated underwater vehicles. IEEE Trans Robot 23(6):1184–1199
37. Najem J, Sarles SA, Akle B, Leo DJ (2012) Biomimetic jellyfish-inspired underwater vehicle actuated by ionic polymer metal composite actuators. Smart Mater Struct 21(9):094026 (11 pp)

38. Palmre V, Hubbard JJ, Fleming M, Pugal D, Kim S, Kim KJ, Leang KK (2013) An IPMC-enabled bio-inspired bending/twisting fin for underwater applications. Smart Mater Struct 22 (1):014003 (11 pp)
39. Pelrine R, Kornbluh R, Pei Q, Joseph J (2000) High-speed electrically actuated elastomer with strain greater than 100 %. Science 287(5454):836–839
40. Peterson SD, Porfiri M, Rovardi A (2009) A particle image velocimetry study of vibrating ionic polymer metal composites in aqueous environments. IEEE/ASME Trans Mechatron 14 (4):474–483
41. Punning A, Anton M, Kruusmaa M, Aabloo A (2004) A biologically inspired ray-like underwater robot with electroactive polymer pectoral fins. In: Proceedings of the 2004 IEEE international conference on mechatronics and robotics, vol 2004, pp 241–245
42. Rosenberger LJ (2001) Pectoral fin locomotion in batoid fishes: undulation versus oscillation. J Exp Biol 204(2):379–394
43. Sader JE (1998) Frequency response of cantilever beams immersed in viscous fluids with applications to the atomic force microscope. J Appl Phys 84(1):64–76
44. Salumäe T (2010) Design of a compliant underwater propulsion mechanism by investigating and mimicing the body of a rainbow trout. MS Thesis, Tallinn University of Technology
45. Shahinpoor M, Bar-Cohen Y, Simpson JO, Smith J (1998) Ionic polymer-metal composites (IPMCs) as biomimetic sensors, actuators and artificial muscles-a review. Smart Mater Struct 7 (6):R15
46. Shahinpoor M, Kim KJ (2001) Ionic polymer-metal composites: I. Fundamentals. Smart Mater Struct 10(4):819–833
47. Shibuya K, Kado Y, Honda S, Iwamoto T, Tsutsumi K (2006) Underwater robot with a buoyancy control system based on the spermaceti oil hypothesis. In: Proceedings of the international conference on intelligent robots and systems, pp 3012–3017
48. Suo Z (2010) Theory of dielectric elastomers. Acta Mech Solida Sin 23(6):549–578
49. Tan X, Kim D, Usher N, Laboy D, Jackson J, Kapetanovic A, Rapai J, Sabadus B, Zhou X (2006) An autonomous robotic fish for mobile sensing. In: Proceedings of the IEEE/RSJ international conference on intelligent robots and systems, pp 5424–5429
50. Trease BP, Lu KJ, Kota S (2003) Biomimetic compliant system for smart actuator-driven aquatic propulsion: preliminary results. In: Proceedings of the ASME international mechanical engineering congress and exposition, pp 43–52
51. Um T, Chen Z, Bart-Smith H (2011) A novel electroactive polymer depth control device for bio-inspired underwater vehicles. In: Proceedings of the IEEE international conference on robotics and automation, pp 172–177
52. Villanueva A, Smith C, Priya S (2011) A biomimetic robotic jellyfish (Ro-bojelly) actuated by shape memory alloy composite actuators. Bioinspir Biomim 6(3):036004 (16 pp)
53. Wang Z, Wang Y, Li J, Hang G (2009) A micro biomimetic manta ray robot fish actuated by SMA. In: Proceedings of the 2009 IEEE international conference on robotics and biomimetics (ROBIO), pp 1809–1813
54. Yeom SW, Oh IK (2009) A biomimetic jellyfish robot based on ionic polymer metal composite actuators. Smart Mater Struct 18(8):085002 (10 pp)
55. Yim W, Lee J, Kim KJ (2007) An artificial muscle actuator for biomimetic underwater propulsors. Bioinspir Biomim 2(2):S31
56. Yu J, Wang L, Tan M (2007) Geometric optimization of relative link lengths for biomimetic robotic fish. IEEE Trans Robot 23(2):382–386
57. Zhao S, Yuh J (2005) Experimental study on advanced underwater robot control. IEEE Trans Robot 21(4):695–703

Macro-Fiber Composite Actuated Piezoelectric Robotic Fish

Alper Erturk

Abstract This chapter is centered on fish-like aquatic robotics using flexible bimorphs made from Macro-Fiber Composite (MFC) piezoelectric laminates. Bimorph propulsors employing MFCs offer a balance between the actuation force and velocity response for performance enhancement in bio-inspired swimming, in addition to noiseless and efficient actuation over a range of frequencies, geometric scalability, and simple design. The experimental component of this work first explores the effect of a passive substrate fin extension on the thrust frequency response of MFCs bimorphs in quiescent water along with measurement procedures. Specifically, it is shown that broadband thrust generation can be achieved in the presence of a passive substrate extension. The second part of the experiments is focused on the characterization of an elastically constrained uniform MFC bimorph propulsor (in-air and underwater) as well as the development of a robotic fish prototype combining a microcontroller and a printed circuit board amplifier to generate high actuation voltage for untethered locomotion. A distributed-parameter electroelastic model including the hydrodynamic effects and actuator dynamics is coupled with the elongated-body theory for estimating the mean thrust in quiescent water. For electroelastically nonlinear actuation levels, experimentally obtained underwater vibration response is coupled with the elongated-body theory to predict the thrust output. The measured mean thrust levels in quiescent water (on the order of ~ 10 mN) compare favorably with thrust levels of biological fish. An untethered robotic fish prototype that employs a single bimorph fin for straight swimming and turning motions is developed and tested in free locomotion. A swimming speed of 0.3 body length/s (7.5 cm/s swimming speed for 24.3 cm body length) is reported at 5 Hz for a nonoptimized main body-propulsor bimorph combination under a moderate actuation voltage level.

A. Erturk (✉)
G.W. Woodruff School of Mechanical Engineering, Georgia Institute of Technology, Atlanta, GA 30332, USA
e-mail: alper.erturk@me.gatech.edu

© Springer-Verlag Berlin Heidelberg 2015
R. Du et al. (eds.), *Robot Fish*, Springer Tracts in Mechanical Engineering,
DOI 10.1007/978-3-662-46870-8_9

1 Introduction

The potential applications for fish-like biomimetic locomotion at different geometric scales range from underwater sensing and exploration for sustainable ecology to drug delivery and disease screening in medicine [1–3]. Other than the use of conventional actuators, such as servomotors and hydraulic actuators employed in conjunction with various mechanisms [4–13], recently, various smart materials have been utilized for fish-like robotic fish development, such as Ionic Polymer-Metal Composites (IPMCs) [14–31], Shape Memory Alloys (SMAs) [32–37], and magnetostrictive thin films [38–40], among other alternatives [41–43]. In particular, the IPMC technology [14–31] has received great interest for biomimetic locomotion primarily due to its low-voltage actuation and large amplitude deflection capabilities. Conventional motor-based actuation involves complex structural design and provides high swimming speeds (per body length); whereas, the use of smart materials enables the geometric scalability option along with simple design and noiseless performance at the expense of reduced swimming speeds.

Piezoelectric materials offer strong electromechanical coupling and actuation forces, high power density, and their fabrication methods at different scales are well established [44–47]. These materials exhibit the so-called direct and converse piezoelectric effects. The direct effect is the process of electric charge development in response to mechanical deformation, while the converse effect is the mechanical deformation resulting from an applied electric field in a piezoelectric material. From the standpoint of multifunctionality, the converse piezoelectric effect can be used for dynamic actuation in biomimetic locomotion over a range of frequencies, while the direct piezoelectric effect can be employed for harvesting underwater energy toward enabling self-powered swimmer-sensor platforms [46]. Similar to IPMCs, the Macro-Fiber Composite (MFC) piezoelectric actuators (introduced by the NASA Langley Research Center in the first decade of this century [48–51]) also exhibit high efficiency in size, reduced energy consumption, and noiseless performance. In addition, unlike IPMCs, the MFCs offer large dynamic stresses in bending actuation as well as high performance for both low-frequency and high-frequency applications. The MFC technology employs piezoelectric (PZT: lead zirconate titanate) fibers of rectangular cross-section in epoxy matrix along with interdigitated electrodes and it leverages the 33-mode of piezoelectricity in bending actuation, i.e., strain and electric field directions are coincident (Fig. 1). The composite architecture is embedded in Kapton film which provides substantial structural robustness and integrity. Furthermore, polyester electrode sheets enable the option of waterproof applications. With these characteristics, MFC-based robotic fish can provide both geometric scalability and high performance swimming, as discussed in detail throughout this chapter.

High-voltage input requirement and low-strain output are the two downsides of piezoelectric transduction limiting the application of previously investigated piezoelectric structures for robotic fish development to use in free locomotion. In order to overcome the shortage of low strain in piezoelectric robotic configurations prior

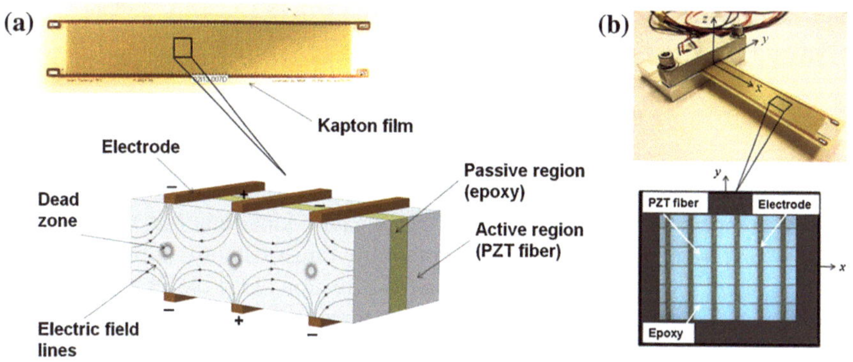

Fig. 1 **a** Picture of a single MFC laminate and close-up schematic of the piezoelectric fibers, epoxy matrix, and interdigitated electrodes; **b** picture of a bimorph cantilever made from two MFC laminates (bonded in a vacuum process using high shear strength epoxy) for bending actuation and close-up view of a small region showing the major components of the composite electromechanical structure

to the MFC technology, various kinematic magnification mechanisms were proposed by others [52–54]. However, the magnification component that is employed for creating larger vibration amplitudes typically creates energy loss and noise. As far as the high-input voltage requirement is concerned, research groups have used tethered configurations to power piezoelectric robotic fish, which restricts the free (unconstrained) locomotion capability [52–55]. Only recently, an untethered piezoelectric robotic fish [56] was reported by using an MFC bimorph as the propulsor with simple electronics and no motion amplification mechanism.

In this chapter, fish-like aquatic robotics using MFC piezoelectric bimorphs is presented based on our recent work [46, 56]. Focusing on quiescent water condition, extensive experiments are conducted to measure the thrust generated by MFCs oscillated under resonant actuation. Performance comparison of two MFC fish configurations with and without as passive caudal fin extension is presented by reviewing the work by Erturk and Delporte [46] in detail. In-air and underwater dynamics of an MFC bimorph cantilever are modeled for linear bending vibrations under dynamic piezoelectric actuation. The in-air electroelastic model is extended to obtain an underwater electrohydroelastic model accounting for the hydrodynamic effects. Underwater dynamics of an elastically constrained bimorph propulsor is then coupled with Lighthill's elongated-body theory to predict the thrust output in quiescent water based on Lighthill's mean thrust expression. In-air and underwater experiments are conducted for model validation and for characterizing a bimorph propulsor. Finally, an untethered robotic fish prototype is reported for straight swimming and turning motions in free locomotion based on the work by Cen and Erturk [56].

2 Underwater Thrust Generation Using MFCs and Effect of a Passive Fin Extension

2.1 Calibration of the Thrust Measurement Setup

Hydrodynamic thrust measurement under piezoelectric actuation is a more involved task as compared to measurements of the dynamic kinematic variables, such as velocity and acceleration. The reason is that the thrust output is a one-directional force resultant achieved during the oscillatory actuation of the piezoelastic structure at steady state. A 254 mm × 25.4 mm × 6.35 mm aluminum cantilever is combined with a laser sensor to obtain an elastic transducer for this purpose. The MFC fish sample with its plexiglass clamp is attached to the tip of the horizontally located *transducer cantilever* as shown in Fig. 2a. A set of small masses are then gradually located at the center of the MFC fish head to emulate the force (thrust) by the help of the gravity. The resulting deflection is recorded by the laser sensor (Fig. 2b) and eventually the linear calibration curve shown in Fig. 2c is obtained. Note that the transducer cantilever responds linearly up to several hundreds of milli-Newtons and the decoder of the laser sensor is sensitive enough to capture the resulting deflection amplitudes.

2.2 Underwater Actuation and Thrust Measurement

After the calibration curve of the transducer cantilever is obtained in air, the transducer cantilever with a clamped MFC fish sample is immersed in water for the hydrodynamic thrust measurements using the experimental setup shown in Fig. 3a. It is important to note that the dimensions of the transducer cantilever are such that its underwater fundamental resonance frequency is sufficiently higher than the underwater actuation frequencies of interest (this is checked by impact hammer

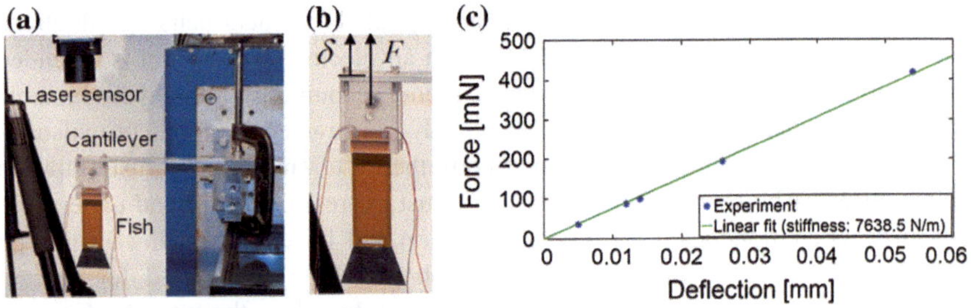

Fig. 2 **a** Calibration of the setup to obtain the force-deflection relationship in the presence of an MFC fish sample and its clamp; **b** close-up view showing the point of applied calibration loads (F) at the head of the MFC fish and the point of deflection (δ) measurement; **c** linear calibration curve with the identified linear stiffness (F/δ) value (© IOP Publishing. Reproduced with permission. All rights reserved)

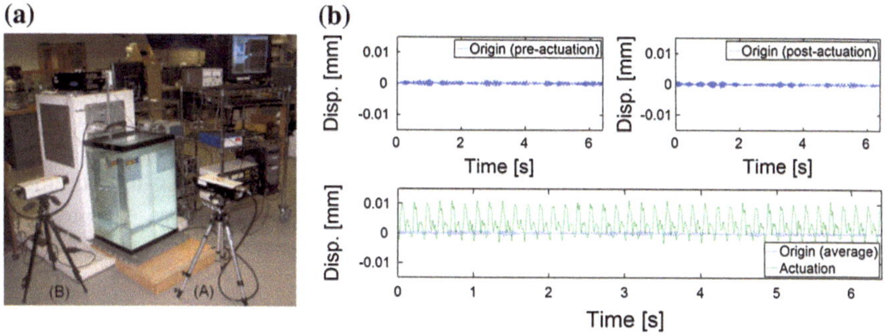

Fig. 3 a Experimental setup used for underwater thrust measurement after the transducer cantilever with a clamped MFC fish sample is submerged in water. Laser A is used for obtaining the transverse tail velocity-to-actuation voltage input FRFs whereas Laser B is used to measure the mean head displacement during the manual frequency sweep for evaluating the hydrodynamic thrust; **b** displacement measurements at the head of the MFC fish sample (in the direction of positive thrust) during pre-actuation, post-actuation, and actuation (along with the average of the pre-actuation and post-actuation histories, which defines the averaged origin) (© IOP Publishing. Reproduced with permission. All rights reserved)

testing). Hence, the measurement is in the quasi-static region of the transducer cantilever in the thrust generation experiments and the tip deflection of the cantilever is due to the dynamics of the MFC fish sample only (i.e., there is no interaction with the dynamics of the transducer cantilever). In addition, the hydrostatic pressure distributions on both faces of the transducer cantilever cancel each other so that the in-air thrust-deflection calibration is valid.

In Fig. 3a, the laser vibrometer pointing from the transverse direction of the MFC fish sample (Laser A) is employed for extracting the modal frequencies of the MFC fish sample (in bending) through the tail velocity-to-actuation voltage FRFs. The second laser vibrometer (Laser B) measures the displacement of the transducer cantilever (as in Fig. 2a) in the perpendicular direction so that the mean displacement can be converted to thrust using the calibration curve (Fig. 2c).

Harmonic actuation is used for hydrodynamic thrust generation. The frequency increment used in the time-domain thrust measurements is 0.5 Hz. At each frequency of voltage actuation, three time-domain displacement measurements are taken (using Laser B in the configuration described by Fig. 3a): Pre-actuation, actuation, and post-actuation. An example is displayed in Fig. 3b for the thrust measurement at 6 Hz under the peak-to-peak actuation input of 1050 V. The first measurement in this scheme is the *pre-actuation* measurement, which is simply the laser reading in the absence of piezoelectric actuation (i.e., noise around the origin). Then, the voltage actuation is started and the data is recorded after the system reaches its steady state, which is the *actuation* measurement. Finally, the actuation is stopped and a last measurement is taken in the absence of any actuation or transients. This is the *post-actuation* measurement. The origin is defined as the average of the pre-actuation and post-actuation measurements. The mean displacement caused by the thrust is the difference between the mean value of the actuation measurement and the mean value of the averaged origin measurement. The mean displacement reading is then used in Fig. 2c to give the mean thrust at the frequency of measurement. Note that the laser signal amplitude is divided by the refractive index of water ($n = 1.333$) in the underwater experiments and the validity of this signal correction is checked through another set of experiments not discussed here. Another important optical consideration when taking laser measurements through a transparent but reflective interface (clean glass in this case) is to make sure that the reflection from the interface is *not* on the lens.

2.3 The Effect of a Passive Caudal Fin Extension on the Thrust Frequency Response

Two bimorph fish samples are fabricated using a 0.127-mm-thick aluminum sheet as the substrate material and MFC piezoelectrics as the active material (MFC-8528-P1 type from the Smart Material Corporation). The active length and width of the MFC layers are 85 and 28 mm, respectively. The MFC layers are bonded onto both faces of the aluminum substructure using high-shear-strength epoxy in a vacuum bonding process. The active region of each bimorph MFC fish is approximately 0.8-mm thick. As shown in Fig. 4, one of the two MFC fish configurations has no caudal fin; whereas, the aluminum substrate of the other bimorph extends outside the active region to give a passive tapered caudal fin of 35 mm length and 48 mm maximum width (at the tail tip). Each MFC layer has a free capacitance of 5.7 nF according to the manufacturer and parallel connection is employed here. Therefore, a resultant capacitance of 11.4 nF is expected for each unclamped MFC fish sample based on the technical specifications. The measured clamped capacitance values (in-air) are 7.4 and 8.6 nF for the configurations without and with a caudal fin, respectively.

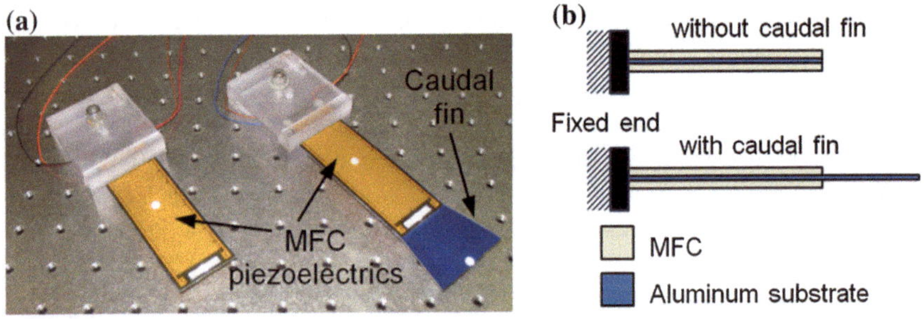

Fig. 4 **a** Bimorph MFC fish samples without and with a passive caudal fin and **b** side-view schematics of the composite bimorph structures (not to scale). Both samples have continuous aluminum substrates (*blue*) of the same thickness but different length (© IOP Publishing. Reproduced with permission. All rights reserved)

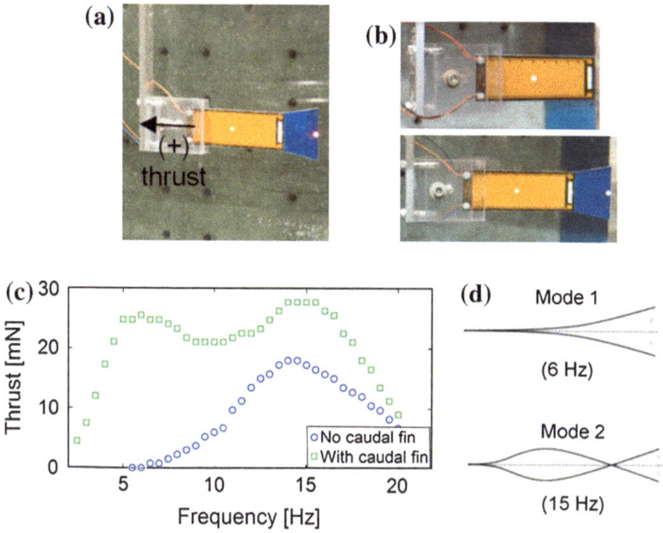

Fig. 5 **a** Underwater configuration of an MFC fish sample; **b** two MFC fish configurations without and with a passive caudal fin; **c** comparison of thrust frequency response for the same actuation input (peak-to-peak voltage: 1050 V) showing the substantial advantage of the MFC fish sample with a caudal fin; **d** mode shapes of the configuration with a passive caudal fin (© IOP Publishing. Reproduced with permission. All rights reserved)

The MFC fish samples are attached to the transducer cantilever for the underwater thrust measurements. Figure 5a shows the alignment of the transducer cantilever with a bimorph MFC fish sample attached at its tip. The underwater configurations of the two MFC fish samples without and with a caudal fin are shown in Fig. 5b. The thrust measurements are conducted for frequencies below 20 Hz (which cover the first two bending modes of the MFC fish sample with a passive caudal fin). For the same harmonic actuation voltage input to each sample (1050 V of peak-to-peak voltage), Fig. 5c shows the thrust FRFs. The fundamental resonance frequency of the MFC fish sample with no caudal fin is around 14.5 Hz and the mean thrust at this frequency is 18 mN. Remarkably, the MFC fish sample with a tapered passive caudal fin exhibits two peaks in the frequency range of 0–20 Hz with much larger thrust output. In addition to its fundamental vibration mode around 6 Hz (the first bending mode), the MFC fish with a passive caudal fin has its second vibration mode around 15 Hz (the second bending mode). The mean thrust readings for the sample with a passive caudal fin in Fig. 5c are 26 mN at 6 Hz (mode 1) and 28 mN at 15 Hz (mode 2). The mode shapes of this configuration are shown in Fig. 5d. Note that the second mode shape (at 15 Hz) has a node near the root of the caudal fin, hence an inflection point close to the head of the MFC fish. It can be concluded from Fig. 5c that the configuration with a passive caudal fin is a *wideband thrust generator* with substantially larger and relatively flat thrust output as compared to the configuration with no caudal fin for the same dynamic actuation input.

It is worth mentioning that the actuation performance for the second vibration mode can be improved by optimizing the surface coverage of the piezoelectric layers. It is known that the second mode shape (in Fig. 5d) has a *strain node* (which is simply an inflection point for a thin cantilever) near the root. Therefore, a significant portion of the actuation input cancels itself with the present surface coverage of the piezoelectric layers since the strain distributions on two sides of the strain node are 180° out of phase. Using segmented piezoelectrics (or segmented electrodes) can improve actuation performance dramatically for swimming with the second mode shape.

2.4 The Effect of Actuation Voltage on the Thrust Output

Further experiments are conducted with both MFC fish samples to investigate the dependence of the thrust output on the actuation voltage. The allowable voltage range of the MFC actuators is −500 to 1500 V and neither of these levels should be exceeded during the dynamic actuation. Therefore, the maximum peak-to-peak voltage input level without imposing any DC offset is 1000 V and this level can be increased up to the peak-to-peak level of 2000 V with sufficient DC offset (i.e., using +500 V DC offset with an oscillatory amplitude of 1000 V so that the maximum is 1500 V and the minimum is −500 V). In this work, the peak-to-peak voltage levels of 800, 1050, and 1300 V are studied and the resulting thrust frequency response curves are shown in Fig. 6. The thrust output increases monotonically with increasing voltage amplitude at every frequency. For all peak-to-peak actuation voltage levels shown in Fig. 6, the configuration with a passive caudal fin results in substantially better thrust generation performance with larger peak thrust as well as wideband behavior. At the highest actuation voltage level (1300 V peak-to-peak), mean thrust values of more than 30 mN are achieved over the frequency range of 4–17 Hz (with the peak values of 40 and 50 mN at the first two resonance frequencies, respectively). Note that the mass of the configuration with a passive caudal fin is only 10 g (excluding the mass of its plexiglass clamp head).

3 Piezohydroelastic Modeling of a Uniform MFC Propulsor Dynamics

3.1 In-Air Dynamics of a Bimorph Propulsor

The linear electroelastic equation of motion for in-air bending vibrations of a thin bimorph cantilever (Fig. 7) with uniform cross-section under dynamic voltage actuation is given by

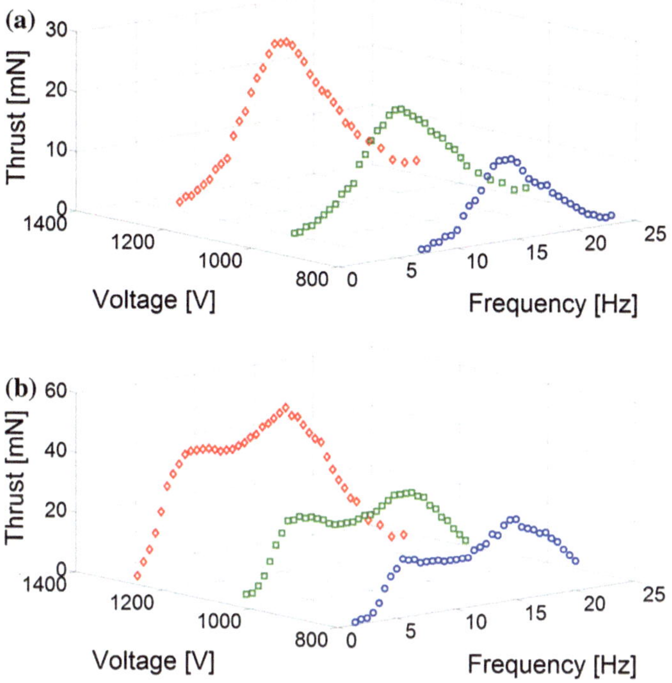

Fig. 6 Mean thrust frequency response curves of different peak-to-peak actuation voltage levels for the configurations **a** without and **b** with a passive caudal fin (© IOP Publishing. Reproduced with permission. All rights reserved)

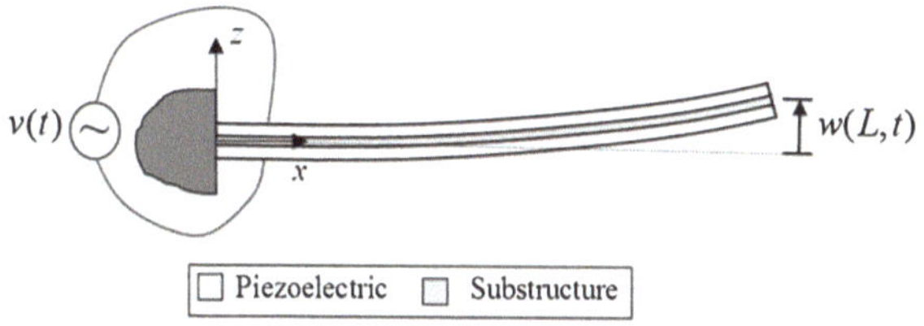

Fig. 7 Schematic of a uniform cantilevered bimorph propulsor under dynamic voltage actuation to create bending vibrations (piezoelectric layers can be combined in series or in parallel) (© IOP Publishing. Reproduced with permission. All rights reserved)

$$D\frac{\partial^4 w(x,t)}{\partial x^4} + m\frac{\partial^2 w(x,t)}{\partial t^2} + c\frac{\partial w(x,t)}{\partial t} = \vartheta\left[\frac{\mathrm{d}\delta(x)}{\mathrm{d}x} - \frac{\mathrm{d}\delta(x-L)}{\mathrm{d}x}\right]v(t) \qquad (1)$$

where D is the flexural rigidity of the composite cross-section, m is the mass per length, c is the damping coefficient, ϑ is the electromechanical coupling term in the physical coordinates, $\delta(x)$ is the Dirac delta function, $v(t)$ is the actuation voltage, and $w(x,t)$ is the deflection of the reference surface in the transverse (z) direction at the longitudinal position x and time t. Here, m and c are altered for underwater vibrations due to the added mass and damping effects of hydrodynamic loads as discussed in Sect. 3.2. While the formulation is given for a uniform cross-section (e.g., the case of no caudal fin extension in the previous section), a similar modeling approach can easily be used in the case of a two-segment configuration with a substrate extension as long as the individual segments are uniform.

Separating the space- and time-domain variables and assuming single-mode (fundamental mode) response yields

$$w(x,t) \cong \phi(x)\eta(t) \tag{2}$$

where $\phi(x)$ and $\eta(t)$ are the mass-normalized eigenfunction and the modal coordinate of the fundamental transverse vibration mode for a clamped-free uniform beam. The mass-normalized eigenfunction for the first mode can be obtained as

$$\phi(x) = \sqrt{\frac{1}{mL}}\left[\cosh\frac{\lambda x}{L} - \cos\frac{\lambda x}{L} - \sigma\left(\sinh\frac{\lambda x}{L} - \sin\frac{\lambda x}{L}\right)\right] \tag{3}$$

and it satisfies

$$\int_0^L m\phi^2(x)\mathrm{d}x = 1, \int_0^L \phi(x)D\frac{\mathrm{d}^4\phi(x)}{\mathrm{d}x^4}\mathrm{d}x = \omega_{n,\mathrm{air}}^2 \tag{4}$$

Here, $\lambda = 1.87510407$, $\sigma = 0.734095514$, L is the length of the bimorph, and $\omega_{n,\mathrm{air}}$ is the in-air natural frequency:

$$\omega_{n,\mathrm{air}} = \lambda^2\sqrt{\frac{D}{m_s L^4}} \tag{5}$$

where m_s is the structural mass per length.

For in-air vibrations, the mass per length in Eq. (1) is merely the structural mass per length:

$$m = m_s \tag{6}$$

while the in-air damping coefficient is

$$c = 2m_s\zeta_s\omega_{n,\mathrm{air}} \tag{7}$$

where ζ_s is the damping ratio which is assumed to be dominated by structural losses for infinitesimal in-air vibrations.

Following the standard analytical modal analysis procedure [57], i.e., substituting Eq. (2) into Eq. (1), multiplying the latter by the eigenfunction and integrating the resulting equation over the beam length, one obtains

$$
\begin{aligned}
&\frac{\mathrm{d}^2\eta(t)}{\mathrm{d}t^2}\int_0^L m\phi^2(x)\mathrm{d}x + \frac{\mathrm{d}\eta(t)}{\mathrm{d}t}\int_0^L c\phi^2(x)\mathrm{d}x + \eta(t)\int_0^L \phi(x)D\frac{\mathrm{d}^4\phi(x)}{\mathrm{d}x^4}\mathrm{d}x \\
&\quad = v(t)\int_0^L \phi(x)\vartheta\left[\frac{\mathrm{d}\delta(x)}{\mathrm{d}x} - \frac{\mathrm{d}\delta(x-L)}{\mathrm{d}x}\right]\mathrm{d}x
\end{aligned} \tag{8}
$$

yielding

$$
\frac{\mathrm{d}^2\eta(t)}{\mathrm{d}t^2} + 2\zeta_s\omega_{n,\mathrm{air}}\frac{\mathrm{d}\eta(t)}{\mathrm{d}t} + \omega_{n,\mathrm{air}}^2\eta(t) = \theta v(t) \tag{9}
$$

where θ is the electromechanical coupling term in the generalized modal coordinate η:

$$
\theta = \int_0^L \phi(x)\vartheta\left[\frac{\mathrm{d}\delta(x)}{\mathrm{d}x} - \frac{\mathrm{d}\delta(x-L)}{\mathrm{d}x}\right]\mathrm{d}x = \vartheta\left.\frac{\mathrm{d}\phi(x)}{\mathrm{d}x}\right|_{x=L} \tag{10}
$$

If the actuation voltage is assumed to be harmonic of the form

$$
v(t) = V_0 e^{j\omega t} \tag{11}
$$

where V_0 is the actuation voltage amplitude, ω is the actuation frequency (in rad/s), and j is the unit imaginary number, then the steady-state response for the modal coordinate is

$$
\eta(t) = \frac{\theta V_0 e^{j\omega t}}{\omega_{n,\mathrm{air}}^2 - \omega^2 + j2\zeta_s\omega_{n,\mathrm{air}}\omega} \tag{12}
$$

The resulting in-air tip velocity amplitude is therefore

$$
\left|\frac{\partial w(L,t)}{\partial t}\right| \cong \left|\phi(L)\frac{\mathrm{d}\eta(t)}{\mathrm{d}t}\right| = \left|\frac{V_0\omega\theta\phi(L)}{\omega_{n,\mathrm{air}}^2 - \omega^2 + j2\zeta_s\omega_{n,\mathrm{air}}\omega}\right| \tag{13}
$$

from which the tip velocity-to-actuation voltage frequency response function (FRF) can be extracted. It is important to note that this solution is valid for excitations around the fundamental natural frequency since higher vibration modes are not used in Eq. (2). In addition, the foregoing derivation neglects the geometric, material, and dissipative nonlinearities [58–60] and is strictly valid for linear vibrations only.

3.2 Underwater Dynamics of a Bimorph Propulsor

Hydrodynamic effects are included to predict the underwater vibrations of the cantilever based on the previous work by Sader and coworkers [61–63] on atomic force microscopy cantilevers as a straightforward approach for slender configurations. Similar efforts are due to Brunetto et al. [23], Mbemmo et al. [24], and Aureli et al. [28] for the underwater dynamics of IPMC propulsors. The following hydroelastic formulation [61–63] assumes geometrically small oscillations (relative to both length and width dimensions) of a uniform cantilever in unbounded fluid. Moreover, Sader's theory [61–63] assumes that the length-to-width ratio (L/b) is large and the accuracy of predictions decay as L and b become comparable [64] due to nonlinear 3-D hydrodynamic effects.

The added mass per length m_a and the hydrodynamic damping ratio ζ_h can be expressed in terms of the hydrodynamic function Γ as [61]

$$m_a = \frac{\pi \rho_w b^2}{4} \Gamma_r \tag{14}$$

$$\zeta_h = \frac{1}{2Q_h} = \frac{\Gamma_i}{2\left(\frac{4m_s}{\pi \rho_w b^2} + \Gamma_r\right)} \tag{15}$$

where ρ_w is the mass density of water, Q_h is the quality factor due to hydrodynamic damping, Γ_r, and Γ_i are the real and imaginary parts of hydrodynamic function Γ, respectively, and b is the width of the bimorph propulsor. The hydrodynamic function Γ can be calculated analytically or numerically. Simplified expressions of the hydrodynamic function Γ are available depending on the range of the Reynolds number [65, 66]. Although the present discussion is given for small oscillations, we note that particularly hydroelastic nonlinearities can easily be pronounced due to large amplitude vibrations and for comparable aspect ratios, requiring correction of the hydrodynamic function [64, 66].

As far as the dissipation mechanisms are concerned, both structural and hydrodynamic damping effects are taken into account for the total damping ratio (ζ_w) of underwater vibrations:

$$\zeta_w = \zeta_s + \zeta_h \tag{16}$$

The mass per length and damping coefficient terms in Eq. (1) for underwater vibrations are

$$m = m_w = m_s + m_a \tag{17}$$

$$c = 2m\zeta_w \omega_{n,\text{water}} = 2m_w(\zeta_s + \zeta_h)\omega_{n,\text{water}} \tag{18}$$

and the mass-normalized eigenfunction in Eq. (2) satisfies

$$\int_0^L m\phi^2(x)\mathrm{d}x = 1, \int_0^L \phi(x)D\frac{\mathrm{d}^4\phi(x)}{\mathrm{d}x^4}\mathrm{d}x = \omega_{n,\mathrm{water}}^2 \tag{19}$$

Since the underwater mass per length is due to Eq. (17), the underwater natural frequency $\omega_{n,\mathrm{water}}$ is obtained from the in-air natural frequency $\omega_{n,\mathrm{air}}$ (which is approximately the in-air resonance frequency for $\zeta_s \ll 1$) through Chu's formula [67] (for large aspect ratio) modified by the real part of the hydrodynamic function Γ_r [61]:

$$\omega_{n,\mathrm{water}} = \omega_{n,\mathrm{air}}\sqrt{\left(1 + \frac{\pi\rho_w b^2}{4m_s}\Gamma_r\right)^{-1}} \tag{20}$$

where $\omega_{n,\mathrm{air}}$ is given by Eq. (5).

The resulting underwater tip velocity response amplitude is therefore

$$\left|\frac{\partial w(L,t)}{\partial t}\right| \cong \left|\phi(L)\frac{\mathrm{d}\eta(t)}{\mathrm{d}t}\right| = \left|\frac{V_0\omega\theta\phi(L)}{\omega_{n,\mathrm{water}}^2 - \omega^2 + j2\zeta_w\omega_{n,\mathrm{water}}\omega}\right| \tag{21}$$

where θ is due to Eq. (10). However, the eigenfuction $\phi(x)$ in Eqs. (10) and (21) is now normalized according to Eq. (19) with m given by Eq. (17), i.e., the mass term in Eq. (3) is due to Eq. (17) for underwater vibrations.

3.3 Hydrodynamic Mean Thrust in Terms of the Underwater Velocity Response

In most cases of robotic fish thrust estimation [27, 28, 38–40], it becomes necessary to identify certain calibration factors, such as the drag coefficient, which removes the possibility of obtaining an a priori estimate of the thrust resultant solely from the underwater vibration response. Lighthill's elongated-body theory was used in previous IPMC-based robotic fish studies to predict the steady-state cruising speed [24, 27] by equating the thrust expressions from the reactive and resistive methods, where the drag coefficient was measured by spring scales while pulling the fish with different velocities. In the present work, Lighthill's theory [68–71] is employed alone to estimate the mean thrust in quiescent water as a first approximation. The mean thrust (T) in Lighthill's theory is given in the presence of an external relative free stream of speed U (which is essentially the swimming speed) as

$$T = \frac{1}{2}m_v\left[\overline{\left(\frac{\partial w}{\partial t}\right)^2} - U^2\overline{\left(\frac{\partial w}{\partial x}\right)^2}\right]_{x=L} \tag{22}$$

where the over-bar stands for the mean value and m_v is the virtual mass density at $x = L$, expressed as

$$m_v = \frac{\pi \rho_w b^2}{4} \beta \tag{23}$$

Here, β is a virtual mass coefficient that is close to unity [70] for the wavelength and body length considered in this work (hence $\beta \cong 1$) with slender body assumption. Once again, aspect ratio effects can alter this term as well [64, 66] due to hydrodynamic nonlinearities which can be accounted for (by an aspect ratio-dependent inertia coefficient as in well-known semi-empirical Morison's equation).

In Eq. (22), we set $U \to 0$ to approximate quiescent water condition:

$$T \cong \frac{1}{2} m_v \overline{\left(\frac{\partial w}{\partial t}\right)^2}\bigg|_{x=L} = \frac{\pi \rho_w b^2}{8} \overline{\left(\frac{\partial w(L,t)}{\partial t}\right)^2} \tag{24}$$

where the mean thrust T depends only on the tip velocity and the virtual mass for quiescent water approximation.

4 Details of the Experimental Setup for Model Validation

4.1 Setup for In-Air Tip Velocity FRF Measurements

The MFC bimorph tested and characterized in the experiments is shown in Fig. 8 along with its clamp and fixture employed for the in-air actuation FRF measurements. The bimorph is made of two custom-made hydrophobic M8528-P1 [51] MFC laminates with no separate substructure layer other than the Kapton and epoxy layers of the MFCs. A vacuum bonding process is employed by using high shear strength epoxy to assemble the piezoelectric laminates (this process is described

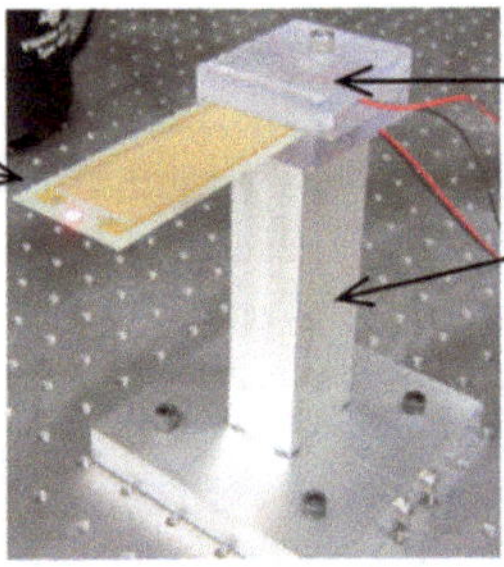

Fig. 8 In-air configuration of the bimorph MFC cantilever for the measurement of its tip velocity-to-actuation voltage FRF (© IOP Publishing. Reproduced with permission. All rights reserved)

elsewhere [72]). The electrode leads of the two MFCs are combined in parallel throughout the experiments discussed in this chapter. A vertically aligned laser vibrometer is used along with the monitored actuation signal in order to obtain the transverse (vertical direction in Fig. 8) tip velocity-to-actuation voltage FRFs of the MFC bimorph in air.

4.2 Setup for Underwater Tip Velocity FRF and Mean Thrust Measurements

The experimental setup used for the underwater tip velocity and thrust measurements is shown in Fig. 9a. As shown in Fig. 9b, the MFC bimorph is fixed with the same clamp in the underwater experiments. Laser (1) measures the transverse tip velocity under dynamic actuation while Laser (2) provides the head displacement (constrained by an aluminum cantilever) through a small mirror that makes a 45° angle with the horizontal plane. The elastically constrained mean head displacement is correlated to the mean thrust as described in the next section. Note that both laser signals (velocity and displacement) are divided by the refractive index of water in the underwater experiments [46] as discussed in Sect. 2.1 (note that the setup and the calibration cantilever are different from those reviewed in more detail in Sect. 2; however, the basic measurement concept is the same).

4.3 Calibration of the Thrust Measurement Setup

The MFC bimorph and its clamp are fixed at the tip of a horizontally located aluminum beam which functions as a *transducer cantilever* along with a vertically pointing laser vibrometer used in the displacement measurement mode (Fig. 10a).

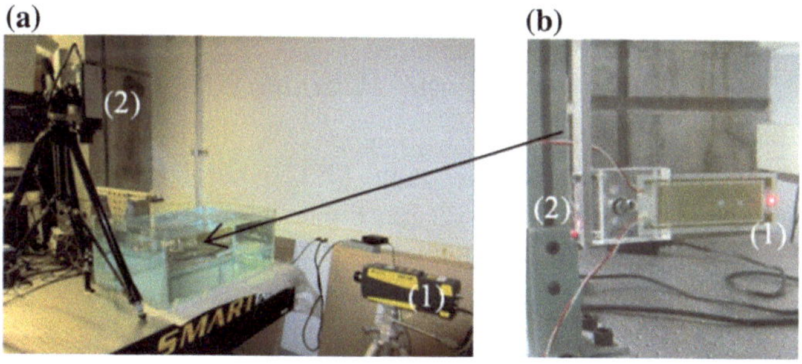

Fig. 9 **a** Experimental setup used for thrust measurement of a bimorph propulsor in quiescent water: Laser *(1)* measures the transverse tip velocity while Laser *(2)* measures the elastically constrained head displacement through a 45° mirror; **b** close-up view showing the measurement points of Lasers *(1)* and *(2)* on the MFC bimorph propulsor (© IOP Publishing. Reproduced with permission. All rights reserved)

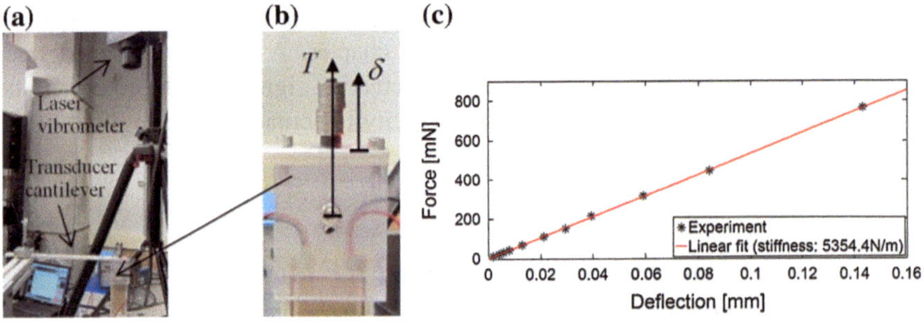

Fig. 10 **a** Setup used for the thrust-displacement calibration experiment with the MFC bimorph, its clamp, and the transducer cantilever; **b** close-up view showing the point of applied loads at the center of MFC bimorph and the deflection measurement point; **c** linear calibration curve with the calculated linear stiffness (T/δ) value (© IOP Publishing. Reproduced with permission. All rights reserved)

This laser vibrometer employed for measuring the head displacement corresponds to Laser (2) in the underwater experiments (Fig. 9a). The purpose of the in-air setup shown in Fig. 10a is to relate the thrust caused by actuation in the underwater experiments to the deflection of the aluminum transducer cantilever as detailed in Sect. 2 (however, this is a separate setup and is not identical to that in Fig. 2). It is assumed that the mean thrust resultant (T) of the bimorph propulsor (in the underwater experiments) acts through the center of its head, causing the deflection of δ at the location of the reflector for Laser (2) in the underwater arrangement given by Fig. 9a. Different values of small masses are gradually located at the center of the top surface to emulate the mean thrust (Fig. 10b). The vertically pointing laser measures the resulting deflection at the reflector, which is employed to obtain the thrust-displacement calibration line shown in Fig. 10c.

5 Experiments and Model Validation

5.1 In-Air Tip Velocity FRF and Parameter Identification

Low-voltage harmonic input is applied to the MFC bimorph for the frequency range of 5–70 Hz with an increment of 0.05 Hz. Figure 11a exhibits the experimentally measured tip velocity-to-actuation voltage FRF and the model prediction using Eq. (13) for the linear actuation regime of the bimorph. The fundamental in-air resonance frequency is 35.5 Hz. The in-air damping ratio (attributed mostly to structural losses for small oscillations) is identified as $\zeta_s = 0.02$ while the identified electromechanical coupling in the physical coordinates [see Eq. (1)] is $\vartheta = 23.03$ μNm/V. The modal electromechanical coupling that depends on the in-air eigenfunction due to Eq. (10) is $\theta = 8.878 \times 10^{-3}$ N/(Vkg$^{-1/2}$).

5.2 Underwater Tip Velocity FRF

The MFC bimorph is submerged in water (as depicted in Fig. 9) along with its clamp and aluminum fixture used in thrust calibration (Fig. 10). Low-voltage harmonic actuation is applied to the MFC bimorph for the frequency range of 2–15 Hz with an increment of 0.05 Hz. The fundamental underwater resonance frequency of linear vibrations is measured as 8.7 Hz. According to Eq. (16), the total underwater damping ratio is due to the structural and hydrodynamic damping effects. Equation (15) yields $\zeta_h = 0.0119$ and $Re \cong 19{,}430$, and eventually, from Eq. (16), one obtains $\zeta_w = 0.0319$. The electromechanical coupling in the physical coordinates is the same as the one obtained from in-air vibration test ($\vartheta = 23.03\ \mu\text{Nm/V}$) while the modal electromechanical coupling becomes $\theta = 2.325 \times 10^{-3}$ N/(Vkg$^{-1/2}$) due to Eq. (10) in which the eigenfunction is normalized according to Eq. (19) by using the underwater mass per length given by Eq. (17). Therefore, one can predict the underwater tip velocity FRF using Eq. (21) as shown in Fig. 11b.

The agreement between the experimental measurement and theoretical prediction is reasonable in the linear actuation regime of the bimorph. Equation (20) predicts the underwater resonance frequency as 8.8 Hz with an error of 1.1 % relative to the experimental value (8.7 Hz). Both the total underwater damping and natural frequency are predicted in terms of the in-air dynamics and fluid properties with good accuracy. Expectedly, the linear model predictions would fail under high actuation voltage levels due to geometric and electroelastic nonlinearities [58–60]. Nonlinear modeling of MFC dynamics under high-voltage actuation and incorporation of nonlinear hydrodynamic effects [66] in such a nonlinear model are of interest for future work.

5.3 Mean Thrust and Tip Velocity Correlation for Different Actuation Voltage Levels

The frequency range covered in the underwater thrust measurements is 0.5–15 Hz with a fine increment of 0.25 Hz in the 6–8 Hz range (resonance region) and a

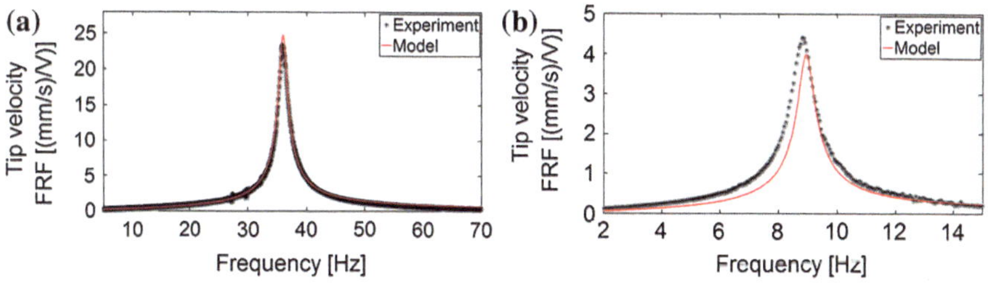

Fig. 11 Measured and calculated **a** in-air and **b** underwater tip velocity-to-actuation voltage FRFs of the MFC bimorph in its linear actuation regime (© IOP Publishing. Reproduced with permission. All rights reserved)

relatively coarse increment of 0.5 Hz outside the resonance region. Three time-domain head displacement measurements are taken at each frequency (preactuation, actuation, and post-actuation as detailed in Sect. 2.2) while the tip velocity in the transverse direction is measured simultaneously. The reference point is calculated as the average of the preactuation and post-actuation values. Based on the previously discussed calibration, the mean thrust is a linear function of the mean displacement of the aluminum cantilever. This mean displacement is the difference between the mean values of the *actuation* displacement and the *reference* point. From this mean displacement, the mean thrust is extracted using the linear calibration curve in Fig. 9c.

The experimental tip velocity and mean thrust measurements for the peak-to-peak actuation voltage levels of 200, 400, 600, and 800 V are shown in Fig. 12a, b, respectively. Clearly, there is a direct correlation between these two independent measurements since the thrust level increases with increasing tip velocity. It should be noted that these high actuation voltage levels fall into the nonlinear regime due to the electroelastic, geometric, and dissipative nonlinear effects. The softening nonlinearity (resulting in the shifting of the resonance frequency to the left) with increased actuation input is a typical behavior of piezoelectric cantilevers under high-voltage actuation [59].

Using Eq. (24) resulting from Lighthill's theory for the quiescent water condition, one can estimate the thrust curves in terms of the tip velocity and the virtual mass. These predictions are shown in Fig. 13a–d for four different actuation voltage levels. Note that, particularly in Fig. 13a (which is the lowest voltage case among these four measurement sets), the frequencies away from the resonance region are

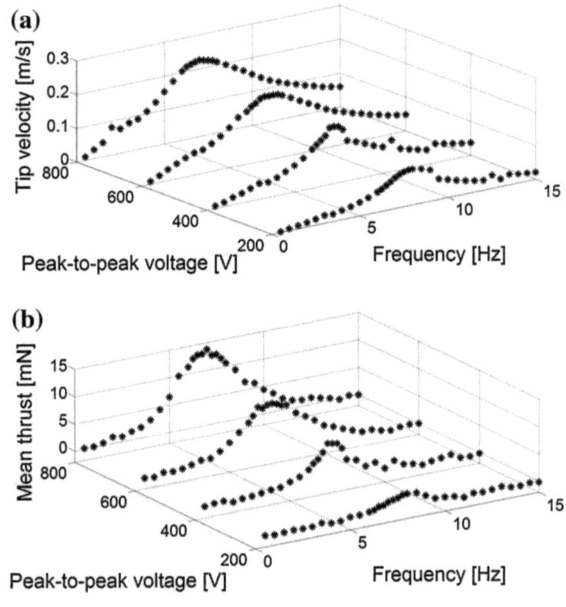

Fig. 12 Experimental **a** tip velocity and **b** mean thrust curves for four different peak-to-peak voltage levels: 200, 400, 600, and 800 V (© IOP Publishing. Reproduced with permission. All rights reserved)

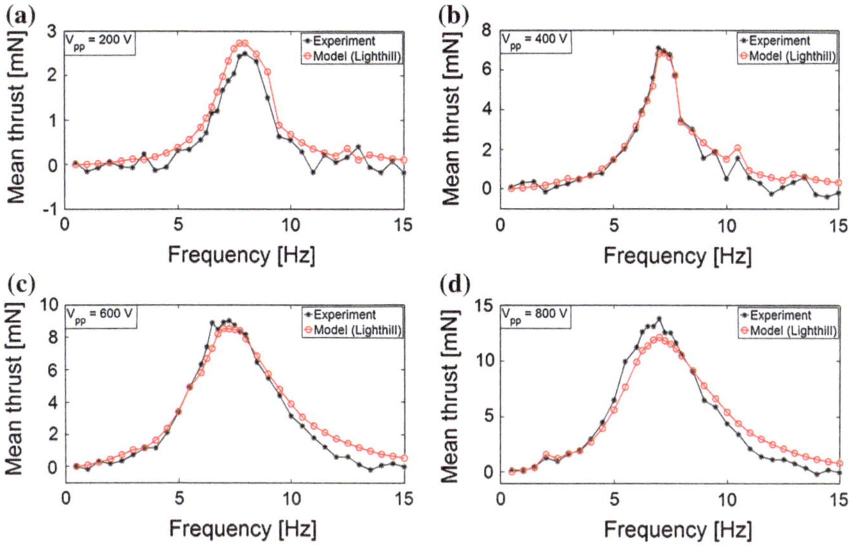

Fig. 13 Measured and predicted thrust curves for the peak-to-peak voltage inputs of **a** 200 V, **b** 400 V, **c** 600 V, and **d** 800 V (© IOP Publishing. Reproduced with permission. All rights reserved)

prone to noise effects in the measurements due to low thrust resultant (caused by low displacement). Based on Fig. 13a–d, it can be concluded that the reduced form of Lighthill's theory [69–71] for quiescent water can predict the mean thrust in terms of the tip velocity with good accuracy. One should recall that Figs. 12 and 13 are electroelastically nonlinear, and therefore the linear derivations given in this chapter do not intent to predict these dynamics quantitatively. However, Lighthill's formula reduced for quiescent water in Sect. 3.3 does correlate the tip velocity-to-mean thrust with reasonable accuracy in Fig. 13 as a first approximation.

5.4 Identification of the Thrust Coefficient

In this section, the hydrodynamic thrust coefficient of the propulsor is extracted based on the mean thrust and transverse tip velocity of the piezoelectric propulsor. First we define the modified Reynolds number as

$$Re_L = \frac{\omega \delta L}{\upsilon}, \tag{25}$$

where $\delta = |w(L,t)|$ is the tip displacement amplitude at frequency ω while υ is the kinematic viscosity of water ($\upsilon = \mu/\rho_w$). Note that the modified Reynolds number introduced at this point uses the underwater vibration response as the geometric scale (rather than the width).

The thrust coefficient C_τ is defined as [73]

$$C_\tau = \frac{\tau}{\frac{1}{2}\rho_w \omega^2 \delta^2 L}, \tag{26}$$

where τ is the thrust per unit width ($\tau = T/b$) and we note from Eqs. (25) and (26) that $\tau \propto Re_L^2$.

Substituting Eq. (24) into Eq. (26) gives

$$C_\tau = \frac{\frac{\pi \rho_w b}{8}\overline{\left(\frac{\partial w(L,t)}{\partial t}\right)^2}}{\frac{1}{2}\rho_w \omega^2 \delta^2 L} = \frac{\frac{\pi \rho_w b}{8}\frac{\omega^2 \delta^2}{2}}{\frac{1}{2}\rho_w \omega^2 \delta^2 L} = \frac{\pi b}{8L} \cong 0.1524 \tag{27}$$

Another manipulation of Eqs. (25) and (26) provides the variation of the mean thrust with modified Reynolds number as

$$T = \frac{\pi \rho_w b^2 \upsilon^2}{16 L^2} Re_L^2 \cong 3.322 \times 10^{-11} Re_L^2 \quad \text{[in Newtons]} \tag{28}$$

Figure 14 displays the variation of the mean thrust with modified Reynolds number based on the experimental thrust measurements at different actuation voltage levels and frequencies along with the prediction of Eq. (28) in log–log scale. The proportionality between mean thrust T and Re_L^2 is observed in the experimental data and is well predicted by the modeling approach based on Lighthill's theory.

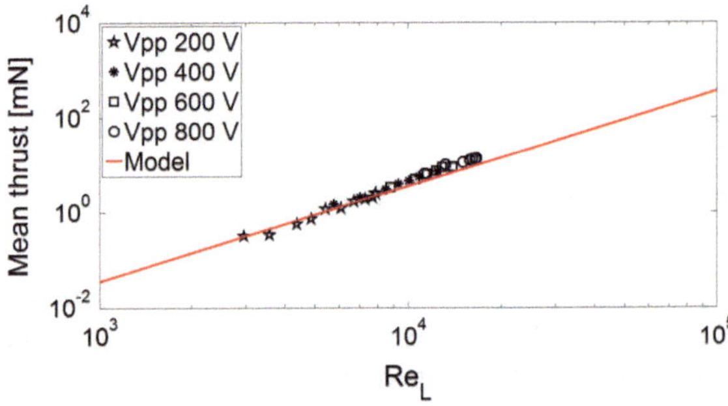

Fig. 14 Experimental and theoretical mean thrust versus the modified Reynolds number. The experimental data belongs to four different actuation voltage levels and different frequencies around the fundamental resonance (© IOP Publishing. Reproduced with permission. All rights reserved)

6 Prototype for Untethered Locomotion

6.1 Electronic Architecture for Untethered Swimming

Despite the advantages of MFCs due to large dynamic actuation stresses, structural flexibility, silent operation, and wide frequency range of effective performance, the requirement of high-voltage input limits its application in free (untethered) locomotion for robotic fish development. In this section, a PCB high-voltage amplifier is implemented along with a microcontroller to overcome this issue in free locomotion. To our knowledge, this section presents the first untethered piezoelectric robotic fish since the configurations in previous efforts [52–55] were actuated through external power through tethers.

A separate bimorph propulsor is fabricated for the free locomotion experiments. An embedded power and actuation system is designed for this prototype, which can generate high-input voltage for the MFC bimorph propulsor. As shown in Fig. 15a, this system consists of two 9 V batteries, a microcontroller (ATmega 128), a wireless device [XBee 1mW Wire Antenna—Series 1 (802.15.4)], and a PCB amplifier (AMD2012-CE3) specially designed for the MFC actuator [51]. In order to obtain smooth sinusoidal voltage for the MFC bimorph, a low-pass filter is added

(a)

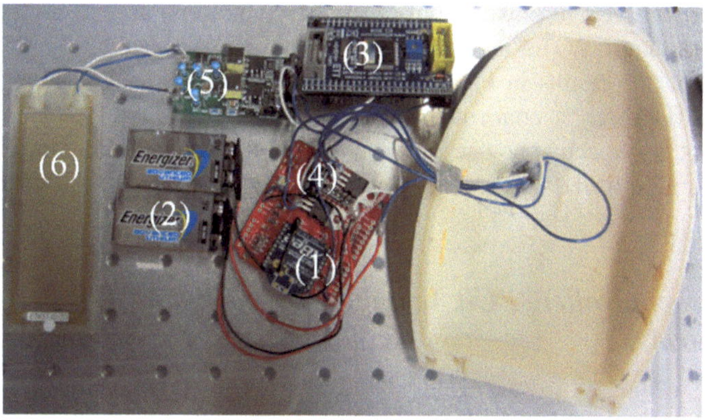

(b)

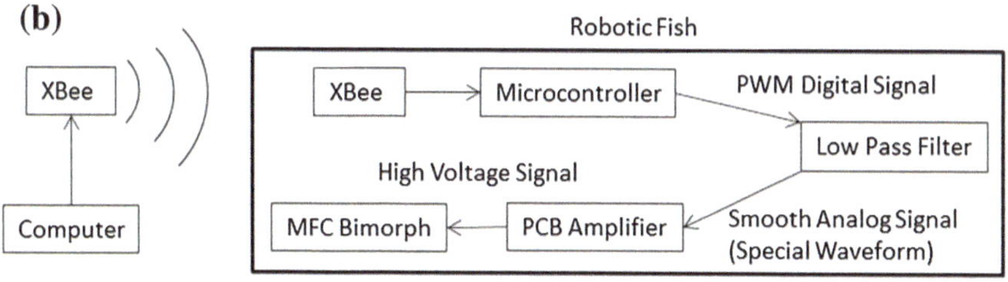

Fig. 15 **a** Components for the untethered piezoelectric robotic fish system: *(1)* XBee radio; *(2)* 9 V batteries; *(3)* microcontroller; *(4)* low pass filter and voltage regulators; *(5)* voltage amplifier; *(6)* MFC bimorph; and **b** electronic schematic of the robotic fish system (© IOP Publishing. Reproduced with permission. All rights reserved)

to filter out the high-frequency noise from the pulse width modulation (PWM) signals. The detailed schematic of the actuation system is shown in Fig. 15b.

6.2 Power Consumption Analysis

In order to enable untethered locomotion of the MFC-based robotic fish, a portable power system has to be designed to generate the high voltage for MFC bimorph actuation. The general maximum output voltage level for the microcontroller is around 5 volts, which is much lower than the requirements for MFC actuation. A specifically designed PCB amplifier (AMD2012-CE3) [51] is utilized in the power system, which is able to generate high voltage from −500 to 1500 V according to the control input signal from 0 to 5 V. This PCB amplifier requires only 12 V input voltage supply. Therefore, one can build the mobile power system by simply using two 9 V batteries, the microcontroller, the PCB amplifier, and the corresponding voltage regulators, as shown in Fig. 15a. These two 9 V batteries can support the continuous operation of the system for almost 30 min. The power consumption of the overall electronic system is around 3–5 W.

The current amplitude and average power plots for MFC bimorph vibration under different frequencies and actuation voltages are shown in Fig. 16. Figure 16b shows that the average power input to the MFC bimorph around its fundamental mode is less than 1 W (for the sinusoidal peak-to-peak voltage of 800 V). It should be noted that the results in Fig. 16b is an overestimation as far as the truly consumed (real) power is concerned since these modulus curves include not only resistive but also reactive power. The overall power consumption of the robotic fish can therefore be further reduced by optimizing the actuation circuit design.

6.3 Microcontroller, PCB Voltage Amplifier, and Wireless Control

In order to provide sinusoidal high voltage for the MFC bimorph, the microcontroller is programmed to generate a special waveform by PWM. Specifically, 0 V

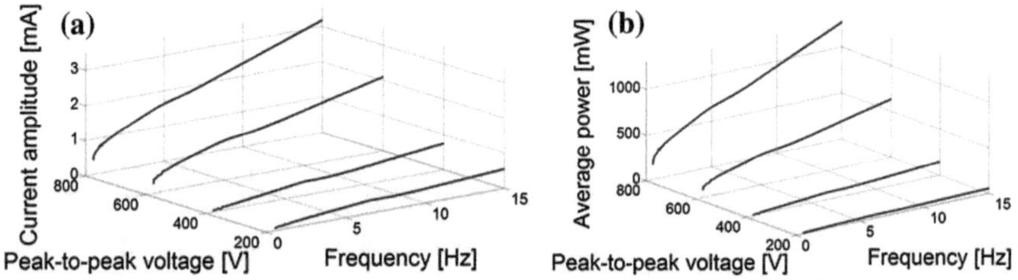

Fig. 16 Experimental **a** current amplitude and **b** average power curves of the MFC bimorph under four different peak-to-peak voltage levels: 200, 400, 600, and 800 V (© IOP Publishing. Reproduced with permission. All rights reserved)

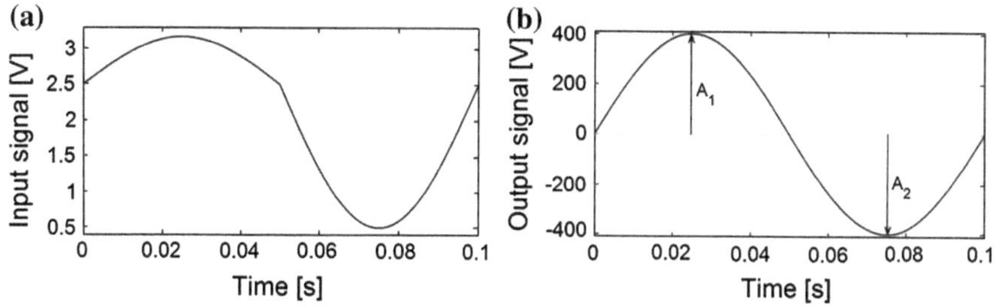

Fig. 17 **a** Input and **b** output signals of the PCB amplifier to generate a sinusoidal peak-to-peak actuation voltage of 800 V at 10 Hz. The microcontroller can provide signal inputs to PCB amplifier to create oscillatory actuation voltage signals with $A_1 \neq A_2$ for turning motion (© IOP Publishing. Reproduced with permission. All rights reserved)

input signal generates −500 V output; 2.5 V input signal generates 0 V output; 5 V input signal generates 1500 V output. These values are the voltage limits of MFCs without depolarization [51]. An example is given in Fig. 17 for the case of generating a sinusoidal peak-to-peak voltage of 800 V at 10 Hz using the PCB amplifier. The microcontroller is able to generate various waveforms by its PWM function and the power system can provide the high-voltage sinusoidal output (up to 2000 V peak-to-peak) for the MFC bimorph propulsor. The frequency, mean voltage, and amplitudes of the PCB amplifier output signal can be controlled by adjusting the rate and duty cycle of the PWM signals. The swimming speed is easily controlled through PCB amplifier output signal frequencies and amplitudes. In addition, turning speed and direction are controlled by setting different values for the amplitudes A_1 and A_2 shown in Fig. 17b. Wireless control is performed through serial communication. Commands are sent by a laptop computer through a USB connected XBee Explorer. An XBee module inside the robotic fish body receives the commands and transfers them to the microcontroller, which can change the PWM waveform. Therefore, wireless communication controls the swimming speed of the robotic fish by setting the rate and duty cycle of the PWM, which affect the vibration frequency and amplitude of the MFC propulsor. The Xbee signal would penetrate the water just a few centimeters so the robotic fish is tested close to surface. The wireless communication is used for testing the change of speed and direction during swimming.

6.4 Fabricated Prototype and Free Locomotion Tests

A picture of the robotic fish prototype is shown in Fig. 18. The fish body is designed to provide a waterproof enclosure for the electronics components and is realized using a Fused Deposition Modeling (FDM) machine. This creates an Acrylonitrile Butadiene Styrene (ABS) plastic shell, that when printed is buoyant in water. An O-ring was integrated into the rim to ensure the interior of the fish stayed

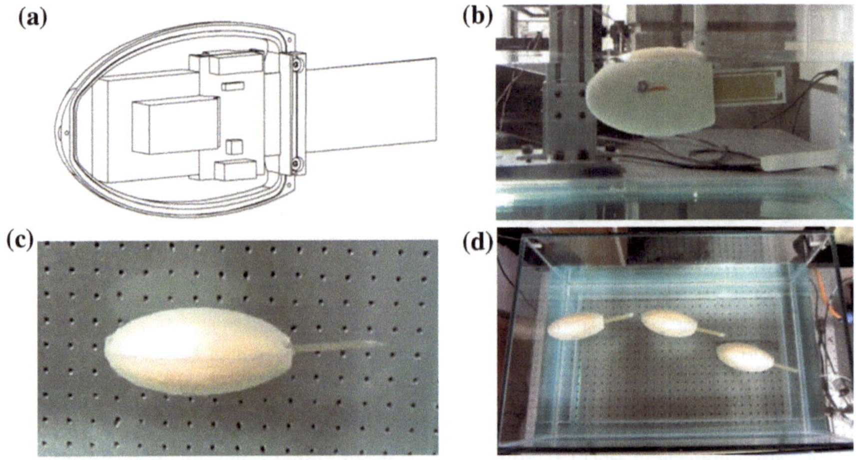

Fig. 18 Untethered robotic fish prototype for free locomotion in **a** modeled view; **b** side-view; **c** top view; and **d** combined motion capture involving turning motion. Swimming speed under peak-to-peak actuation voltage of 1000 V at 5 Hz is approximately 7.5 cm/s (© IOP Publishing. Reproduced with permission. All rights reserved)

dry in order to protect the electronics. This prototype is intended merely for proof of concept, as the nonoptimized volume (and hence buoyant force) requires a significant amount of counterweight.

The swimming speed for a peak-to-peak actuation voltage of 1000 V at 5 Hz is measured as 7.5 cm/s (measured by video recording).[1] This is equivalent to almost 0.3 body length/s and it compares favorably with several smart actuator-based aquatic robots reported in the literature (including tethered ones) [3] even though the present prototype excludes the additional passive caudal fin extension and volumetric optimization. The passive caudal fin extension is known to improve not only the thrust amplitude but also the bandwidth of effective excitation frequencies (see Figs. 5 and 6). Further improvements can be made by increasing the actuation voltage to a larger level with a DC offset (since MFC laminates have asymmetric voltage limits: −500 to 1500 V).

In general, motor-based robotic fish has larger speed per body length than smart material-based counterparts, whereas the latter offers ease of fabrication and geometric scalability. According to the comparative study by Cen and Erturk [56], the preliminary MFC robotic fish design presented in this work is near the intersection of smart material-based and motor-based robotic fish. Therefore, MFC-based robotic fish provides geometric scalability (as compared to motor-based robotic fish) and high performance swimming (as compared to IPMC-based robotic fish).

[1] Note that the boundary conditions in free (unconstrained) locomotion are no longer clamped-free due to the hardware attachment at the head with finite translational and rotary inertia (which can be accounted for in structural dynamic modeling). Therefore, the resonance frequencies of Fig. 9 (clamped-free) and Fig. 18 (free-free with hardware) cases are different.

7 Conclusions

Bio-inspired aquatic robotics using MFC piezoelectric bimorphs is investigated for fish-like locomotion. First, the effect of a passive substrate fin extension on the thrust frequency response of MFC bimorphs in quiescent water is presented along with measurement procedures. Specifically it is shown that broadband thrust generation can be achieved in the presence of a passive substrate extension. In-air and underwater dynamics of an MFC bimorph cantilever are then modeled for small amplitude bending vibrations under piezoelectric actuation. Underwater dynamics of the bimorph propulsor is coupled with Lighthill's elongated-body theory to predict the thrust output in quiescent water. In-air and underwater experiments are conducted for model validation and for the characterization of a bimorph propulsor. Simplified hydrodynamics is assumed to estimate the underwater electrohydroelastic behavior. For future work, nonlinear electrohydroelastic modeling (combining nonlinear electroelastic dynamics [58–60] with nonlinear hydrodynamic effects [66]) is required to predict the dynamics of the propulsor for large oscillations, comparable length-to-width aspect ratios, and under high electric field levels.

The underwater experiments resulted in mean thrust levels as high as 14 mN around 7 Hz for the peak-to-peak actuation voltage of 800 V using a 90 mm × 35 mm × 0.67 mm cantilever in the absence of a passive fin extension (note that the MFCs can perform without depolarization up to peak-to-peak actuation voltage of 2000 V). In the presence of a passive caudal fin extension, not only the frequency bandwidth but also the thrust levels can be enhanced substantially. The passive caudal fin brings the second vibration mode close to the first one and makes the smart fish a wideband thrust generator. The effect of actuation voltage on the thrust output is also investigated for the fish configuration with a passive caudal fin. For the highest actuation voltage level, mean thrust values of more than 30 mN are obtained in the frequency range of 4–17 Hz (with the peak values of 40 and 50 mN at the first two resonance frequencies, respectively). Overall, fish-like propulsors made from MFCs can successfully imitate thrust levels of biological fish (see Lauder and Drucker [74] for typical values).

An untethered robotic fish prototype that incorporates a microcontroller and a printed-circuit-board amplifier is developed and tested in free locomotion. A swimming speed of 0.3 body length/s (7.5 cm/s swimming speed for 24.3 cm body length) is achieved for a nonoptimized main body–propulsor combination. This swimming speed can be improved substantially by increasing the actuation voltage, optimizing the volume, and using a passive fin extension. According to normalized swimming speed vs. body length comparison by Cen and Erturk [56] with the literature of untethered robotic fish, the MFC-based robotic fish provides geometric scalability and high performance swimming, located at the intersection of smart material-based and motor-based swimmers with these favorable characteristics.

Acknowledgements The author acknowledges contributions of his previous undergraduate visiting researcher Mr. Ghislain Delporte (formerly at Virginia Tech) and of his previous master's student Mr. Lejun Cen (at Georgia Tech) to the original work that is reviewed in this chapter. The author also acknowledges support from the NSF Grant CMMI-1254262 for the ongoing research efforts.

References

1. Bandyopadhyay PR (2005) Trends in biorobotic autonomous undersea vehicles. IEEE J Ocean Eng 30(1):109–139
2. Roper D et al (2011) A review of developments towards biologically inspired propulsion systems for autonomous underwater vehicles. Proc Inst Mech Eng [M] J Eng Marit Environ 225(2):77–96
3. Chu WS et al (2012) Review of biomimetic underwater robots using smart actuators. Int J Precis Eng Manufact 13(7):1281–1292
4. Barrett DS (1994) The design of a flexible hull undersea vehicle propelled by an oscillator foil. MS thesis, Department of Ocean Engineering, Massachusetts Institute of Technology, Boston, 177 pp
5. Barrett DS (1996) propulsive efficiency of a flexible hull underwater vehicle. PhD thesis, Department of Ocean Engineering, Massachusetts Institute of Technology, Boston, 215 pp
6. Anderson JM, Chhabra V (2002) Maneuvering and stability performance of a robotic tuna. Integr Comp Biol 42(1):118–126
7. Low K, Willy A (2006) Biomimetic motion planning of an undulating robotic fish fin. J Vib Control 12(12):1337–1359
8. Valdivia y Alvarado P (2007) design of biomimetic compliant devices for locomotion in liquid environments. PhD thesis, Department of Ocean Engineering, Massachusetts Institute of Technology, Boston, 164 pp
9. Crespi A et al (2008) Controlling swimming and crawling in a fish robot using a central pattern generator. Auton Robot 25(1):3–13
10. Liu F, Lee KM, Yang CJ (2012) Hydrodynamics of an undulating fin for a wave-like locomotion system design. IEEE/ASME Trans Mech 17(3):554–562
11. Zhou C, Low K (2012) Low, design and locomotion control of a biomimetic underwater vehicle with fin propulsion. IEEE/ASME Trans Mech 17(1):25–35
12. Wen L et al (2012) Quantitative thrust efficiency of a self-propulsive robotic fish: experimental method and hydrodynamic investigation. IEEE/ASME Trans Mechatron 18(3):1027–1038
13. Serchi FG, Arienti A, Laschi C (2012) Biomimetic vortex propulsion: toward the new paradigm of soft unmanned underwater vehicles. IEEE/ASME Trans Mechatron 18(2):484–493
14. Shahinpoor M (1992) Conceptual design, kinematics and dynamics of swimming robotic structures using ionic polymeric get muscles. Smart Materials and Structures 1:91–95
15. Mojarrad M, Shahinpoor M (1996) Noiseless propulsion for swimming robotic structures using polyelectrolyte ion-exchange membrane. Smart Mater Technol Biomim Smart Struct Mater 1996(2716):183–192
16. Laurent G, Piat E (2001) Efficiency of swimming microrobots using ionic polymer metal composite actuators. In: Proceedings IEEE international conference on robotics and automation, vol I–IV, pp 3914–3919
17. Kim KJ (2003) Fabrication and development of electroactive ionic polymer-metal composites and their applications as smart materials. THERMEC'2003, pts 1-5, vol 426–424, pp 2249–2254
18. Kim B et al (2005) A biomimetic undulatory tadpole robot using ionic polymer-metal composite actuators. Smart Mater Struct 14(6):1579–1585

19. Chung CK et al (2005) A novel fabrication of ionic polymer-metal composites (IPMC) actuator with silver nano-powders. Transducers 05, digest of technical papers, vol 1–2, pp 217–220
20. Guo SX et al (2006) Underwater swimming micro robot using IPMC actuator. In: Proceeding of the 2006 IEEE international conference on mechatronics and automation, IEEE ICMA 2006, vol 1–3, pp 249–254
21. Chung CK et al (2006) A novel fabrication of ionic polymer-metal composites (IPMC) actuator with silver nano-powders. Sens Actuators B 117(2):367–375
22. Tan XB et al (2006) An autonomous robotic fish for mobile sensing. In: IEEE/RSJ 2006 international conference on intelligent robots and systems, vol 1–12, pp 5424–5429
23. Brunetto P et al (2008) A model of ionic polymer-metal composite actuators in underwater operations. Smart Mater Struct 17(2):12
24. Mbemmo E et al (2008) Modeling of biomimetic robotic fish propelled by an ionic polymer-metal composite actuator. In: IEEE international conference on robotics and automation, vol 1–9, pp 689–694
25. Jeon JH, Yeorn SW, Oh IK (2008) Fabrication and actuation of ionic polymer metal composites patterned by combining electroplating with electroless plating. Compos A Appl Sci Manufact 39(4):588–596
26. De Witt BJ, Hugo RJ (2009) A preliminary study of the transition of an in-line pipe vortex to slug flow using particle image velocimetry. In: Proceedings of the ASME fluids engineering division summer conference-2008, vol 1, pt a and b, pp 637–646
27. Chen Z, Shatara S, Tan XB (2010) Modeling of biomimetic robotic fish propelled by an ionic polymer-metal composite caudal fin. IEEE-ASME Trans Mechatron 15(3):448–459
28. Aureli M, Kopman V, Porfiri M (2010) Free-locomotion of underwater vehicles actuated by ionic polymer metal composites. IEEE-ASME Trans Mechatron 15(4):603–614
29. Takagi K et al (2006) Development of a Rajiform swimming robot using ionic polymer artificial muscles. In: IEEE/RSJ international conference on intelligent robots and systems, vol 1–12, pp 1861–1866
30. Chen Z, Um TI, Zhu J (2011) Bio-inspired robotic cownose ray propelled by electroactive polymer pectoral fin. In: ASME 2011 international mechanical engineering congress and exposition
31. Chen Z, Um TI, Bart-Smith H (2011) Ionic polymer-metal composite enabled robotic manta ray. In: Proceedings of SPIE 7976:797637
32. Shinjo N, Swain GW (2004) Use of a shape memory alloy for the design of an oscillatory propulsion system. IEEE J Ocean Eng 29(3):750–755
33. Wang ZL et al (2008) A micro-robot fish with embedded SMA wire actuated flexible biomimetic fin. Sens Actuators A 144(2):354–360
34. Wang Z et al (2008) Embedded SMA wire actuated biomimetic fin: a module for biomimetic underwater propulsion. Smart Mater Struct 17:025039
35. Cho KJ et al (2008) Design, fabrication and analysis of a body-caudal fin propulsion system for a microrobotic fish. In: IEEE international conference on robotics and automation, vol 1–9, pp 706–711
36. Rossi C et al (2011) Bending continuous structures with SMAS: a novel robotic fish design. Bioinspir Biomim 6:045005
37. Rossi C et al (2011) A motor-less and gear-less bio-mimetic robotic fish design. IEEE International Conference on Robotics and Automation, pp 3646–3651
38. Zhang YS, Liu GJ (2005) Design, analysis and experiments of a wireless swimming micro robot. In: IEEE international conference on mechatronics and automations, vol 1–4, pp 946–951
39. Zhang YS, Liu GJ (2005) Wireless micro biomimetic swimming robot based on giant magnetostrictive films. IEEE Int Conf Rob Biomim 2006:195–200
40. Zhang YS, Liu GJ (2009) Wireless swimming microrobot: design, analysis, and experiments. J Dyn Syst Meas Control Trans ASME 131(1):014502

41. Zhang Z (2007) Design and control of a fish-like robot using an electrostatic motor. In: IEEE Proceedings of ICRA, Rome
42. Zhang Z, Philen M, Neu W (2010) A biologically inspired artificial fish using flexible matrix composite actuators: analysis and experiment. Smart Mater Struct 19:094017
43. Philen M, New W (2011) Hydrodynamic analysis, performance assessment, and actuator design of a flexible tail propulsor in an artificial alligator. Smart Mater Struct 20:094015
44. Uchino K (2008) Piezoelectric actuators 2006—expansion from IT/Robotics to ecological/ energy applications. J Electroceram 20(3–4):301–311
45. Cook-Chennault KA, Thambi N, Sastry AM (2008) Powering MEMS portable devices—a review of non-regenerative and regenerative power supply systems with special emphasis on piezoelectric energy harvesting systems. Smart Mater Struct 17(4):043001
46. Erturk A, Delporte D (2011) Underwater thrust and power generation using flexible piezoelectric composites: an experimental investigation toward self-powered swimmer-sensor platforms. Smart Mater Struct 20:125013
47. Trolier-Mckinstry S, Muralt P (2004) Thin film piezoelectrics for MEMS. J Electroceram 12 (1–2):7–17
48. Wilkie WK et al (2000) Low-cost piezocomposite actuator for structural control applications. In: Proceedings of SPIE, 3991:323
49. High JW, Wilkie WK (2003) Method of fabricating NASA-standard macro-fiber composite piezoelectric actuators, National Aeronautics and Space Administration, Langley Research Center
50. Bryant RG (2007) Overview of NASA langley's piezoelectric ceramic packaging technology and applications, National Aeronautics and Space Administration, Langley Research Center
51. Smart Material Corp. http://www.smart-material.com
52. Fukuda T et al (1995) Steering mechanism and swimming experiment of micro mobile robot in water. In: IEEE proceedings on micro electro mechanical systems, pp 300–305
53. Heo S et al (2007) Effect of an artificial caudal fin on the performance of a biomimetic fish robot propelled by piezoelectric actuators. J Bionic Eng 4(3):151–158
54. Wiguna T et al (2009) Design and experimental parameteric study of a fish robot actuated by piezoelectric actuators. J Intell Mater Syst Struct 20(6):751–758
55. Ming AG et al (2009) Development of underwater robots using piezoelectric fiber composite. In: IEEE international conference on robotics and automation ICRA, vol 1–7, pp 3435–3440
56. Cen L, Erturk A (2013) Bio-inspired aquatic robotics by untethered piezohydroelastic actuation. Bioinspir Biomim 8:016006
57. Meirovitch L (2001) Fundamentals of vibrations, vol XVIII. Mcgraw-hill, Boston, p 806
58. Wolf K, Gottlieb O (2002) Nonlinear dynamics of a noncontacting atomic force microscope cantilever actuated by a piezoelectric layer. J Appl Phys 91(7):4701–4709
59. Usher T, Sim A (2005) Nonlinear dynamics of piezoelectric high displacement actuators in cantilever mode. J Appl Phys 98(6):064102
60. Stanton SC et al (2012) Nonlinear nonconservative behavior and modeling of piezoelectric energy harvesters including proof mass effects. J Intell Mater Syst Struct 23(2):183–199
61. Sader JE (1998) Frequency response of cantilever beams immersed in viscous fluids with applications to the atomic force microscope. J Appl Phys 84(1):64–76
62. Van Eysden CA, Sader JE (2007) Frequency response of cantilever beams immersed in viscous fluids with applications to the atomic force microscope: arbitrary mode order. J Appl Phys 101(4):044908
63. Chon JWM, Mulvaney P, Sader JE (2000) Experimental validation of theoretical models for the frequency response of atomic force microscope cantilever beams immersed in fluids. J Appl Phys 87(8):3978–3988
64. Aureli M, Porfiri M (2010) Low frequency and large amplitude oscillations of cantilevers in viscous fluids. Appl Phys Lett 96(16):164102
65. Maali A et al (2005) Hydrodynamics of oscillating atomic force microscopy cantilevers in viscous fluids. J Appl Phys 97(7):074907

66. Aureli M, Basaran ME, Porfiri M (2012) Nonlinear finite amplitude vibrations of sharp-edged beams in viscous fluids. J Sound Vib 331(7):1624–1654
67. Chu W (1963) Technical report no. 2. DTMB, contract NObs-86396 (X), Southwest Research Institute, San Antonio, Texas
68. Lighthill MJ (1960) Note on the swimming of slender fish. J Fluid Mech 9(2):305–317
69. Lighthill MJ (1969) Hydromechanics of aquatic animal propulsion. Annu Rev Fluid Mech 1:413–446
70. Lighthill MJ (1970) Aquatic animal propulsion of high hydromechanical efficiency. J Fluid Mech 44:265–301
71. Lighthill MJ (1971) Large-amplitude elongated-body theory of fish locomotion. Proc R Soc Lond B Biol Sci 179(1055):125–138
72. Anton SR, Erturk A, Inman DJ (2010) Multifunctional self-charging structures using piezoceramics and thin-film batteries. Smart Mater Struct 19(11):115021
73. Abdelnour K et al (2009) Hydrodynamics of underwater propulsors based on ionic polymer-metal composites: a numerical study. Smart Mater Struct 18(8):085006
74. Lauder GV, Drucker EG (2002) Forces, fishes, and fluids: hydrodynamic mechanisms of aquatic locomotion. News Physiol Sci 17:235–240

A Multifunctional Underwater Biomimetic Microrobot

Shuxiang Guo and Liwei Shi

Abstract Robots play an important role in underwater monitoring and recovery operations, such as pollution detection, submarine sampling and data collection, video mapping, and object recovery in dangerous places. However, regular-sized robots may not be suitable for applications in some restricted underwater environments. Accordingly, in previous research we designed several novel types of bio-inspired microrobots, using ionic polymer metal composite (IPMC) and shape memory alloy (SMA) actuators. These microrobots possess some of the attributes of compact structure, multifunctionality, flexibility, and precise positioning. However, they lack the attributes of long endurance, stable high speed, and large load capacity necessary for real-world applications. To overcome these disadvantages, we propose a mother–son robot system, composed of several microrobots as sons and a newly designed amphibious spherical robot as the mother. In this system, the mother robot is actuated by four water-jet propellers and eight servomotors, capable of providing stable high speed and carrying the microrobots to the desired target location where tasks are to be performed. Generally speaking, compact structure, multifunctionality, and precise positioning are considered incompatible characteristics for underwater microrobots. To realize the necessary multifunctionality for adapting to complex underwater environments, we introduce a walking biomimetic microrobot with two kinds of motion attitudes: a lying state and a standing state. The microrobot uses eleven IPMC actuators to move and two SMA actuators to change its motion attitude. In the lying state, the microrobot implements stick-insect-inspired walking/rotating motion, fishlike swimming motion, horizontal grasping motion, and floating motion. In the standing state, it implements inch-worm-inspired crawling motion in two horizontal directions and grasping motion in

S. Guo (✉) · L. Shi
School of Life Science, Beijing Institute of Technology, Beijing, China
e-mail: guo@eng.kagawa-u.ac.jp

L. Shi
e-mail: shiliwei@bit.edu.cn

S. Guo
Faculty of Engineering, Kagawa University, 2217-20 Hayashi-cho, Takamatsu, Kagawa, Japan

© Springer-Verlag Berlin Heidelberg 2015

R. Du et al. (eds.), *Robot Fish*, Springer Tracts in Mechanical Engineering,
DOI 10.1007/978-3-662-46870-8_10

the vertical direction. We constructed a prototype of this biomimetic microrobot and evaluated its walking, rotating, and floating speeds experimentally. The experimental results indicated that the robot could attain a maximum walking speed of 3.6 mm/s, a maximum rotational speed of 9 °/s, and a maximum floating speed of 7.14 mm/s. Obstacle-avoidance and swimming experiments were also carried out to demonstrate its multifunctionality.

Keywords Ionic polymer metal composite (IPMC) actuators · Biomimetic underwater microrobot · Motion attitudes · Micromechanism

1 Introduction

Robots are now widely used to implement underwater tasks considered by humans to be dangerous, dull, or dirty, primarily because of their long endurance, stable high speeds, and large load capabilities. This trend has continued into underwater monitoring and recovery operations, including pollution detection, submarine sampling and data collection, video mapping, exploration of unstructured underwater environments, object recovery in dangerous places, and other tasks [1, 2]. Various configurations, shapes, and sizes of underwater robots are required for different applications or tasks. For underwater environmental detection or observation, a compact structure with multifunctionality and flexibility enables a robot to work in limited spaces. When a large range of motions and large load capacity are required, a traditional motor-actuated electromagnetic structure is essential. When large interior space and flexible multidirectional rotation in a restricted space are required, a spherical robot body is recommended. When high-speed cruising is required, a streamlined robot body may be the best choice [3, 4].

If a robot is to be used in a complicated underwater environment, such as a narrow pipeline or a region filled with reefs, it should be endowed with the combined attributes of endurance, stable high speed, large load capability, flexibility, compact structure, and multifunctionality. Many types of underwater robots have been developed in recent years. While the use of some of these robots involves changing the angles of rudders or adjusting the differential propulsive forces of thrusters, a number of vectored propeller-actuated underwater robots have also been introduced [5]. A multi-channel Hall-effect thruster has also been reported, involving vector composition of underwater robots [6]. Moreover, we have developed a spherical underwater robot equipped with three vectored water-jet-based thrusters [3]. However, most of these robots are steered by traditional electromagnetic thrusters, which are difficult to miniaturize [4].

Accordingly, motors are rarely found in microrobot applications [7, 8] and special actuator materials are used instead. A variety of smart materials, such as ionic polymer metal composite (IPMC), piezoelectric elements, pneumatic actuators, and shape memory alloy (SMA), have been investigated for use as artificial muscles in new types of microrobots [9–15]. In this research, IPMC is used as actuator material to develop a microrobot with a compact structure, multifunctionality, and flexibility. The actuation characteristics of IPMC, which include suitable response time, high bending deformation and long life, show a significant potential for the propulsion of underwater microrobots [16–21]. Although many biomimetic microrobots with smart actuators have been introduced in recent years, developing a single microrobot with compact structure, flexibility, and multifunctions remains a challenge, due to conflicts between the three characteristics [4]. For this reason, many microrobot designers have abandoned compact structure in favor of biomimetic multi-jointed configurations to improve flexibility and obtain multifunctions. Others have sacrificed flexibility and multifunctions in pursuit of miniaturization.

For real-world applications, an underwater robot should possess the attributes of endurance, stable high speed, large load capacity, flexibility, compact structure, and multifunctionality. To implement these characteristics, we propose a mother–son robot system, which includes several microrobots as sons and a newly designed amphibious spherical robot as the mother. This is an original concept and is inspired by the design of aircraft carrier systems [4]. In this system, the mother robot is actuated by four water-jet propellers and eight servomotors, capable of providing stable high speed and carrying the microrobots to the desired target location where tasks are to be performed. When the mother robot reaches the desired location, or encounters a narrow channel that is difficult to navigate, it assumes a stable position and acts as a base station for the microrobots. Then, the microrobots exit the mother robot, proceed to the target position, and carry out their tasks.

Compared with a single large robot, when the final tasks are carried out by microrobots, it is easier to adapt to narrow environments and implement relatively high positioning precision. In addition, compared with individual microrobots, the mother–son system offers the following advantages.

(1) The range of motions of the overall system is expanded, owing to the relatively high speed and endurance of the mother robot.
(2) The microrobots can obtain a relatively stable, high power supply via cables.
(3) Since the microrobots are all controlled by the mother robot, communications between microrobots can be implemented by the mother when cooperation is needed.
(4) Since the power supply and control units are installed in the mother robot, the microrobots can be designed with a more compact structure, suitable for restricted spaces such as narrow pipelines or channels.

We introduced a newly designed spherical amphibious mother robot in [22–24]. A spherical body has both a compact structure and maximum interior space, compared to a streamlined body. It can rotate and change direction more easily than a streamlined design, which is very important for microrobots in restricted spaces. Following the mother robot in previous researches, this chapter will mainly focus on the microrobots. At the end of this chapter, we will show a schematic diagram to illustrate the mother-son system.

Aside from fishlike and manta-ray-like swimming locomotion, we have developed several microrobots that employ biomimetic locomotion to implement walking, floating and swimming motions [1, 2, 25–27]. However, each of these units implements only some of these motions and none of them are able to carry out simple tasks such as grasping and carrying objects to a desired position, detecting an object, or avoiding an obstacle. To realize the necessary multifunctionality for adapting to different environments, a hybrid underwater microrobot with two motion attitudes is introduced in this paper. The microrobot uses eleven IPMC actuators to implement three-dimensional underwater motions, and two SMA actuators for attitude change. The robot can change between two attitudes: a lying attitude and a standing attitude. It uses the standing attitude to cross a high, narrow gap, and changes to the lying attitude while walking through a low, wide tunnel. We have constructed a prototype microrobot and carried out a series of experiments to evaluate its performance.

The remainder of this chapter is divided into four parts. First, we describe the characteristics of IPMC actuators and summarize the implemented techniques of underwater locomotion. Second, we review the feasibility results for several previously developed microrobots. Third, based on several types of biomimetic locomotion, we introduce the new hybrid microrobot design, including the structural design and motion mechanisms in the two attitudes. Fourth, we discuss the construction of a prototype of this hybrid microrobot, together with a series of experiments to evaluate its walking, rotating, floating, and swimming speeds. Attitude change and obstacle-avoidance experiments are also included. Finally, we present our conclusions.

2 Biomimetic Locomotion

2.1 IPMC Actuators

(1) IPMC is an innovative material made of an ionic polymer membrane, chemically plated with gold electrodes on both sides. When an electrical stimulus is applied across the thickness of the IPMC, bending deformation will be generated for the water molecules' redistribution in the ionic polymer membrane, which expands the cathode side of the membrane and contract the anode side. Its actuation characteristics show significant potential for the propulsion of underwater microrobots. It is lightweight and has a suitable response time, high bending deformation, and long life. IPMC is widely used in soft robotic actuators such as artificial muscles, as well as on dynamic sensors [16]. The IPMC adopted for this research consists of Au deposited on Nafion™ film with a thickness of 0.22 mm.

An IPMC actuator can be regarded as equivalent to a cantilever beam. Figure 1 shows the mechanical configuration and relevant geometrical parameters, which are as follows: L_c denotes the length of the clamped part of the IPMC, L_f is the total free

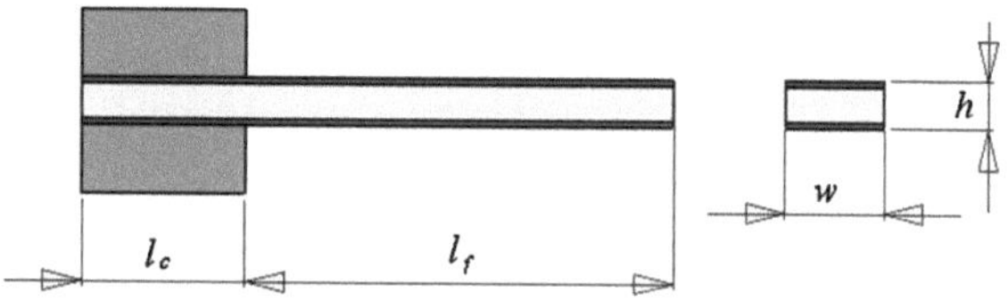

Fig. 1 Mechanical configuration of the actuator and relevant parameters

length of the IPMC, and w and h denote the width and the height of the IPMC cross-section. The pinned end is used to apply electrical voltages across the thickness.

According to mechanical analysis, bending deformation of an IPMC actuator results from the redistribution of internal water molecules. Under the influence of an applied stimulus, the water molecules in the actuator are redistributed in the following two stages [28].

(2) When an electrical stimulus is applied across the thickness of the IPMC, each hydrated sodium ion moves to the cathode side, accompanied by four hydrated water molecules. Bending deformation is then generated by the expansion of Nafion117 near the cathode side and contraction near the anode side.

(3) After a short time, the self-diffusion of water molecules causes free water molecules to gradually flow to the anode side, reducing the concentration of water molecules at the cathode and demonstrating the deformation recovery potential of the IPMC actuator.

We measured the free-end deflection of an IPMC actuator in a water tank for different applied square-wave signals. The sample IPMC actuator was 20 mm long, 4 mm wide, and 0.22 mm thick. It was driven by a personal computer (PC) equipped with a digital-to-analog converter card, and the deflection of the IPMC was measured via a laser displacement sensor. The laser sensor was used to translate the displacement to a voltage and then the voltages were recorded and translated to the PC by an analog-to-digital converter card. We used the square-wave signals to drive the IPMC actuator. Figure 2 indicates that the displacement was inversely proportional to the frequency of the input signal, and proportional to the input voltage at a low frequency. However, the displacement variation with respect to voltage was greatly reduced at a high frequency [29].

We also measured the bending force generated at the free end of the equivalent cantilever beam by applying different signals. In this experiment, the sample IPMC actuator was 24 mm long, 18 mm wide, and 0.22 mm thick. The actuator was driven by a PC equipped with a digital-to-analog converter card, and the bending force of the IPMC was measured by an electronic balance. To reduce the torque on the electronic balance, we used a needle to transfer the press force from the IPMC actuator. The initial distance between the IPMC and the needle tip is set as 3 mm. Figure 3 shows the experimentally recorded tip-bending force of the actuator for different voltages. The experimental results indicate that the tip-bending force increases as the driving voltage increases [30].

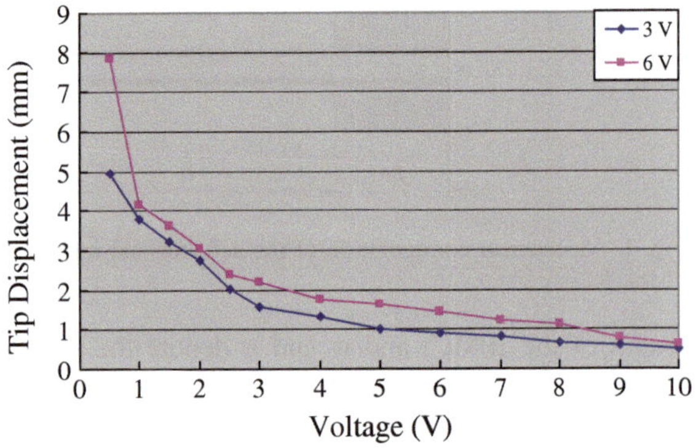

Fig. 2 Tip displacements of the IPMC actuator [29]

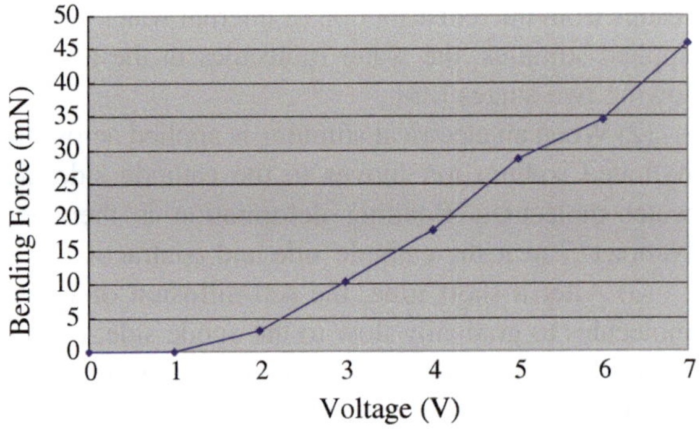

Fig. 3 Bending forces of the IPMC actuator [30]

2.2 Bio-inspired Locomotion

IPMC actuators can be used as oscillating or undulating fins for swimming microrobots when a fast response is required [16–18, 31, 32]. However, this type of swimming motion cannot ensure precise positioning of the robot. Furthermore, fishlike propulsion mechanisms simply mimic the undulating and oscillatory body/fin motions of a fish. Some simple underwater tasks are not easily carried out without hands or fingers. Therefore, in addition to swimming, other types of biomimetic locomotion are required for microrobots with compact structure, multi-functions, and flexibility.

2.2.1 Stick Insect-Inspired Walking Locomotion

Nature provides perfect models for robots. Biomimetic robots borrow their senses and structure from animals, such as insects, fish, and birds. In the case of the stick insect, each leg is composed of the coxa, femur, tibiae, and tarsus. The tarsus is also

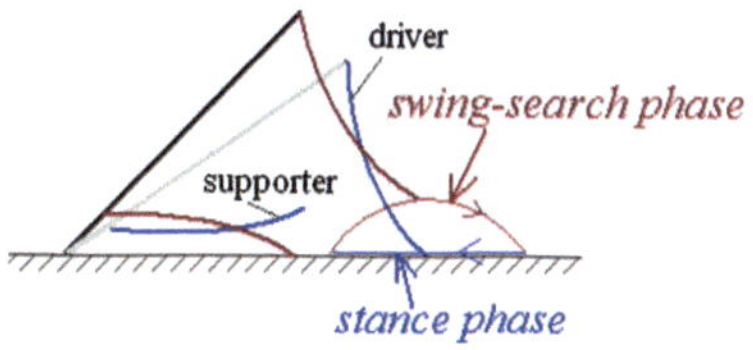

Fig. 4 Two-phase driving locomotion with IPMC actuators [25]

called the foot and does not contribute to its movements. The coxa offers the foot one degree of freedom (DOF) in the direction of movement. The femur and the tibiae offer the foot two DOF to enable it to find a reliable foothold during the swing–search phase to touch the ground and support the body during the stance phase.

A stick insect-inspired biomimetic locomotion prototype using two IPMC actuators was introduced in [1]. As Fig. 4 shows, the actuator in the vertical direction is called the driver, while the actuator in the horizontal direction is called the supporter. The free end of the driver is the foot. Each step cycle can be separated into four periods. First, the supporter lifts the body up, and the driver is off the ground. Second, the driver bends forward. Third, the supporter bends upward far enough so that it is off the ground, while the driver in contact with the ground supports the body. Fourth, the driver bends backward in the propulsion stroke, and the body is pushed forward. The driver and supporter are controlled by two channels of square waves, each with the same frequency. The phase of the supporter lags 90° behind that of the driver [1, 2, 25].

2.2.2 Jellyfish-like Floating Locomotion

Jellyfish movement is dependent on floatation, ocean currents, and winds and is accomplished via a form of jet propulsion. Specifically, jellyfish move by squeezing their bodies so that jets of water are ejected from underneath, propelling them forward.

A jellyfish-inspired biomimetic locomotion prototype with SMA actuators was introduced in [1, 33]. The jellyfish-inspired body uses SMA actuators to imitate the circular muscles of a real jellyfish, as shown in Fig. 5. The body shrinks when voltage is applied and water is squeezed out of it. This changes the buoyancy and produces an upward force. The body floats upward when the force reaches a certain value. The upward force can be changed by controlling the frequency of the actuator shrinkage and the voltage between its two ends. This means that the microrobot can be induced to float upward, remain neutrally buoyant, or sink as required.

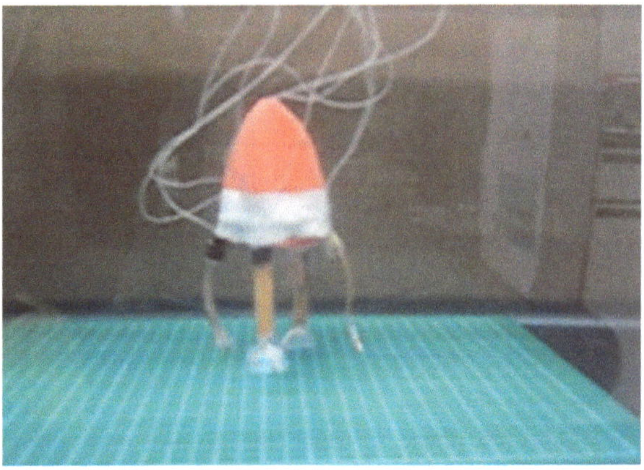

Fig. 5 A jellyfish-inspired biomimetic locomotion prototype [1]

2.2.3 Butterfly-Inspired Swimming Locomotion

Butterfly movement is accomplished by the counter force of air. Specifically, they flap their wings to push the air at different frequencies and speeds. In a single flapping period, there are two motions, folding and unfolding. Movement results from the fact that the folding motion pushes the air between their wings at a higher speed than the unfolding motion.

A butterfly-inspired biomimetic locomotion prototype with SMA actuator was also introduced to implement fast swimming. We use two pectoral fins to imitate the flapping motions of a butterfly. The fins are installed perpendicular to the horizontal plane, and the initial angle between them is set at $\pi/3$. They squeeze the water between them to create a counter force in the horizontal direction. In this way, the microrobot can be induced to implement a swimming motion. The horizontal propulsion can be changed by controlling the frequency of the actuator shrinkage and the driving voltage. The two pectoral fins are driven by separate SMA actuators, and thus the robot can swim or make turns via the cooperation of the fins, as shown in Fig. 6 [27].

2.2.4 Inchworm-Inspired Crawling Locomotion

Inchworms have smooth, hairless bodies, usually about 25 mm long. Also known as measuring worms, spanworms, or loopers, they lack appendages in their midsections, causing them to have a characteristic looping gait. They have three pairs of true legs at the front end, like other caterpillars, but only two or three pairs of prolegs at the rear end. An inchworm moves by drawing its hind end forward while holding on with its front legs, and then advancing its front section while holding on with its prolegs [26, 34, 35].

An inchworm-inspired biomimetic locomotion prototype with two IPMC actuators was introduced to implement fast creeping. The design was based on a one

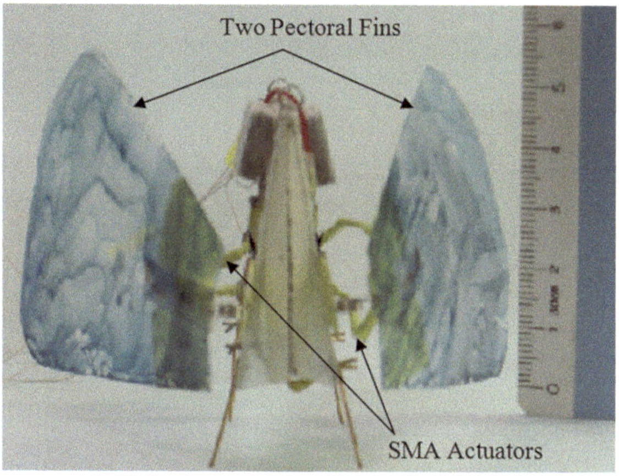

Fig. 6 A butterfly-inspired biomimetic locomotion prototype [27]

Fig. 7 An inchworm-inspired biomimetic locomotion [35]

DOF leg. The structure of the one DOF walking mechanism is described in [35]. This mechanism can only implement crawling motion, as shown in Fig. 7.

3 Developed Microrobots

Swimming motion is a universal motion in water for the underwater species. Till date, the majority of research work has been focused on fish-like propulsion mechanisms, fin materials, remote operation, multi-agent cooperation, and

Fig. 8 Fish-inspired microrobot [32, 36]. **a** Fish-like microrobot. **b** Living with the real fish

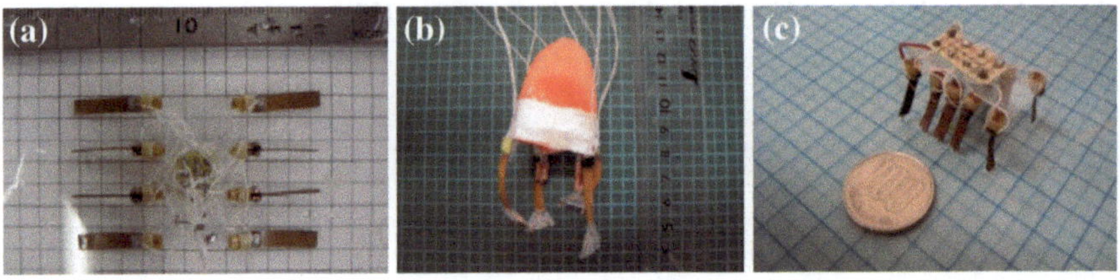

Fig. 9 Prototype microrobots. **a** Stick insect-inspired [2]. **b** Jellyfish-inspired [1]. **c** Inchworm-inspired [26, 35]

mechanical structures. We have developed a robotic fish by using one IPMC actuator, which was reported by CCTV 10 in China. Figure 8 shows this fishlike microrobot, which can live with real fish in the same water tank.

However, the swimming motion cannot ensure the position precision for the robot. Also, the fishlike propulsion mechanisms just mimic the undulating and oscillatory body/fin motions. It is hard to implement some simple underwater tasks without hands or fingers. So, besides the swimming, we proposed some other biomimetic locomotion.

Based on stick insect-inspired walking locomotion, a prototype of an eight-legged microrobot was developed, as shown in Fig. 9a [2]. It was 33 mm long, 56 mm wide, and 9 mm high. Four legs were used as drivers and the other four actuators were used as supporters. It was capable of walking, rotating, and diving/surfacing. However, the floating efficiency of this microrobot was not high. To improve the floating motion, a prototype of a jellyfish-type microrobot was constructed, based on jellyfish-inspired locomotion, as shown in Fig. 9b [1]. It was 68 mm high, with a weight of 4.81 g in air. This biomimetic microrobot consisted of a two-ring body and four legs. The body was designed to imitate a jellyfish's diving/surfacing motions. Additionally, four IPMC actuators were fixed on the body to implement walking motion in two directions. Although the floating motion was improved, the prototype was unable to rotate, and the walking motion was unsatisfactory because the center of gravity was located in one of the two halves of the body, causing an imbalance in the overall body and a large amount of slippage.

For the purpose of creating a microrobot with a compact structure and multi-functions, an inchworm-inspired microrobot with ten IPMC actuators was developed, as shown in Fig. 9c. It was 33 mm long, 14 mm wide, and 14 mm high. Four outside actuators were used as legs to implement walking, rotating, and floating motions. The other six actuators were used as fingers to grasp small objects [26, 34, 35]. Figure 10 shows the hybrid motion of the inchworch-inspired microrobot. Compared with the jellyfish-like robot, this design offered the advantages of stability, compact structure, less water resistance, and grasping motion implementation. However, because the rotating radii were not the same for the outside four legs, a large amount of slippage

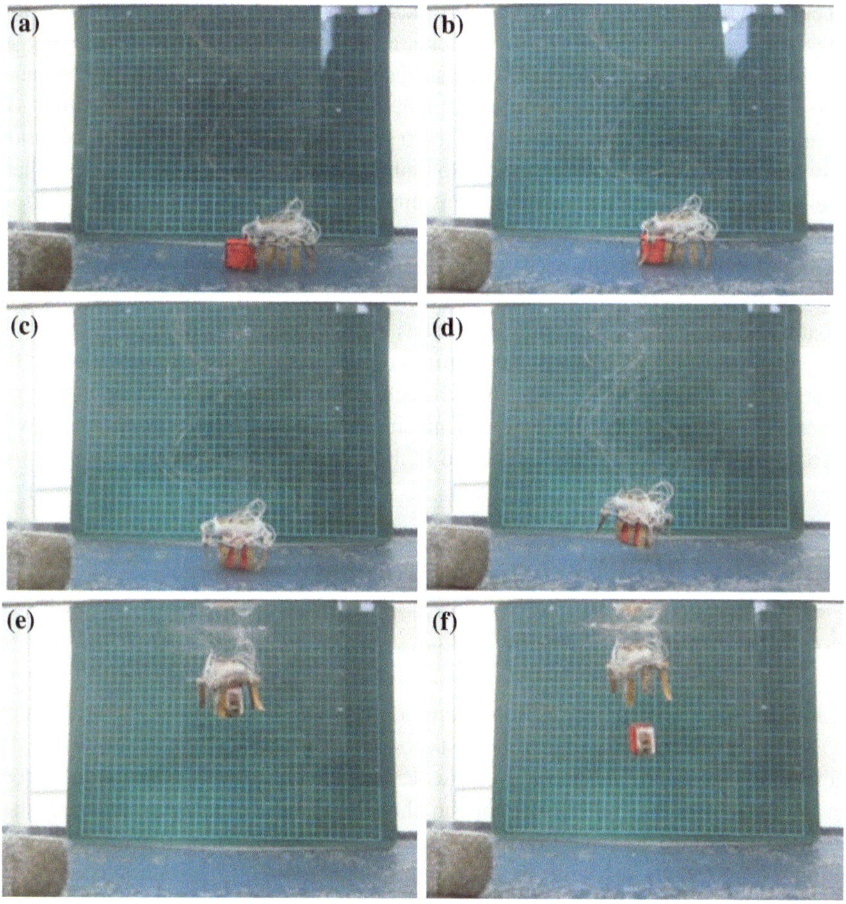

Fig. 10 Hybrid motion of the inchworch-inspired microrobot. **a** Initial position. **b** Walking motion. **c** Grasping motion. **d** Floating motion. **e** Reaching the surface. **f** Opening motion

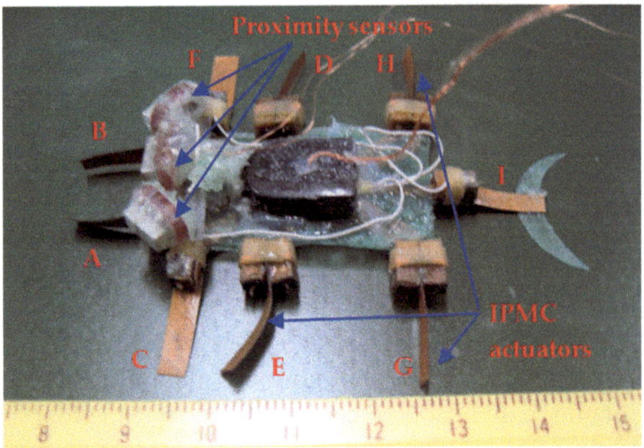

Fig. 11 Prototype lobster-like microrobot

occurred while rotating and the rotating efficiency was not high. Only the outside four legs were used to electrolyze the water around the IPMC surface, generating air bubbles, which became attached to the surfaces of the legs, increasing the buoyancy and implementing the floating motion. Due to the limitations of the structure, the inside six legs were used solely as fingers to grasp an object and could not contribute any buoyancy to the floating motion, so that the floating speed was slow. To inherit the multifunctions of the inchworm-inspired microrobot and overcome its disadvantages, we introduced a new lobster-like microrobot, intended for underwater exploration in a restricted space, as shown in Fig. 11. It uses nine IPMC actuators as legs or claws. Seven actuators are used as legs to implement walking, rotating, and floating. The other two actuators are used as claws to grasp small objects. To imitate the antennae of actual lobsters, three infrared proximity sensors are installed on the head of the microrobot to detect an obstacle. To realize the necessary multifunctionality for adapting to different environments, a hybrid underwater microrobot with two motion attitudes will be introduced in the following sections.

4 Proposed Multifunctional Microrobot

4.1 Structure of the Microrobot

Based on the above types of biomimetic locomotion, we propose a hybrid underwater microrobot, consisting of a plastic body, eleven IPMC actuators, two SMA actuators, a passive tail fin, and two plastic sheets. With the SMA actuators affixed to the plastic sheets, the microrobot can change its attitude between the lying state and the standing state, as illustrated in Fig. 12. The body of the microrobot is 35 mm long and 20 mm wide, as determined by the motion functions and balance of the overall body. It is 3 mm high in the lying state and 21 mm high in the standing state. The eleven actuators are all 17 mm long, 3 mm wide, and 0.2 mm thick.

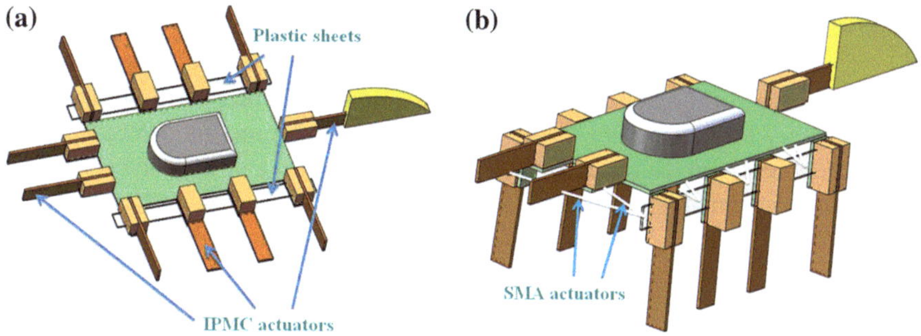

Fig. 12 Proposed hybrid microrobot. **a** Lying attitude. **b** Standing attitude

Fig. 13 Leg sequence and dimensions of the proposed microrobot

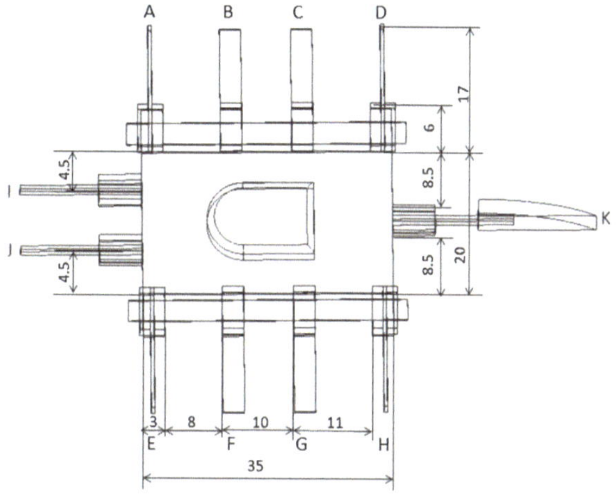

The microrobot uses eleven 1-DOF IPMC actuators to realize walking, rotating, grasping, swimming, and floating motions [37]. Figure 13 shows the leg sequence of these actuators. In the lying attitude, actuators I and J are used as fingers, and are designed for grasping. Actuators B, C, F, and G are called supporters, while actuators A, D, E, and H are called drivers. By changing the bending directions of the four drivers, the robot can walk forward or backward, and rotate clockwise or counterclockwise. In the standing attitude, actuators B, C, F, and G are used as fingers for grasping. Legs A and E are used as leading legs, while legs D and H are used as following legs to implement walking and rotating motions. In both attitudes, actuator K is used to actuate the passive tail fin for swimming.

4.2 Force Analysis of the Attitude Change

The SMA actuators are used to change the attitude of the proposed microrobot. It was necessary to calculate the force required for standing motion before attaching the SMA actuators to the robot body. We then constructed a physical mechanism to transform horizontal forces into vertical forces that could be measured with a spring dynamometer. Figure 14 shows a diagram of the force transition mechanism. We first inserted two fishing lines through the points A–D and B–C, respectively, and then connected the four ends of the two lines at the point O. The vertical force F was measured via a spring dynamometer at point O. The force F_n required to pull the plastic sheet from the horizontal to the vertical direction is given by

$$F_n = 2F_3 \tag{1}$$

where F_3 denotes the tensile force in either of the lines (AD or BC). According to Fig. 14, the tensile forces F_3 and F_1 and the resultant force F_2 can be obtained from the following equations:

$$F_3 = -F_1 \sin\theta_1 \tag{2}$$

$$F_1 = \frac{F_2}{2\cos\theta_1} \tag{3}$$

$$F_2 = -\frac{F}{2\cos\alpha_1} \tag{4}$$

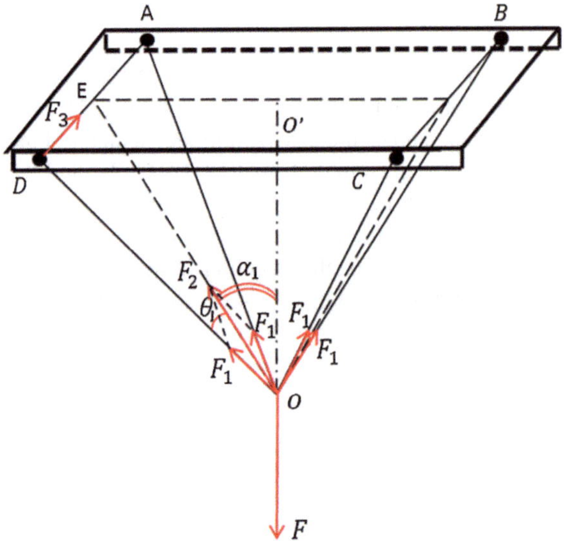

Fig. 14 The scheme of tensile force measurement for the attitude change

where F denotes the measured vertical force.

Utilizing (1)–(4), F_n is given by

$$F_n = \frac{F \sin\theta_1}{2\cos\theta_1 \cos\alpha_1} \tag{5}$$

We used this formula to calculate the force F_n required for our proposed structure.

4.3 Mechanism of the Walking/Rotating Motion in the Lying Attitude

In the lying attitude, the proposed microrobot can implement stick-insect-inspired walking motions using supporters B, C, F, and G and drivers A, D, E, and H. The drivers provide the propulsion for the motion, and the supporters are employed to raise the drivers off the ground and reduce the resistance. The drivers and supporters are controlled by two square-wave channels, each with the same frequency. The phase of the four supporters lags 90° behind that of drivers. Figure 15 shows a single step cycle of the forward motion. Each cycle is divided into four periods as follows [1].

(1) The four supporters lift the body to raise the drivers off the ground.
(2) As the supporters lift the body, the drivers bend forward.
(3) The four supporters bend upward, causing the four drivers to contact the ground.
(4) The four drivers bend backward to push the body forward.

The walking speed is determined by the displacements of the four drivers and the frequency of the control signal. Since the drivers are distributed symmetrically on both sides of the body, and have the same size and deflection characteristics, they bear equivalent loads and drag forces. Therefore, all four drivers provide the same tip displacement for a given applied input voltage. Assuming that the robot is moved by a fixed driving voltage and current, the tip displacement of the actuator in one direction is $d/2$, and the distance the robot advances is d, as shown in Fig. 15c, d. The walking speed can then be obtained from

$$v = d \times f = (d_0 - \Delta d) \times f \tag{6}$$

where v denotes the average walking speed, d_0 denotes the tip displacement of a driver without a payload, Δd is the reduction in the actual displacement of a driver due to friction, and f is the frequency of the input signal.

By changing the bending directions of the four drivers, forward and backward walking motions and clockwise and counterclockwise rotations can be implemented. Figure 16 shows a single step cycle of the rotational motion, which can also be divided into four periods. When the four supporters lift the body, the two left drivers bend backward and the two right drivers bend forward. When the four

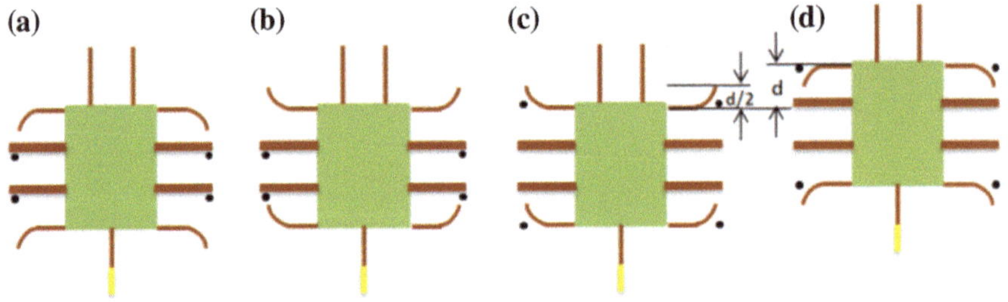

Fig. 15 One step cycle of moving forward motion in lying structure (The marks • indicate which actuator contacts the ground). **a** The supporters lift the body. **b** The drivers bend forward. **c** The supporters bend upward. **d** The drivers bend backward

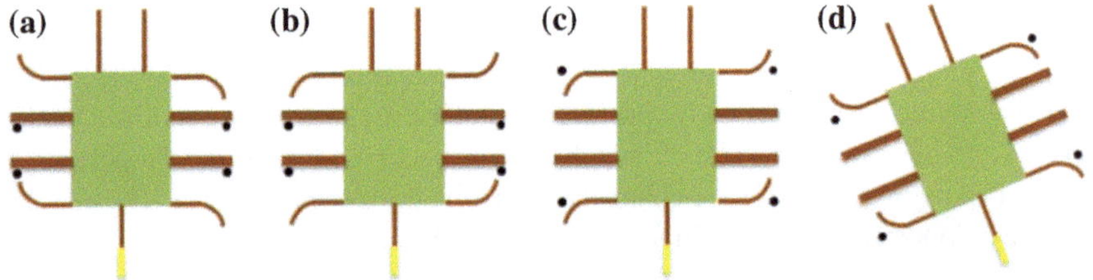

Fig. 16 One step cycle of rotating motion in lying structure (The marks • indicate which actuator contacts the ground). **a** The supporters lift the body. **b** The two left drivers bend backward and the two right drivers bend forward. **c** The supporters bend upward. **d** The two left drivers bend forward and the two right drivers bend backward

supporters bend upward, the four drivers contact the ground and bend in the reverse direction.

When the rotational direction of drivers E and H is opposite to that of drivers A and D, the microrobot can implement clockwise rotation or counterclockwise rotation. The robot rotates through the angle θ in a single step cycle, as shown in Fig. 17a. Here, θ is given by

$$\theta = \frac{L}{R} \tag{7}$$

where L denotes the length of the rotational arc and R denotes the radius of rotation with center-point O. From Fig. 17b, we have

$$r\cos\alpha = r - \frac{d}{2} \tag{8}$$

$$l = \alpha \times r \tag{9}$$

$$h = \sqrt{\frac{d}{2} \cdot \left| 2r - \frac{d}{2} \right|} \tag{10}$$

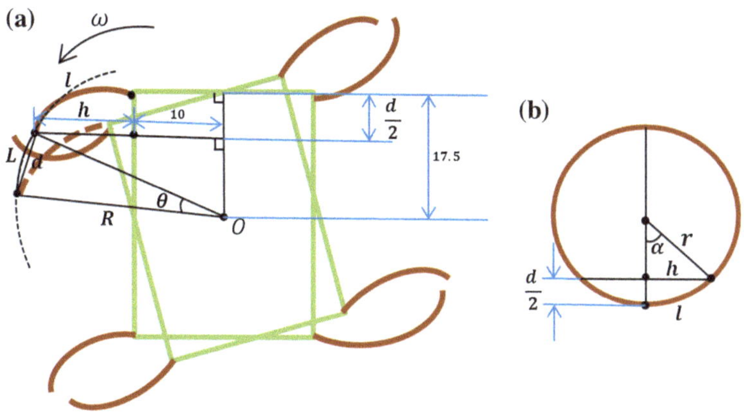

Fig. 17 **a** The rotating angle in one step cycle. **b** The calculation of the value of h (only drivers are drawn)

where r is the bending radius of the IPMC actuator, α denotes the central angle of the IPMC bending arc, l denotes the length of the IPMC actuator, and h denotes the semifocal chord length of the IPMC bending arc. The radius R can be calculated using the equation

$$R = \sqrt{(h + 10)^2 + (17.5 - \frac{d}{2})^2} \tag{11}$$

When d is very small, we can approximate the arc length L by d, the linear distance between the initial and final robot position. According to Eqs. (7) and (11), the theoretical rotational speed can then be calculated from

$$\omega = \theta * f = \frac{2d}{\sqrt{(h + 10)^2 + (17.5 - \frac{d}{2})^2}} f \tag{12}$$

4.4 *Mechanism of the Walking/Rotating Motion in the Standing Attitude*

In the standing attitude, the microrobot can implement inchworm-inspired crawling motions in two directions (longitudinal and transverse) using the eight legs A–H. Unlike the motions in the lying attitude, legs A and E are used as leading legs, while legs D and H are used as following legs. This allows the robot to implement walking motion in the longitudinal direction. When the robot walks forward, the phase of the leading leg lags 90° behind that of the following leg, as shown in Fig. 18 [35]. In this attitude, the robot can fold all legs below its body to get across

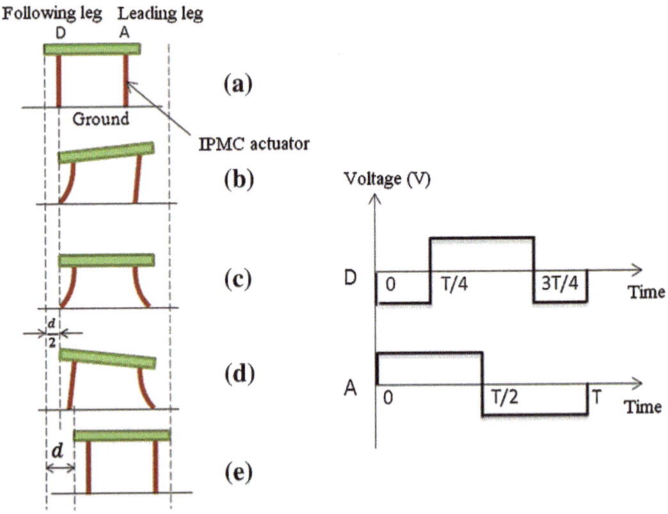

Fig. 18 One step cycle of crawling motion in standing attitude [35]

high narrow gaps. The crawling speed in the standing attitude is determined by the same parameters as in the lying attitude.

Based on this walking mechanism, when one side of the microrobot moves forward and the other side moves backward, or remains stationary, the robot can rotate in either the clockwise or counterclockwise direction. The rotational speed of the robot is determined by the rotational angle in a single step and the frequency [35].

4.5 Mechanism of the Grasping Motion

In the lying attitude, the microrobot can grasp small objects and carry them to a specified location using fingers I and J. First, the microrobot moves close to the object using legs A–H. Second, fingers I and J bend toward each other to grasp the object. Then the microrobot carries the object to the desired destination. In the standing attitude, the microrobot can also grasp small objects using the leg pairs B–F and C–G for this purpose, while legs A, D, E, and H provide the crawling and rotational motions.

4.6 Mechanism of the Floating Motion

When the frequency of the driving voltage is decreased to 0.3 Hz, the water around the IPMC actuators is electrolyzed. Air bubbles are generated and become attached to the leg surfaces, and the buoyancy of the microrobot is increased. In the lying attitude, four drivers and four supporters are used to electrolyze the water and implement floating motion. In the standing attitude, leg pairs A–E and D–H are used to implement floating motion. The tail fin can also be used to provide buoyancy, and to adjust the balance of the overall body while floating.

4.7 Mechanism of the Swimming Motion

In a similar manner to the BCF and MPF locomotion of fish, robots can be classified into body and/or caudal actuator (BCA) types, and median and/or paired actuator (MPA) types [38]. The proposed microrobot utilizes the BCA mode, which generates thrust by bending the caudal fin K, as shown in Fig. 13. The bending of the caudal fin provides oscillatory motion, and is triggered by a single IPMC actuator. A passive fin is attached to the free end of this actuator to increase the thrust.

5 Prototype Microrobot and Experiments

5.1 Prototype Microrobot

Based on the proposed structure, a prototype hybrid underwater microrobot with two motion attitudes was constructed, as shown in Fig. 19. The body was composed of two layers, to which eleven IPMC actuators were attached with wooden clips. Two IPMC fingers and a tail fin were attached to the first layer, while eight IPMC legs were attached to the second layer. Two SMA actuators were affixed to two sheets attached to the second layer. The prototype microrobot employed eight legs to walk, rotate, and float in two attitudes. Two fingers were utilized to implement grasping, and the tail fin was used for swimming. The control signals of the IPMC actuators were all square waves, in order to drive the actuators more efficiently [29].

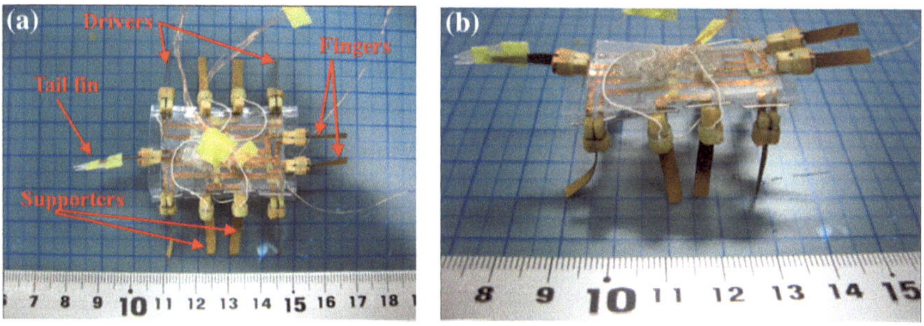

Fig. 19 The prototype microrobot (in air). **a** The lying attitude. **b** The standing attitude

In addition, two SMA actuators were employed to pull the two sheets and fold the eight legs below the body, to implement the attitude change. The prototype driving system consisted of an AVR atmega16 and twelve Omron G6 K-2P electric relays that were used as circuit changers to vary the input voltages. The microrobot received its control signals through enamel-covered wires with a diameter of 0.03 mm. The wires were soft enough for their resistance to be ignored [29].

5.2 Walking, Rotating, and Grasping Experiments in the Lying Attitude on an Underwater Flat

The walking experiments were conducted on a flat underwater surface. In these experiments, we varied the applied signals, and calculated the walking speed by recording the time required to cover a distance of 50 mm. The experiment was repeated five times for each set of control signals to obtain an average speed.

At a fixed current of 0.7 A, we carried out two groups of experiments with different applied voltages and frequencies. Figure 20 shows the experimental results for voltages of 4 and 6 V, which indicated that the walking speed was proportional to the input voltage, and that the walking motion was highly efficient in the control frequency range from 2 to 6 Hz. We can see that: (1) the walking speed increased with the input voltage; (2) at 6 V, a maximum speed of 3.6 mm/s was attained at 2.5 Hz; and (3) at 4 V, a maximum speed of 2.1 mm/s was attained at 2.5 Hz. When the frequency was higher than 7 Hz, the walking speed approached 0.

At a fixed frequency of 1 Hz, we also carried out three groups of experiments with applied voltages 3, 5 and 8 V. We obtained an average speed for every set of signals, varying the current as shown in Fig. 21. From the results, the walking speed was proportional to the applied current and input voltage. The microrobot required only low current and voltage for walking motion in the lying attitude.

Fig. 20 Experimental walking speeds with different frequencies

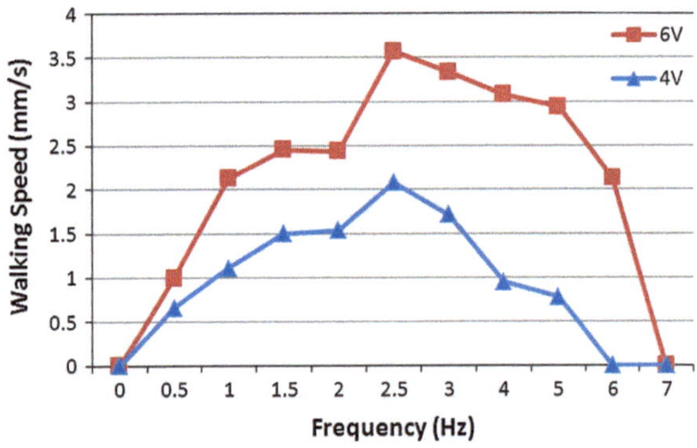

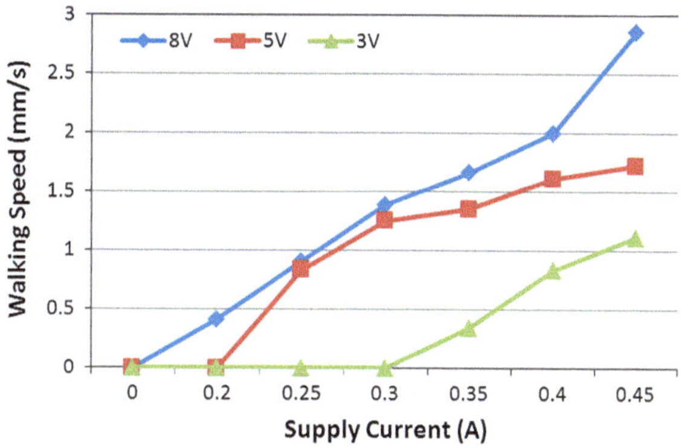

Fig. 21 Experimental walking speeds with different currents

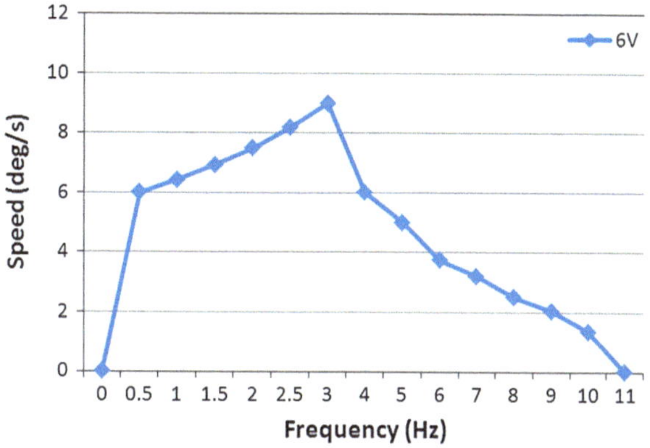

Fig. 22 Experimental rotating speeds

In the rotating experiments, we varied the control frequency from 0.5 to 11 Hz at a fixed voltage of 6 V and a fixed current of 1 A, and calculated the average rotational speeds. Figure 22 shows the experimental results, which indicated that the microrobot had a higher rotational speed in the frequency range from 0.5 to 4 Hz, and a maximum rotational speed of 9 °/s. When the control frequency was lower than 3 Hz, the rotational speed was proportional to the frequency, since the oscillatory amplitude was relatively large. However, when the control frequency was higher than 3 Hz, the rotational speed was inversely proportional to the frequency, since the rotational angle in a single step cycle became small, and the decreased displacement became a primary factor affecting the rotational speed.

In the lying attitude, the microrobot was able to use its two fingers to implement grasping motion. A hybrid walking, rotating, and grasping motion is shown in Fig. 23. First, the robot walked forward. Second, it rotated clockwise and opened its two fingers. Then it closed its fingers and rotated counterclockwise. Finally, it walked backward.

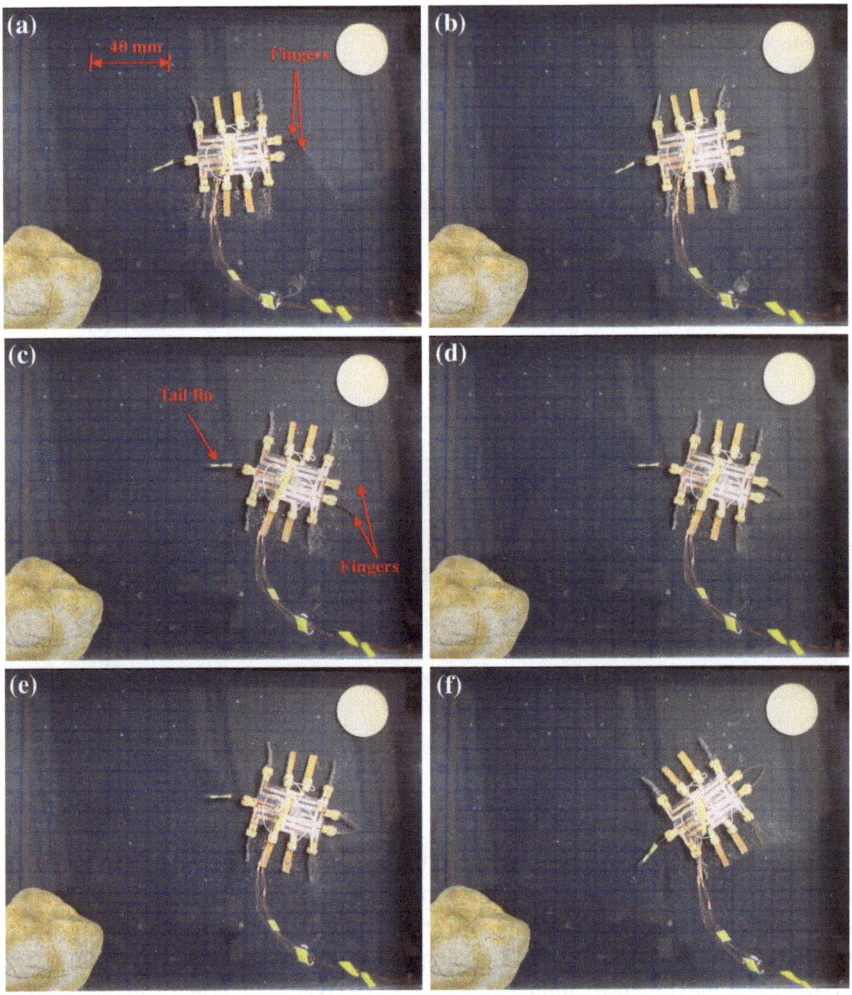

Fig. 23 Walking, rotating, and grasping motions. **a** Walking forward. **b** *Right* turning. **c** Two fingers open. **d** Grasping motion. **e** *Left* turning. **f** Walking backward

5.3 Floating Experiments Without Payloads

Legs A–H were used to electrolyze the water and implement floating motion. In the floating experiments, we varied the frequencies of the driving voltages and calculated the floating speed by recording the time required to float through a vertical distance of 100 mm. Figure 24 shows a video sequence of the floating motion.

At a fixed voltage of 6 V, we varied the control frequencies from 0.05 to 0.5 Hz. The experiment was repeated 5 times for each set of control signals to obtain an average speed. Figure 25 shows the experimental floating speeds for different

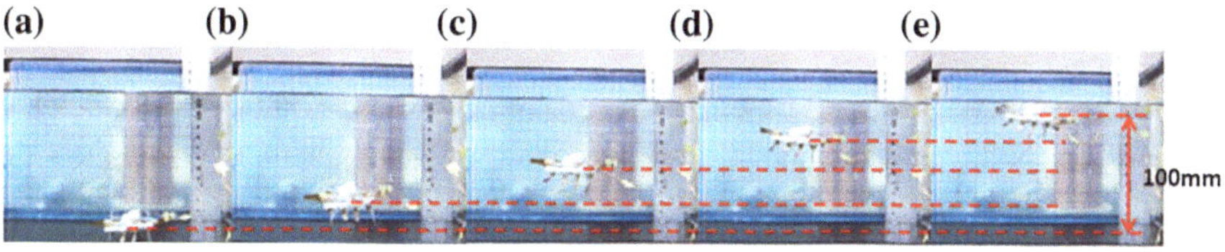

Fig. 24 Floating experiment. **a** Phase 1. **b** Phase 2. **c** Phase 3. **d** Phase 4. **e** Phase 5

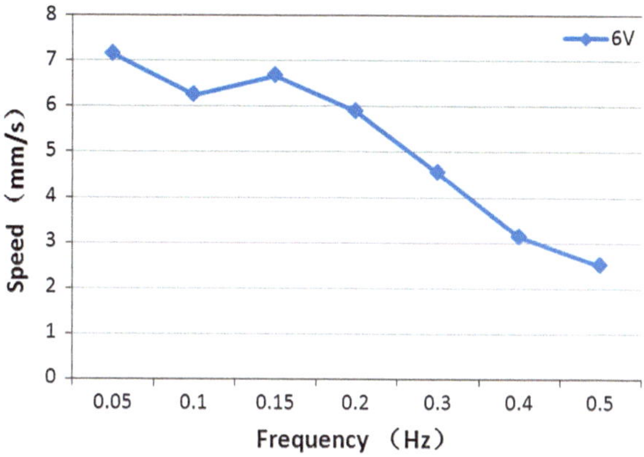

Fig. 25 Experimental floating speeds

frequencies. From the results, the average floating speed was inversely proportional to the control frequency, and the maximum speed was achieved with a frequency of 0.05 Hz.

5.4 Standing Experiments

In the standing experiments, we used the two SMA actuators to make the microrobot stand up, both in air and on the underwater flat. Figure 26 shows video sequences of the standing motion on the underwater flat, from the front and left-side perspectives. We carried out the experiments with a control voltage of 8 V and a maximum current of 1 A. An initially deformed SMA actuator can recover its predetermined low-temperature shape during heating, demonstrating the shape memory effect [39]. Therefore, thermal insulation is important for SMA actuators, especially in water. Accordingly, we sealed the two SMA actuators with elastic adhesive tape to achieve a better heating effect when they were triggered to shrink.

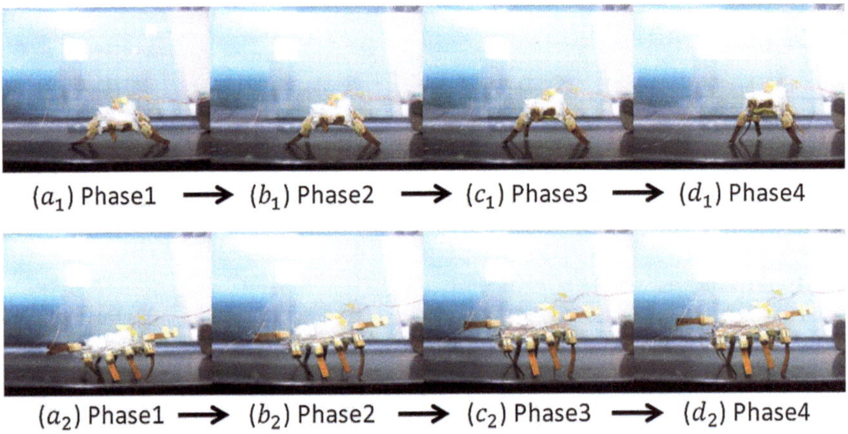

Fig. 26 Standing experiments on the underwater flat

5.5 Obstacle-Avoidance Experiment

To implement closed-loop control, we installed one short-range proximity sensor on the microrobot to detect an object or avoid an obstacle while walking or swimming. The proximity sensor used in the present research was 8 mm long and 5 mm wide, with a weight of 0.5 g. The distance measurement range was 0–60 mm, and the output voltage ranged from 150 mV to the power voltage [29]. The sensor signals were transmitted to a micro-AD board, which converted the voltages to digital values and sent them to the AVR. By utilizing the proximity sensor, the microrobot was able to detect an obstacle in front of it without any physical contact, and avoided it automatically. In the previous experiments, the microrobot avoided an obstacle by changing its walking direction. However, due to the low rotating efficiency of this unit while in a standing attitude, a long time was required to avoid a very wide obstacle via rotation. Therefore, the hybrid robot avoided the obstacle by floating instead. Figure 27 shows the object-avoidance experiment in the standing attitude. First, the microrobot walked toward the obstacle using legs A, D, E, and H driven by an input voltage of 6 V at a frequency of 1 Hz. When the distance between the microrobot and the obstacle decreased to about 10 mm, the proximity sensor detected the obstacle. The microrobot then stopped and floated upward.

5.6 Swimming Experiments in the Standing Attitude

The swimming experiments were carried out in the same water tank. To increase the oscillatory thrust, the swimming motion was evaluated in the standing attitude.

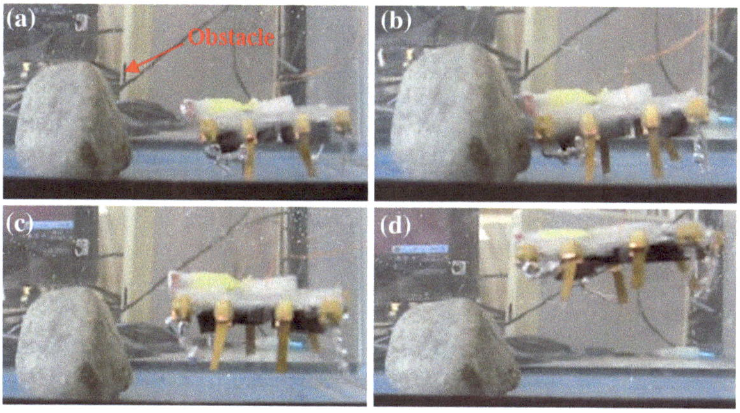

Fig. 27 *Obstacle* avoidance experiment. **a** Walking forward. **b** Detected the obstacle. **c** Floating upward. **d** Floating state

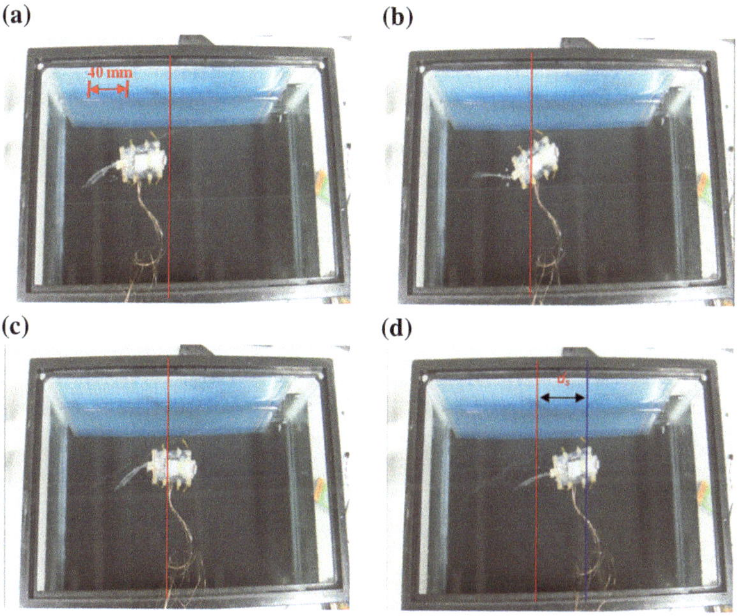

Fig. 28 Swimming experiment. **a** Initial position (*red line*). **b** *Left* bending. **c** *Right* bending. **d** Final position (*blue line*)

Water resistance increases in proportion to the cross-sectional area of the robot body, reducing the oscillatory amplitude of the body. On the other hand, increasing the oscillatory amplitude can reduce the effect of water resistance and increase the swimming speed. The IPMC actuator was actuated by a square-wave signal with a frequency of 0.5 Hz and an input voltage of 6 V. The swimming motion for one oscillatory cycle is shown in Fig. 28.

6 Results and Discussion

Generally speaking, compact structure, multifunctionality, flexibility, and precise positioning are considered incompatible characteristics in underwater microrobots [39, 40]. We have already designed several bio-inspired underwater robots with compact structures using IPMC and SMA actuators. These robots employ biomimetic locomotion to implement walking/rotating, surfacing/diving, grasping, and swimming motions. However, each of the units implements only some of these motions. To design a robot with multifunctionality, we need to integrate the above motions in a single robot. There are three types of underwater walking/rotating motions: inchworm-inspired, stick-insect-inspired, and lobster-inspired. Since the position precision of IPMC legs has not been high, in the present research we proposed an electromechanical model of an IPMC leg for position control [29]. Also, a novel hybrid structure with two motion attitudes was developed to adapt to different environments. Floating can be achieved via the electrolysis characteristics of IPMC, or via jellyfish-inspired or fish-bladder-inspired designs. Since the floating speeds are adjustable in all three of these methods, the first is the best choice to realize a compact structure. Swimming can be achieved via fish-inspired, snake-inspired, butterfly-inspired, or manta-ray-inspired designs. However, due to mechanism limitations, only a caudal actuator was suitable for our hybrid design. Accordingly, we used a single IPMC actuator to drive a passive fin in an oscillatory motion. Human-inspired, inchworm-inspired, and lobster-inspired finger locomotion have been proposed for grasping. Our new design not only inherited lobster-inspired finger locomotion, but also implemented inchworm-inspired grasping motion by changing its attitude from lying to standing. Also, a schematic diagram of the Father–son robot system is shown in Fig. 29.

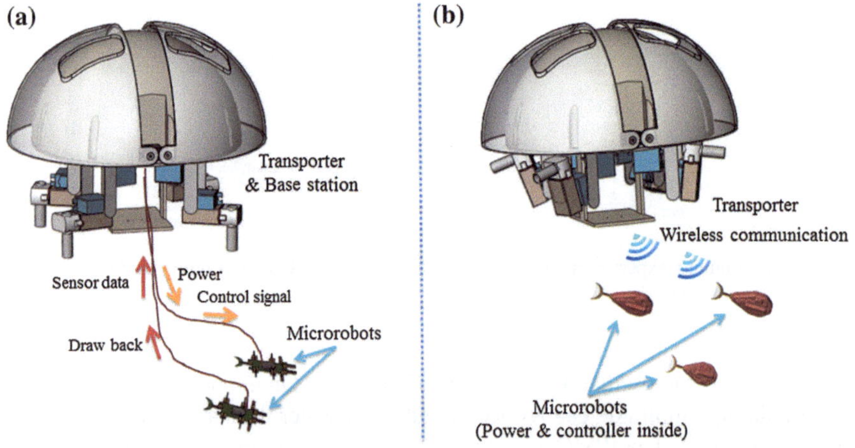

Fig. 29 A schematic diagram of the father–son robot system. **a** Wire communication. **b** Wireless communication

7 Conclusion

In this chapter, stick insect-inspired two-phase walking locomotion, jellyfish-inspired floating/diving locomotion, and inchworm-inspired crawling and grasping locomotion were discussed. The feasibility results for four previously developed prototype microrobots were then reviewed. The floating efficiency of a stick insect-inspired robot was not high. A jellyfish-like robot could not rotate and its walking motion was unsatisfactory, although its floating motion was improved. For an inchworm-inspired robot, there were large differences in the rotational radii of the outside four legs, leading to a large amount of slippage while rotating and low rotating efficiency. In addition, the inside six legs were used solely as fingers to grasp an object and could not contribute any buoyancy to the floating motion, so that the floating speed was slow. To inherit the multifunctions of the inchworm-inspired microrobot and overcome its disadvantages, we introduced a new lobster-like microrobot, intended for underwater exploration in a restricted space.

Then, we introduced a hybrid biomimetic microrobot with two motion attitudes to implement microrobot multifunctionality and flexibility for adaptation to complex underwater environments. In the lying attitude, the new robot implemented stick-insect-inspired walking/rotating motions using eight IPMC legs. These legs were also used to electrolyze the water for floating. Two lobster-inspired IPMC fingers were used to grasp small objects. According to the results of the walking experiments, the robot reached a maximum walking speed of 3.6 mm/s at a control frequency of 2.5 Hz and a fixed current of 0.7 A. The results of the floating experiments indicated that the robot could achieve a maximum floating speed at a control frequency of 0.05 Hz and a control voltage of 6 V. Driven by two SMA actuators, the robot could change its attitude from lying to standing on an underwater flat. In the standing attitude, the microrobot could implement inchworm-inspired walking/rotating using the four outside IPMC legs. The four inside legs were utilized as fingers to grasp large objects. While suspended in the water, the IPMC caudal fin actuated a passive fin to implement oscillatory motion, which provided propulsion for swimming. When equipped with a proximity sensor, the robot could detect and avoid obstacles automatically, either by rotating or by floating.

As the next step in our research, we will introduce the father–son robot system to solve microrobot design problems requiring as low speeds and short operating times in restricted operating areas.

Acknowledgments This research is partly supported by National Natural Science Foundation of China (61375094), Key Research Program of the Natural Science Foundation of Tianjin (13JCZDJC26200) and National High Tech. Research and Development Program of China (No.2015AA043202).

References

1. Shi L, Guo S, Asaka K (2011) Development of a new jellyfish-type underwater microrobot. Int J Robot Autom 26(2):229–241
2. Guo S, Shi L, Asaka K, Li L (2009) Experiments and characteristics analysis of a bio-inspired underwater microrobot. In: Proceedings of the 2009 IEEE international conference on mechatronics and automation, Changchun, China, 9–12 August 2009
3. Lin X, Guo S (2012) Development of a spherical underwater robot equipped with multiple vectored water-jet-based thrusters. J Intell Rob Syst 67(3–4):307–321
4. Shi L, Guo S, Mao S, Li M, Asaka K (2013) Development of a lobster-inspired underwater microrobot. Int J Adv Rob Syst 10:1–15. doi:10.5772/54868
5. Cavallo E, Michelini R, Filaretov V (2004) Conceptual design of an AUV equipped with a three degrees of freedom vectored thrusters. J Intell Robot Syst 39(4):365–391
6. Duchemin O, Lorand A, Notarianni M, Valentian D, Chesta E (2007) Multi-channel hall-effect thrusters: mission applications and architecture trade-offs. In: Proceedings of the 30th international electric propulsion conference, Florence, Italy, 17–20 September 2007
7. Behkam B, Sitti M (2006) Design methodology for biomimetic propulsion of miniature swimming robots. J Dyn Syst Meas Contr 128(1):36–43
8. Zhang W, Guo S, Asaka K (2006) A new type of hybrid fish-like microrobot. Int J Autom Comput 3(4):358–365
9. Heo S, Wiguna T, Park H, Goo N (2007) Effect of an artificial caudal fin on the performance of a biomimetic fish robot propelled by piezoelectric actuators. J Bionic Eng 4(3):151–158
10. Villanueva A, Joshi K, Blottman J, Priya S (2010) A bio-inspired shape memory alloy composite (BISMAC) actuator. Smart Mater Struct 19(025013):1–17
11. Wang Z, Hang G, Li J, Wang Y, Xiao K (2008) A micro-robot fish with embedded SMA wire actuated flexible biomimetic fin. J Sens Actuators A Phys 144(2):354–360
12. Lee S, Kim K, Park I (2007) Modeling and experiment of a muscle-like linear actuator using an ionic polymer–metal composite and its actuation characteristics. J Smart Mater Struct 16(3):583–588
13. Liu S, Lin M, Zhang Q (2008) Extensional ionomeric polymer conductor composite actuators with ionic liquids. Electroact Polym Actuators Devices (EAPAD) 6927:69270H
14. Nakadoi H, Sera A, Yamakita M, Asaka K, Luo Z, Ito K (2007) Integrated actuator-sensor system on patterned IPMC film: consideration of electronic interference. In: Proceedings of the 2007 4th IEEE international conference on mechatronics, 8–10 May 2006
15. McGovern ST, Spinks GM, Xi B, Alici G, Truong V, Wallace GG (2008) Fast bender actuators for fish-like aquatic robots. Electroact Polym Actuators Devices (EAPAD) 6927:69271L
16. Kamamichi N, Yamakita M, Asaka K, Luo Z (2006) A snake-like swimming robot using IPMC actuator/sensor. In: Proceedings of the 2006 IEEE international conference on robotics and automation, Orlando, Florida, USA, 15–19 May 2006
17. Kim B, Kim D, Jung J, Park J (2005) A biomimetic undulatory tadpole robot using ionic polymer–metal composite actuators. J Smart Mater Struct 14:1579–1585
18. Ye X, Su Y, Guo S, Wang L (2008) Design and realization of a remote xontrol centimeter-scale robotic fish. In: Proceedings of the 2008 IEEE/ASME international conference on advanced intelligent mechatronics, Xi'an, China, 2–5 July 2008
19. Yim W, Lee J, Kim KJ (2007) An artificial muscle actuator for biomimetic underwater propulsors. J Bioinspiration Biomimetics 2(2):S31–S41
20. Ye X, Hu Y, Guo S, Su Y (2008) Driving mechanism of a new jellyfish-like microrobot. In: Proceedings of 2008 IEEE International Conference on Mechatronics and Automation, Takamatsu, Japan, 5–8 Aug 2008
21. Kamamichi N, Kaneda Y, Yamakita M, Asaka K, Luo ZW (2003) Biped walking of passive dynamic walker with IPMC linear actuator. In: Proceedings of SICE Annual Conference in Fukui, Fukui University, Japan, 4–6 Aug 2003

22. Shi L, Guo S, Mao S, Yue C, Li M, Asaka K (2013) development of an amphibious turtle-inspired spherical mother robot. J Bionic Eng 10(4):446–455
23. Guo S, Mao S, Shi L, Li M (2012) Development of an amphibious mother spherical robot used as the carrier for underwater microrobots, In: Proceedings of the 2012 ICME International Conference on Complex Medical Engineering, Kobe, Japan, 1–4 July 2012
24. Guo S, Mao S, Shi L, Li M (2012) Design and kinematic analysis of an amphibious spherical robot. In: Proceedings of 2012 IEEE International Conference on Mechatronics and Automation, Chengdu, China, 5–8 August 2012
25. Zhang W, Guo S, Asaka K (2006) Development of an underwater biomimetic microrobot with both compact structure and flexible locomotion. J Microsyst Technol 13(8):883–890
26. Shi L, Guo S, Asaka K (2010) A novel multifunctional underwater microrobot. In: Proceedings of the 2010 IEEE International Conference on Robotics and Biomimetics, Tianjin, China, 14–18 Dec 2010
27. Shi L, Guo S, Asaka K (2012) A novel jellyfish- and butterfly-inspired underwater microrobot with pectoral fins. Int J Robot Autom 27(3):276–286
28. Gong Y, Fan J, Tang C, Tsui C (2011) Numerical simulation of dynamic electro-mechanical response of ionic polymer-metal composites. J Bionic Eng 8:263–272
29. Shi L, Guo S, Li M, Mao S, Xiao N, Gao B, Song Z, Asaka K (2012) A novel soft biomimetic microrobot with two motion attitudes. Sensors 12(12):16732–16758
30. Shi L, Guo S, Kudo H, Asaka K (2012) Development of a venus flytrap-inspired robotic flytrap. In: Proceedings of the 2012 IEEE international conference on robotics and biomimetics, Guangzhou, China, 11–14 Dec 2012
31. Gao B, Guo S (2010) Development of an infrared ray controlled fish-like underwater microrobot. In: Proceedings of the 2010 IEEE International Conference on Automation and Logistics, Hong Kong and Macau, China, 16–20 Aug 2010
32. Gao B, Guo S, Ye X (2011) Motion-control analysis of ICPF-actuated underwater biomimetic microrobots. Int J Mechatron Autom 1(2):79–89
33. Shi L, Guo S, Asaka K (2011) A novel butterfly-inspired underwater microrobot with pectoral fins. In: Proceedings of the 2011 IEEE International Conference on Mechatronics and Automation, Beijing, China, 7–10 Aug 2011
34. Shi L, Guo S, Asaka K (2011) A bio-inspired underwater microrobot with compact structure and multifunctional locomotion. In: Proceedings of 2011 IEEE/ASME International Conference on Advanced Intelligent Mechatronics (AIM 2011), Budapest, Hungary, 3–7 July 2011
35. Guo S, Shi L, Xiao N, Asaka K (2012) A biomimetic underwater microrobot with multifunctional locomotion. Robot Auton Syst 60(12):1472–1473
36. Gao B, Guo S (2011) Dynamic mechanics and electric field analysis of an ICPF actuated fish-like underwater microrobot. In: Proceeding of the IEEE International Conference on Automation and Logistics, Chongqing, China, 15–18 Aug 2011
37. Guo S, Li M, Shi L, Mao S (2012) A smart actuator-based underwater microrobot with two motion attitudes. In: Proceedings of 2012 IEEE International Conference on Mechatronics and Automation, Chengdu, China, 5–8 Aug 2012
38. Chu W, Lee K, Song S, Han M, Lee J, Kim H, Kim M, Park Y, Cho K, Ahn S (2012) Review of biomimetic underwater robots using smart actuators. Int J Precis Eng Manuf 13(7):1281–1292
39. Shi L, Guo S, Asaka K (2012) Modeling and experiments of IPMC actuators for the position precision of underwater legged microrobots. In: Proceedings of the 2012 IEEE International Conference on Automation and Logistics, Zhengzhou, China, 15–17 Aug 2012
40. Park I, Kim S, Kim D, Kin K (2007) The mechanical properties of ionic polymer-metal pomposities. J Electroact Polym Actuators Devices (EAPAD) 6524:65241R

Multiple Autonomous Robotic Fish Collaboration

Guangming Xie, Long Wang and Yonghui Hu

Abstract Most of previous work on multiple robots collaboration is focused on terrestrial robots and seldom deals with underwater applications due to the uncertainties and complexity in a hydroenvironment. In this chapter, three typical collaboration problems with multiple autonomous robotic fish are investigated. The three problems includes target tracking and collision avoidance, formation control, and cooperative transportation. For the first problem, a situated-behavior-based decentralized control is employed on each robotic fish according to its visual data. On dealing with motion planning of the fish during target tracking and collision avoidance, a control law by a combination of an attractive force toward a target and a repulsive force for collision avoidance is utilized. For the formation control problem, leader-following based framework is adopted. Each follower robot estimates the position and orientation angle of its leader with a fast color-tracking vision system, and establishes a Bezier trajectory between its current position and the position of its leader robot. For the cooperative transportation problem, an underwater box-pushing task is designed in which three autonomous robotic fish that sense, plan and act on its own move an elongated box from some initial location to a goal location. The whole task is decomposed into three subtasks and assigned to capable robotic fish. The robotic fish coordinate through explicit communications and distribute the subtasks with a market-based dynamic task allocation method.

G. Xie (✉) · L. Wang
Intelligent Control Laboratory, Department of Mechanics and Space Technologies, College of Engineering, Peking University, Beijing 100871, People's Republic of China
e-mail: xiegming@pku.edu.cn

L. Wang
e-mail: longwang@pku.edu.cn

Y. Hu
School of Control and Computer Engineering, North China Electric Power University, Beijing 102206, People's Republic of China
e-mail: huyhui@gmail.com

© Springer-Verlag Berlin Heidelberg 2015

R. Du et al. (eds.), *Robot Fish*, Springer Tracts in Mechanical Engineering,
DOI 10.1007/978-3-662-46870-8_11

1 Introduction

Multirobot systems are being used increasingly in highly dynamic or adversarial environments to address complex tasks, such as planetary exploration [1], monitoring and surveillance [2], search and rescue [3], mapping of unknown or partially known environments [4], and transportation of large objects [5]. Like humans working in a team to achieve a common goal or a swarm of ants foraging for and hauling food together, a group of cooperating robots can perform certain tasks better than a single robot. By decomposing the task into subtasks and executing them concurrently, multi-robots can accomplish the task in a more efficient and robust manner. Moreover, many tasks not executable by a single robot can be tackled by a robot team by taking advantages of distributed sensing and actuation. However, the design and deployment of multirobot systems in real-world applications represent a formidable scientific challenge. Many problematic issues like group architecture, resource conflict, dynamic and unpredictable environments, noisy perception and limited communication bandwidth and range have to be dealt with in order to achieve effective teamwork.

Majority of multirobot systems are implemented in terrestrial or aerial environment, and few results have been obtained on underwater robots. Unprecise motion control due to the disturbance of waves and unknown currents, the lack of effective acoustic and optical sensors, unreliable underwater communication and high operational costs make it difficult to realize multirobot cooperation in hydro-environment. With the increasing human demand for exploitation and utilization of ocean resources, more research efforts should be devoted to the development of cooperating underwater robots. In recent years, the biomimetic robotic fish, as a novel miniature underwater vehicle, has progressed considerably [6, 7]. By emulating the swimming mechanisms of fish in nature, robotic fish can obtain enhanced locomotion performances over conventional screw-propelled underwater vehicles, such as high efficiency, great agility, increased noise reduction and station-keeping ability. Robotic fish can play an important role in various underwater tasks, especially those that require operations in cluttered environments and in unsteady flow. Most studies of robotic fish focus on the hydrodynamic modelling of swimming fish [8, 9] and building of artificial fish-like devices [10–12], cooperative control of multiple robotic fish has seldom been investigated. The significance of the study of multiple robotic fish cooperation is twofold. From the engineering perspective, multiple cooperating robotic fish provide a feasible solution to a variety of complex underwater missions, which are intractable for a single robotic fish or difficult to be executed by other underwater robots. For example, in naval reconnaissance task multiple robotic fish can improve the performance of the task execution by sharing collected information while reduce the possibility of detection by pretending to be a real fish school. From the scientific perspective, the schooling behaviors of fish in nature can be recorded and better understood with the help of multiple robotic fish. The self-organizing mechanisms of fish can be emulated and verified with multiple robotic fish governed by a localized control regimen, and perhaps the grouping behaviors of fish might be deliberately harnessed to produce certain global patterns via multiple fish-like robots swimming together with live fish.

Research in the field of multirobot systems has resulted in a number of canonical task domains, such as box-pushing [13], formation control [14, 15] and foraging [16]. This chapter presents three typical collaboration tasks carried out by multiple autonomous robotic fish. The first one deals with vision-based target tracking and collision avoidance, in which two robotic fish are required to track one target and at the same time adopt some collision-avoiding strategies without colliding with each other. The aquatic environment poses two major difficulties for the fulfillment of the task. Firstly, the target drifts everywhere due to disturbances, therefore it is quite difficult to predict the position of the target. Secondly, the waves produced by fish lead to uncertainty and inaccuracy to precise locomotion control of the robotic fish. Considering the difficulties in the task, a situated-behavior method is employed to divide the environment into a set of complete and exclusive situations according to the camera data. For each situation, a specific behavior is designed. In designing the situation associated behaviors, a control law that is a combination of an attractive force toward a target and a repulsive force for collision avoidance is utilized. The second task is about formation control, in which multiple robotic fish achieve predetermined formations with a leader-following approach. A Bezier curve between leader and follower robot, which is tangent to current paths of both leader and follower robots, is introduced based on the pose estimation of the leader with a fast color-tracking vision system of the follower. The Bezier curve is optimized by minimizing the curvature's changing rate. Using the penalty function method, the problem of nonlinear constrained optimization is converted into unconstrained, and one-dimensional search is adopted to solve this unconstrained optimization problem. Based on optimized scale factor, the optimal Bezier curves allow the robotic fish to follow a smooth and stable trajectory, satisfying the nonholonomic constraint of the robotic fish. The desired angular velocity of the follower robot can be obtained with the optimal Bezier trajectory. Both numerical simulations and physical robot experiments are performed to validate the proposed approach. The third task is concerned with cooperative underwater box-pushing, in which a decentralized transportation strategy is employed. Without any global information about the posture of the box and themselves, the robotic fish cooperate by sharing visual information and execution ability in order to move the box from its initial location to the designated goal location. The mission is decomposed into the following three subtasks that can be executed each by a single robotic fish: an observing subtask for pose perception of the box at the goal location and two pushing subtasks for pushing of the left and right ends of the box. The robotic fish fulfill the subtasks by executing consecutively a series of behaviors. To determine which robotic fish should execute which subtask, a market-based task allocation method is used. The robotic fish submit bids to compete for the auctioned subtask. To cope with unexpected changes in the environment and the limited sensing range of the robotic fish, subtasks assigned to the robotic fish can be auctioned again in case the robotic fish is no longer competent for the subtask it has committed to.

The remainder of this chapter is structured as follows. Section 2 presents the prototype development and motion control of the robotic fish. The tasks of target tracking and collision avoidance, formation control, and cooperative transportation are addressed in Sects. 3, 4 and 5, respectively. Finally, conclusions are provided in Sect. 6.

2 Autonomous Robotic Fish Prototype

2.1 Mechanical Design

Modeled after boxfish that is characterized by inflexible body and utilizes MPF mode for propulsion and caudal oscillations as auxiliary locomotion means, the robotic fish consists of a rigid main body, a pair of pectoral fins and a caudal fin. The main body, which is a roughly rectangular, waterproofed hull, provides housings for the power, electronics and actuators. Each propulsor can perform 1-DOF movements and is actuated by a servomotor (Hitec HS-5955TG) that is fixed on the bottom chassis. The reciprocatory rotation of the servomotors are transmitted to the outside through dynamic sealing structure filled with grease. The pectoral fins with the approximate shape of NACA-0012 profile are attached to the joints on the side, and the lunate tail fin is linked to the vertical joints with connecting pieces. The rotatory range of the tail fin is limited to $\pm 90°$, while that of the pectoral fins is expanded to $\pm 180°$ through transmission of gear sets of $2:1$ ratio. A pinhole CMOS camera, as the only exteroceptive sensor is installed at the mouth position with a transparent window glued to the hull for waterproof purpose. For most fishes, the center of mass is located above the center of buoyancy and as a result they are hydrostatically unstable [9, 17]. However, it is hard for robotic fish to generate the necessary trimming forces and powered correction forces to stabilize and hold posture, therefore the robotic fish is designed to be hydrostatically stable through lower placement of the mass center. The density of the robotic fish has been designed to be close to that of water through careful calculations, so that little trimming weight or foam can be added to accomplish neutral buoyancy. Figure 1a shows the photograph of the robotic fish prototype and its mechanical configuration is illustrated in Fig. 1b.

2.2 Electronics and Sensor

The robotic fish is designed for autonomous operation such that it is equipped with onboard power, embedded processor, image sensor (OV7620 from OmniVision) and a duplex wireless communication module (GW100B from Unitel Pty Ltd) as the user interface for human-robot interaction. Four rechargeable Ni–Cd cells of 2700 mAh capacity provide the robotic fish about 1 h power autonomy. The control unit is a microcontroller S3C2440 that incorporates a high-performance 32-bit

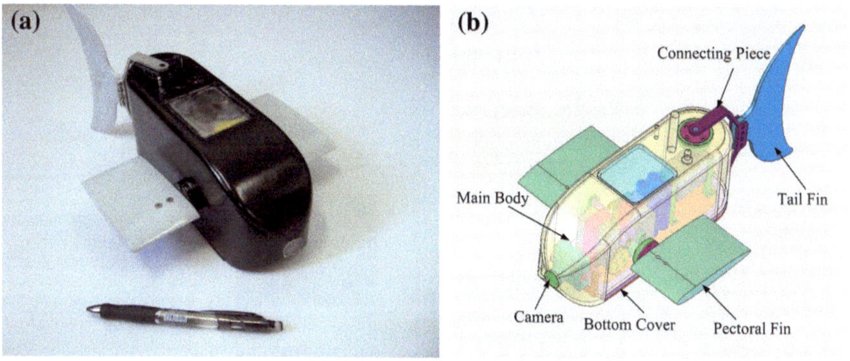

Fig. 1 Prototype and mechanical configuration of autonomous robotic fish. **a** Prototype, **b** mechanical configuration

RISC, ARM920T CPU core running at 400 MHz and a wide range of peripherals from Samsung Electronics. The onboard memory includes 64 MB SDRAM used during program execution and 64 MB Nand Flash for permanent data and code storage. The microcontroller captures image data in YCrCb 4:2:2 format at 320×240 resolution and does real-time image processing for perception of the environment. Three PWM (Pulse Width Modulation) signals are generated by the microcontroller to control the motion of the joints. Figure 2 illustrates hardware architecture of the control system.

2.3 Software

The software running on the robotic fish is comprised of two parts: a boot program named U-boot (universal bootloader) and the application code. U-boot runs immediately when the robotic fish powers up, and then waits a few seconds for the user to update application code through the wireless communication module. By default, U-boot will transfer the control to the application code that is already stored in Nand Flash memory if the user didn't interrupt. The use of U-boot greatly simplifies and accelerates the Flash burning process, which conventionally involves taking apart the bottom chassis, pulling out cables and resealing of the fish body.

2.4 Swimming Locomotion Control

The robotic fish swims by oscillatory movements of the tail and pectoral fins. Since sinusoidal signals can generate smooth oscillations and allow flexible and easy adjustment of joint angles, we model the swimming locomotion as sinusoidal variation of the robot's joint angles. Each joint of the robotic fish oscillates in a harmonic manner according to the following equation:

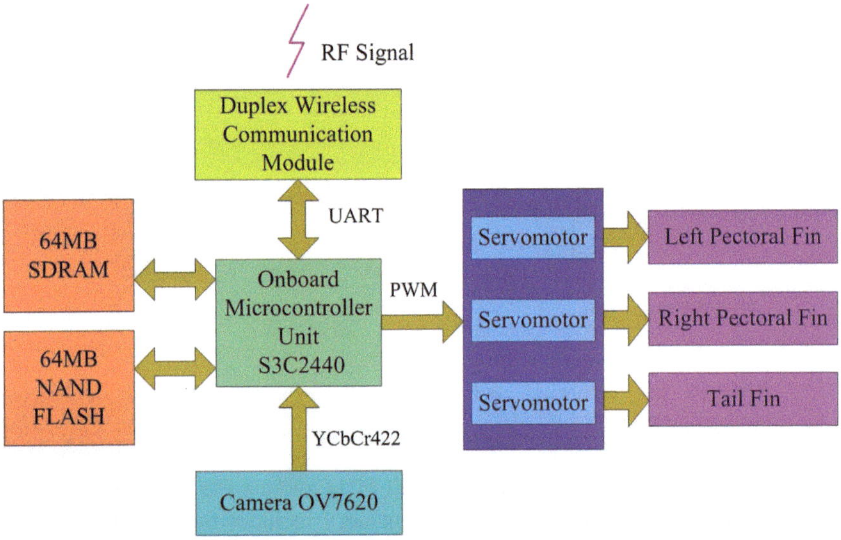

Fig. 2 Hardware architecture of the control system

$$\theta_i(t) = \bar{\theta}_i + A_i \sin(2\pi f_i t + \phi_i) \quad (i = 1, 2, 3) \tag{1}$$

where $\theta_i(t)$ is the angular position of the ith joint at time t, $\bar{\theta}_i$ denotes the angular offset, A_i represents the oscillatory amplitude of the joint angle, and f_i indicates the frequency. The swimming speed of the robotic fish can be adjusted by modulating the value of the frequency f and the amplitude A. The angular offset $\bar{\theta}$ can be used as a strategy for maneuvering and three-dimensional swimming of the robotic fish. The 1st joint, 2nd joint and the 3rd joint are used to drive the left fin, the right fin and the tail fin of the robotic fish, respectively.

Both the tail fin and the pectoral fins of the robotic fish can generate propulsion and maneuvering forces and through coordinated control of the propulsors, a great diversity of swimming gaits can be realized. Based upon the propulsors used, the swimming can be classified into two basic modes: **BCF** (body and/or caudal fin) mode and **MPF** (median and/or paired fin) mode, although the combined use of tail and pectoral fins can produce more complex movements. Having analyzed the governing dynamics of the robot's motion, several typical swimming gaits can be empirically designed. Figure 3 illustrates some typical swimming gaits and the realization of each pattern is described as follows:

- **BCF forward swimming**, by oscillating only the tail fin $\left(\bar{\theta}_3 = 0\right)$, with the pectoral fins held parallel to the horizontal plane functioning to enhance stability $\left(\bar{\theta}_1 = \bar{\theta}_2 = 0,\ A_1 = A_2 = 0\right)$.
- **BCF turning in advancing**, by superimposing a non zero angular offset $\bar{\theta}_3$ on the oscillation of the tail joint.

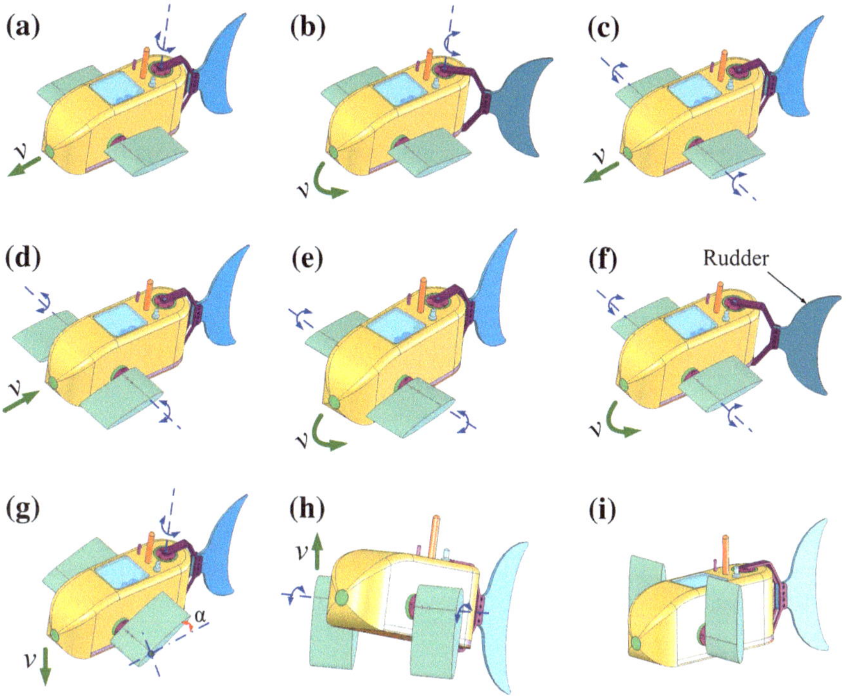

Fig. 3 Illustrations of typical swimming gaits designed for the robotic fish. **a** BCF forward swimming, **b** BCF turning in advancing, **c** MPF forward swimming, **d** MPF backward swimming, **e** MPF turning, **f** hybrid turning, **g** submerging, **h** ascending, **i** braking

- **MPF forward and backward swimming**, by the synchronized oscillations of the paired pectoral fins around the horizontal plane (i.e., $\bar{\theta}_1 = \bar{\theta}_2 = 0$ for forward, $\bar{\theta}_1 = \bar{\theta}_2 = \pi$ for backward), with caudal fin held straight $(\bar{\theta}_3 = 0,\ A_3 = 0)$.
- **MPF turning**, by the differentiation of hydrodynamic forces between the pectoral fins, typically producing anteriorly directed force on one side and posteriorly directed force on the other side ($\bar{\theta}_1 = 0, \bar{\theta}_2 = \pi$, or $\bar{\theta}_1 = \pi, \bar{\theta}_2 = 0$) and stopping the oscillations of the caudal fin ($A_3 = 0$).
- **Hybrid turning**, using synchronized pectoral fins flapping for thrust generation and tail fin as a rudder $(\bar{\theta}_3 \in [-\pi/3, \pi/3],\ A_3 = 0)$.
- **Submerging and ascending**, by adjusting the attack angle α of the pectoral fins when attaining a higher swimming speed [e.g. submerging while $\bar{\theta}_1 = \bar{\theta}_2 \in (0, \pi/2), A_1 = A_2 = 0$], or actuating the pectoral fins' synchronized oscillations around the vertical plane functioning to generate lift forces (e.g. ascending while $\bar{\theta}_1 = \bar{\theta}_2 = -\pi/2$).
- **Braking**, by sudden rotation of the pectoral fins to a position perpendicular to the body $(\bar{\theta}_1 = \bar{\theta}_2 = \pi/2,\ A_1 - A_2 = 0)$.

3 Underwater Target Tracking and Collision Avoidance

3.1 Task Description

The task is depicted as follows: two autonomous robotic fish are required to track a target (water polo) without colliding with each other respectively in a rectangular swimming tank. This task will be decomposed into two subtasks of water-polo-tracking and collision-avoiding.

3.2 Vision Processing

The vision module is responsible for extracting interested information from the camera of the robotic fish. The vision processing is based on color information. The underwater images from the camera are digitized in YCbCr color space, and color thresholds that were learned by sampling offline are then applied to the images. Inspired by [18–22], fast image segmentation, object recognition and localization are processed on the underwater images captured from the onboard camera of the robotic fish.

3.2.1 Color-Based Image Segmentation

A YCbCr image, as captured from the camera, is segmented into regions representing one color class each. Since color is the key feature used for object recognition, all important objects in our task have distinct and unique colors (e.g. water polo with red color and robotic fish with black).

Without regard to the luminance component Y that changes dramatically with ambient light, a 2-D lookup table is used to perform the mapping from CbCr pixel values to symbolic color class [19]. The lookup table is indexed by the raw Cb and Cr values of the pixel. Each entry of the lookup table stores the index number for the symbolic color to assign to the pixel, otherwise set to "white" color if the pixel is background. The thresholds are learned from example images offline. The color segmentation process uses the threshold table on each pixel of the image to classify the image.

After the color thresholding, the vision system then carries out the following steps: pixel connection and region merging [23]. The connection procedure scans adjacent rows and merges runs (horizontal neighboring pixels of the same color) to make a region. The statistics gathered for a region include: bounding box, centroid, and area. The region merging process is done with the calculation of the region's statistic characteristics. The criterion for deciding to merge the regions is the resulting density of the combined region. Two regions are combined into one region if the combined pixel area has a density occupying the area of new bounding box above a threshold for that color class. The purpose of region merging is to combine several nearby smaller regions of the same color into a single larger region in order to identify object and remove noise.

3.2.2 Object Recognition and Feature Extraction

The objects, including the water polo and the other fish, need to be recognized. We recognize these objects with several filters according to their characteristics. And the interested information of the objects in the image can be extracted. To illustrate the process, the recognition and feature extraction of the water polo are presented below.

The water polo is found by scanning through the candidates of red regions and selecting the one with the largest area. A filter is used to check if the region is large enough. If the region's area does not reach a threshold, we abandon it, otherwise we regard it as the target. Figure 4 shows the result of the object recognition.

The X_i'-axis and Y_i'-axis of image coordinate system are illustrated in Fig. 4b. The origin of the image coordinate system is the center of the image. Once the target has been detected, the coordinates of its lowest point in the image should be extracted for its localization. As shown in Fig. 4b, the coordinates $({}^i x'_{\mathbf{m}}, {}^i y'_{\mathbf{m}})$ of the target's lowest point $\mathbf{m}$ in the processed image can be obtained by searching the target region.

3.2.3 Object Localization

In object localization, we use the 2-D regions in the camera image to estimate the position of relevant objects such as the target or the other fish in the 3-D coordinate system of the observing robotic fish. With the help of the intrinsic camera parameters, image coordinates can be converted into robot coordinates. The robot

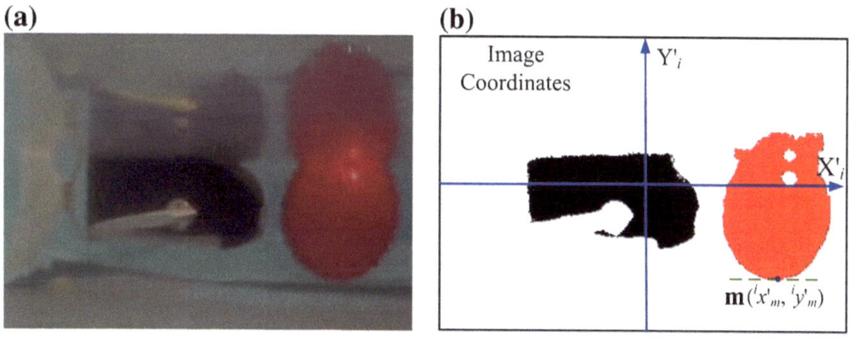

Fig. 4 Object recognition for a water polo and robotic fish. **a** Sample underwater image. **b** Image after processing [water polo (*red*), robotic fish (*black*)]

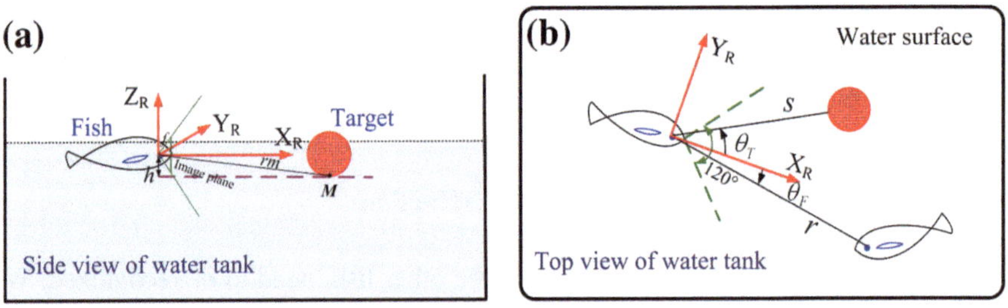

Fig. 5 **a** Estimate of the position of water polo relative to the robotic fish. **b** Estimate of the distance and angular direction in robot coordinate system

coordinate system (X_R, Y_R, Z_R) is fixed to the observing robotic fish, shown in Fig. 5. Notice that the mounted camera has a wide viewing angle of 120°.

Taking the target for example, in order to estimate the location of the water polo, we detect its lowest point **m** in the corresponding processed image [see Fig. 4b]. Since both the robotic fish and the target are slightly buoyant, the vertical distance h from the observing camera to the lowest point of the water polo is fixed. This point **m** and the optical center of the camera define a viewing ray **rm**. Its intersection with the lowest horizontal plane of the water polo yields an estimate of the position of water polo relative to the robotic fish, as illustrated in Fig. 5a. The 3-D point **M** in robot coordinate system corresponding to the observed image point $\mathbf{m} = [{}^i x'_{\mathbf{m}}, {}^i y'_{\mathbf{m}}]^T$ in Fig. 4b is given by

$$\mathbf{M} = [{}^R x_{\mathbf{M}}, {}^R y_{\mathbf{M}}, {}^R z_{\mathbf{M}}]^T = \frac{-h}{{}^i y'_{\mathbf{m}}} [f, {}^i x'_{\mathbf{m}}, {}^i y'_{\mathbf{m}}]^T \tag{2}$$

where f denotes the focal length of the camera. The point **m** given in image coordinate system can easily be expressed in the coordinate system of the robot, as long as the pose of the robot's camera relative to the target is given in robot coordinate system. Based on the estimation of 3-D point **M**, as shown in Fig. 5b, the distance s and angular direction θ_T to the target in $X_R Y_R$-plane of the robot coordinate system can then be computed by

$$s = \sqrt{{}^R x_{\mathbf{M}}^2 + {}^R y_{\mathbf{M}}^2} \tag{3}$$

$$\tan \theta_T = \frac{{}^R y_{\mathbf{M}}}{{}^R x_{\mathbf{M}}} \tag{4}$$

In the same way of target localization, the distance **r** and angular direction θ_F to the other fish in robot coordinate system can also be computed.

3.3 Situated-Behavior Based Decentralized Control

To simplify the difficulty of precise motion control in underwater environment, decentralized control of multiple robotic fish is designed with a situated-behavior method [24]. The situated-behavior mechanism is a design methodology which divides the environment into a set of complete and exclusive situations, and for each situation, an associated behavior is applied. The designed behaviors can also be decomposed into several low-level behaviors, such as avoid-collision behavior, which can be shared by the situation associated behaviors directly. Next, taking fish A for example, the situated-behavior methodology for two robotic fish will be described.

3.3.1 Situations

According to its visual data, four primary situations are defined for robotic fish A:

- **NFT (no fish or target) situation**: No fish or target appears in fish A's sight.
- **OF (only fish) situation**: Only fish B appears in fish A's sight.
- **OT (only target) situation**: Only the target appears in fish A's sight.
- **BFT (both fish and target) situation**: Both fish B and the target appear in fish A's sight.

Figure 6 illustrates all situations in the environment.

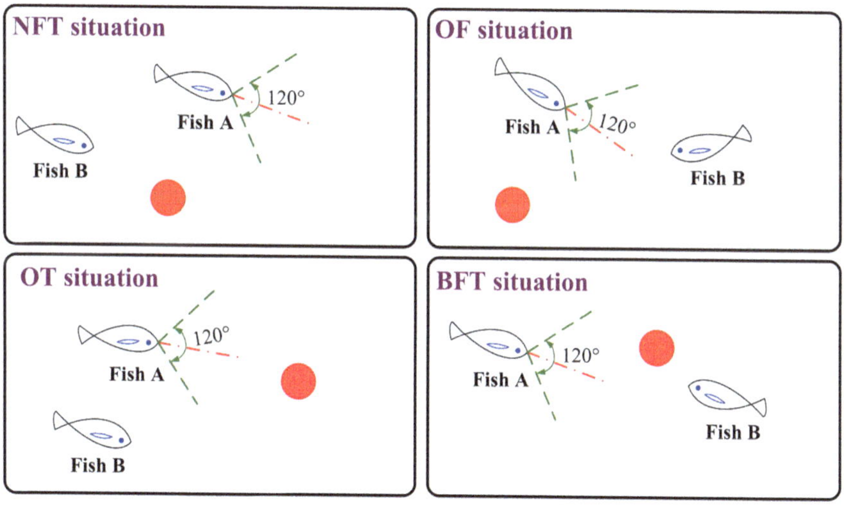

Fig. 6 Illustration of the situations

3.3.2 Situation Associated Behavior Design

The design of the behaviors is required to drive the robotic fish to its target and to force the robotic fish away from the other one. An associated behavior is designed for each situation.

- **B_{NFT}**: In NFT situation, fish A searches for the target by adopting *MPF-turning* gait clockwise or counterclockwise, according to the previous position of the target in its field of view (FOV), namely search-for-target behavior.
- **B_{OF}**: In OF, fish A searches for the target by adopting *MPF-turning* gait clockwise or counterclockwise, to the opposite side of the other fish's position. The associated behavior with OF situation can be decomposed into two low-level behaviors: search-for-target behavior and avoid-collision behavior.
- **B_{OT}**: In OT, fish A goes to the target by adjusting its orientation towards the target direction, namely go-to-target behavior.
- **B_{BFT}**: In BFT, fish A goes to the target and avoids collision with the other fish by integrating the attractive force from its target and the repulsive force from the other fish. The associated behavior with BFT situation can be decomposed into two low-level behaviors: go-to-target behavior and avoid-collision behavior.

The decision-making process is shown in Fig. 7, which illustrates the behaviors associated with the situations. As for each autonomous robotic fish, the inputs of the decision-making module are the target and the other fish's information, including the locations and distance of the target and the other fish to itself. Then the current situation is identified and the situated behavior is selected.

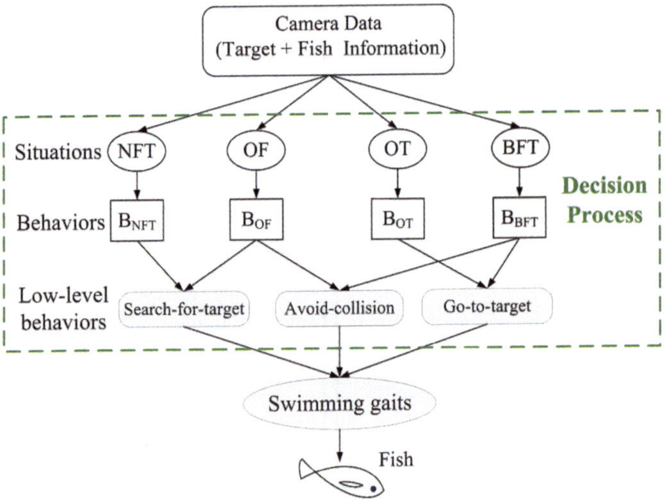

Fig. 7 Schematic diagram of decision-making process based on situated-behavior method

3.3.3 Control Law Design for B_{BFT}

In BFT situation, a control law is designed to determine the attractive force in go-to-target behavior and the repulsive force in avoid-collision behavior. As described in the object localization, the robotic fish has the following noisy measurements available to it:

- $\mathbf{r}$: Distance to the other robotic fish.
- s: Distance to the target.
- θ_F: Angular direction to the other robotic fish in the observing robotic fish coordinate system.
- θ_T: Angular direction to the target in the observing robotic fish coordinate system.

The task is performed in the horizontal plane, and Fig. 8 illustrates the definitions of $\mathbf{r}$, s, θ_F, and θ_T for fish A in its coordinate system (X_A, Y_A). As shown in Fig. 8, let θ_T denote the angle, measured counterclockwise, from X_A to the vector that points from robotic fish A to the centroid of the target. Let θ_F be the angle from X_A to the vector that points from robotic fish A to robotic fish B.

An attractive force in proportion of s is introduced to pull the robotic fish in the direction θ_T of the target. This attractive force, employed in go-to-target behavior, will be given by

$$k_a s \begin{bmatrix} \cos\theta_T \\ \sin\theta_T \end{bmatrix} \tag{5}$$

where the constant k_a is a scaling coefficient and the vector in (5) is expressed in the coordinate system of robotic fish A.

The collision avoidance is obtained by using a $1/\mathbf{r}^3$ repulsive force exerted along the line from the other fish to the robotic fish whose control is under consideration. The direction of this repulsive force is obtained through measurement of θ_F. This repulsive force, utilized in avoid-collision behavior, is expressed in the coordinate system of robotic fish A by

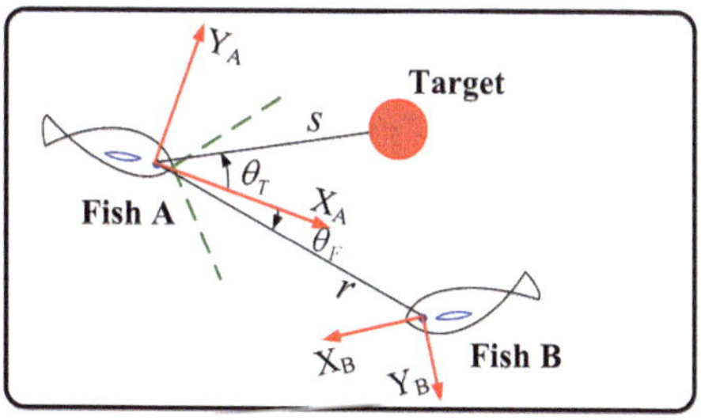

Fig. 8 Schematic illustration of $\mathbf{r}$, s, θ_F, and θ_T for fish A

$$-\frac{k_{\mathbf{r}}}{\mathbf{r}^3}\begin{bmatrix}\cos\theta_F\\ \sin\theta_F\end{bmatrix} \tag{6}$$

where the value of $k_{\mathbf{r}}$ is chosen large enough that any pair of robotic fish will maintain a safe distance apart.

Therefore, the approach will be to integrate the repulsive and attractive forces to determine the velocity of fish A. The absolute velocity of fish A, expressed in the coordinate system of robotic fish A, can be shown to be

$${}^{A}v_A = k_a s\begin{bmatrix}\cos\theta_T\\ \sin\theta_T\end{bmatrix} - \frac{k_{\mathbf{r}}}{\mathbf{r}^3}\begin{bmatrix}\cos\theta_F\\ \sin\theta_F\end{bmatrix} \tag{7}$$

In order to illustrate the rationality of the control law (7) based on a combination of an attractive force toward a target and a repulsive force for collision avoidance, the cost function J is introduced by

$$J \equiv \frac{1}{2}\{k_a[(x_A - x_T)^2 + (y_A - y_T)^2] + \frac{k_{\mathbf{r}}}{(x_A - x_F)^2 + (y_A - y_F)^2}\} \tag{8}$$

where (x_A, y_A) is the position of the robotic fish A, (x_T, y_T) is the position of the target, and (x_F, y_F) is the position of the other fish (fish B), all expressed in the coordinate system of robotic fish A.

Differentiating (8) with respect to $[x_A, y_A]^T$, we obtain

$$\nabla J = k_a\begin{bmatrix}x_A - x_T\\ y_A - y_T\end{bmatrix} - \frac{k_{\mathbf{r}}}{[(x_A - x_F)^2 + (y_A - y_F)^2]^2}\begin{bmatrix}x_A - x_F\\ y_A - y_F\end{bmatrix} \tag{9}$$

It is noted that $\mathbf{r} = \sqrt{(x_A - x_F)^2 + (y_A - y_F)^2}$, $s = \sqrt{(x_A - x_T)^2 + (y_A - y_T)^2}$, then (9) becomes

$$\nabla J = k_a\begin{bmatrix}-s\cos\theta_T\\ -s\sin\theta_T\end{bmatrix} - \frac{k_{\mathbf{r}}}{\mathbf{r}^4}\begin{bmatrix}-\mathbf{r}\cos\theta_F\\ -\mathbf{r}\sin\theta_F\end{bmatrix} \tag{10}$$

Then

$${}^{A}v_A = -\nabla J = k_a s\begin{bmatrix}\cos\theta_T\\ \sin\theta_T\end{bmatrix} - \frac{k_{\mathbf{r}}}{\mathbf{r}^3}\begin{bmatrix}\cos\theta_F\\ \sin\theta_F\end{bmatrix} \tag{11}$$

It is observed that the first term of (11) is the same as our attractive-force component in (7), and the second term is the same as our repulsive-force component in (7).

3.4 Experiments and Results

Experiments with the robotic fish were carried out in an indoor swimming tank with the size of 2250 mm × 1250 mm and with still water of 400 mm in depth. We verify the proposed control methods in several experimental trials of target-tracking and collision-avoiding task.

In experiment I, fish A is tested for its performance of vision-based target tracking and collision avoidance, shown in Fig. 9. Here fish A is required to track the target and at the same time avoid colliding with fish B. The target (red water polo) initially locates at the center of the swimming tank, and robotic fish A starts from the top right corner of the tank while robotic fish B is set in front of fish A. At the beginning, fish A detects no target and carries out search-for-target behavior, shown in Fig. 9a. Then fish A detects that it is obstructed by fish B in the direction to the target, so fish A performs avoid-collision behavior when approaching the target by integrating the attractive force from its target and the repulsive force from fish B, shown in Fig. 9b, c. At 6 s, the fish A has passed fish B without collision successfully, and begins to move toward the target direction. To control the orientation of the fish towards the target, fish A adjusts its angular offset of the tail fin. Shown in Fig. 9d–f, the orientation of fish A is gradually changed toward the target direction. At 10 s, fish A gets to the target location successfully. The trajectory of fish A is given in the sequences of overhead images. In this experiment, the situated-behavior method is verified to be effective.

Experiment II involves a scenario in which two robotic fish are required to track the target from opposite direction, shown in Fig. 10. In this experiment, each robotic fish adjusts the orientation towards the target by decentralized control. The two fish respectively arrive at the target location in a short time and their trajectories are shown in the sequences of overhead images. The experimental results also demonstrate the high efficiency of the proposed control strategies.

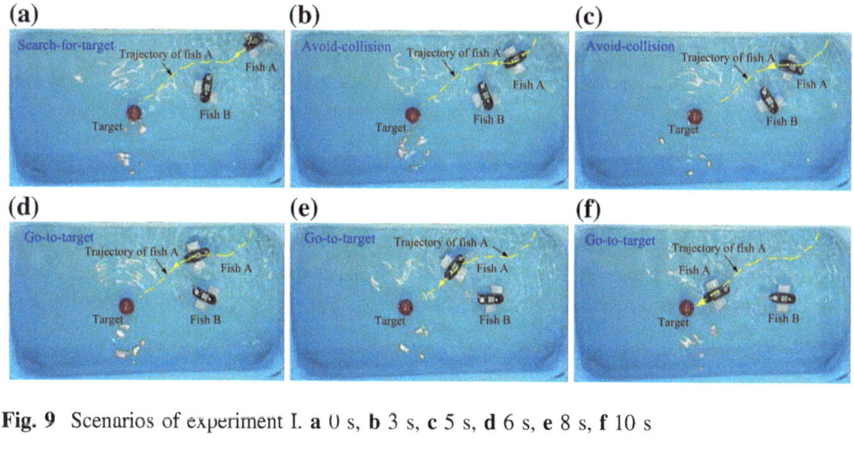

Fig. 9 Scenarios of experiment I. **a** 0 s, **b** 3 s, **c** 5 s, **d** 6 s, **e** 8 s, **f** 10 s

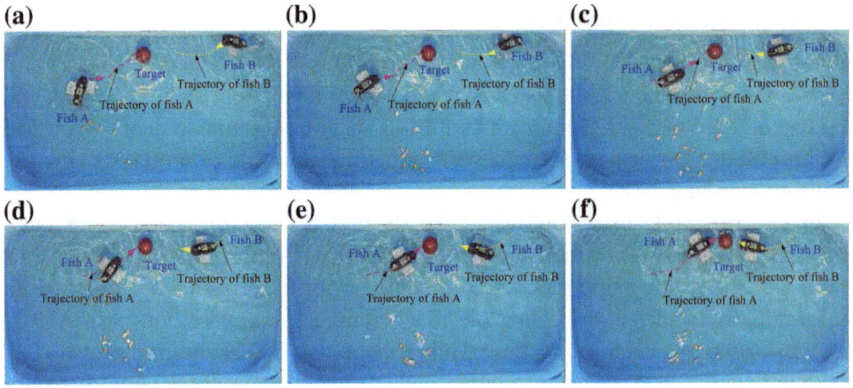

Fig. 10 Scenarios of experiment II. **a** 0 s, **b** 1 s, **c** 3 s, **d** 4 s, **e** 5 s, **f** 6 s

The performance of the experiments is evaluated by the consuming time before the fish reach their target location. In experiments I and II, if the robotic fish can reach the target location without any collision within a limited time, the experiment is regarded as successful. It has been observed that even starting from the same initial states including initial position and heading direction of the fish, and the position of the target, the experiments consume different amounts of time to finish and each robot implements different behaviors. This is because there are more uncertainties in underwater manipulation than in ground operation. Although the experimental results are influenced by the uncertainties and complexity of the underwater environment, they are still successful and promising.

4 Leader-Following Based Formation Control

A framework for the deployment of multiple autonomous robotic fish to achieve leader-following formations with Bezier trajectory is presented. Each follower robot estimates the position and orientation angle of its leader with a fast color-tracking vision system, and establishes a Bezier trajectory between its current position and the position of its leader robot. Considering the nonholonomic properties of the robotic fish, the optimization of Bezier curves curvature to choose appropriate scale factor is conducted by combination of penalty function and one-dimensional search methods to perform a smooth and stable trajectory. The optimal trajectories are accurate enough to estimate the angular velocity of the follower robot, while a fuzzy controller is used to adjust its linear velocity. By the introduction of virtual leaders, formations of different shapes can be generated. Simulations and experimental results show the effectiveness of our approach.

With leader-following method, each robotic fish takes another neighboring fish as a reference point to determine its motion. The referenced robotic fish is called a

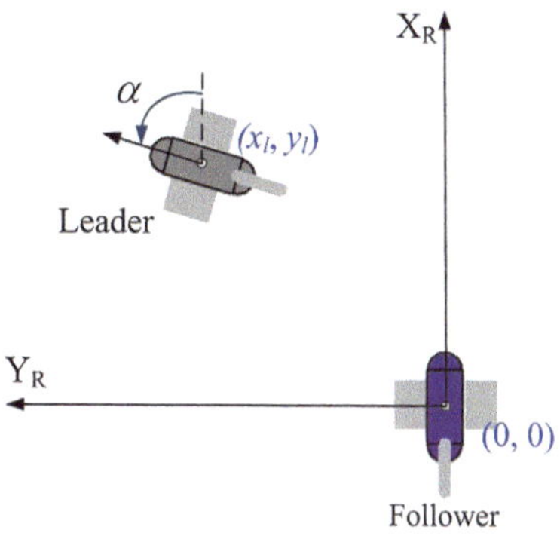

Fig. 11 Position (x_l, y_l) and orientation angle α of the leader robotic fish in the follower's frame of reference (X_R, Y_R)

leader, and the fish following it called a follower. In a group of robots, there are many pairs of leaders and followers and complex formations can be achieved by controlling relative positions of these pairs of robots respectively. This approach has been adopted widely in formation control by the characteristics of simplicity, reliability and no need for global knowledge and computation.

To achieve leader-following formations, a Bezier trajectory between leader and follower robotic fish is introduced for the deployment of multiple autonomous robots. Each follower fish estimates the position and orientation angle of its leader with a real-time color-tracking vision system, and builds a Bezier curve that generates the trajectory from its current position to the position of its leader. The Bezier-trajectory-based approach can be extended to more follower robots for keeping a column formation. By the introduction of virtual leaders, formations of different shapes can also be generated.

As illustrated in Fig. 11, only the position (x_l, y_l) and the orientation angle α of the leader robot need to be estimated with respect to the follower robot by its vision module. The corresponding vision processing for pose estimation is described as follows.

4.1 Pose Estimation

The leader robotic fish is marked with specified color pattern and its pose estimation is based on the tracking of its attached color regions. Hence the vision processing is similar to that in Sect. 3.

Figure 12 depicts the color pattern attached to the leader robot, consisting of a red rectangle and a golden rectangle with the same size. The red rectangle provides an estimate to the position based on its lower-right corner. The difference between the perceived heights of the red and golden rectangles provides an estimate of the orientation angle of the pattern with respect to the observing follower robot.

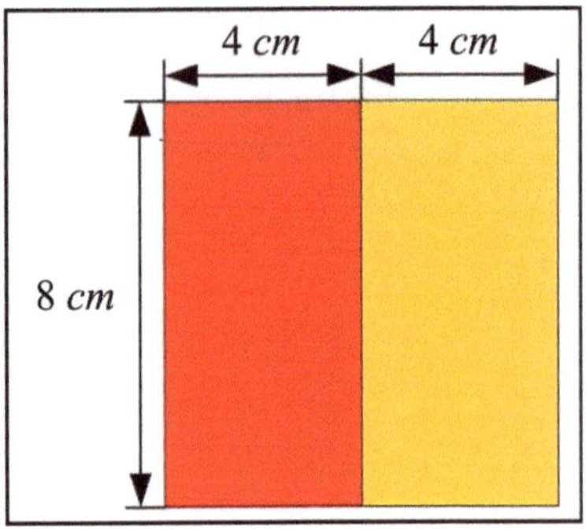

Fig. 12 *Color* pattern of the leader robot used for pose estimation

4.1.1 Color-Based Image Segmentation

Without regard to the luminance component Y that changes dramatically with ambient light, a 2-D threshold table is used to perform the mapping from CbCr pixel values to symbolic color class [19]. After the color thresholding, the vision module then carries out the following steps: pixel connection and region merging [23]. The connection procedure scans adjacent rows and merges runs (horizontal neighboring pixels of the same color) to make a region. The statistics gathered for a region include: bounding box, centroid, and area. The region merging process is done with the calculation of the region's statistic characteristics. The criterion for deciding to merge the regions is the resulting density of the combined region. Two regions are combined into one region if the combined pixel area has a density occupying the area of new bounding box above a threshold for that color class. The purpose of region merging is to combine several nearby smaller regions of the same color into a single larger region in order to identify pattern and remove noise.

4.1.2 Pattern Detection and Feature Extraction

The color pattern, including the red rectangle and the golden rectangle, needs to be recognized. We detect these color regions with several filters according to their characteristics. And the interested information of the regions in the image can be extracted.

The color regions of the pattern are found by scanning through the candidates of red and golden regions and selecting ones with the largest area. A filter is used to check if the region is large enough. Figure 13 shows the result of the color-pattern recognition.

The X_i'-axis and Y_i'-axis of image coordinate system are illustrated in Fig. 13b. The origin of the image coordinate system is the center of the image. Once the color pattern has been detected, we extract the coordinates of the lower-right corner **m** of the red region which is closest to the golden region in the image for its localization. The height h_1 of the red rectangle and h_2 of the golden rectangle in the image are

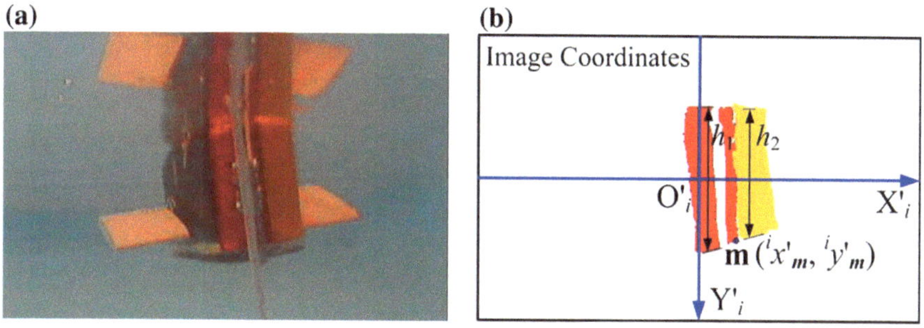

Fig. 13 *Color*-pattern recognition for follower robotic fish. **a** Sample underwater image. **b** Processed image

also extracted for an estimate of the orientation angle. As shown in Fig. 13b, all the interested information, including the perceived heights h_1, h_2, and the coordinates $({}^i x'_{\mathrm{m}}, {}^i y'_{\mathrm{m}})$ of point **m**, can be obtained by searching the color-pattern region.

4.1.3 Localization of Leader Robot

In localization, we use the 2-D regions in the camera image to estimate the position of the leader robotic fish in the 3-D coordinate system of the observing follower robotic fish. With the help of the intrinsic camera parameters, image coordinates can be converted into robot coordinates. The robot coordinate system (X_R, Y_R, Z_R) is fixed to the observing follower robotic fish, and the camera coordinate system (X_C, Y_C, Z_C) is attached to the camera of the follower robotic fish. Figure 14 shows the relation of the robot coordinate system and the camera coordinate system. Notice that Δl denotes the distance between the origin of the robot coordinate system and the origin of the camera coordinate system in the fish body direction, and the mounted camera has a wide viewing angle of 120°.

In order to estimate the location of the leader, we detect the lower-right corner **m** of the red region of its color pattern in the corresponding processed image (see Fig. 13b). Since all the robotic fish are slightly buoyant, the vertical distance h from the observing camera to the lowest point of the color pattern is fixed. This point **m** and the optical center of the camera define a viewing ray **rm**. Its intersection with the lowest horizontal plane of the color pattern attached to the leader yields an estimate of the position of the pattern relative to the observing follower, as illustrated in Fig. 14. The 3-D point **M** in camera coordinate system corresponding to the observed image point $\mathbf{m} = [{}^i x'_{\mathrm{m}}, {}^i y'_{\mathrm{m}}]^T$ in Fig. 13b is given by

$$\mathbf{M_C} = [{}^C x_{\mathbf{M}}, {}^C y_{\mathbf{M}}, {}^C z_{\mathbf{M}}]^T = \frac{h}{{}^i y'_{\mathrm{m}}} [{}^i x'_{\mathrm{m}}, {}^i y'_{\mathrm{m}}, f]^T \tag{12}$$

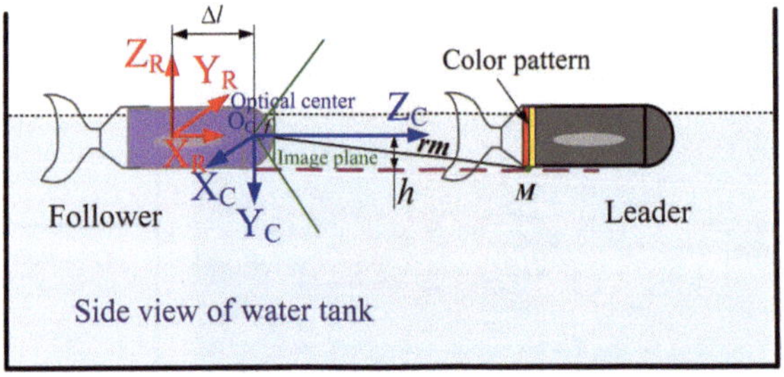

Fig. 14 Estimate of the position of leader robot relative to the follower

where f denotes the focal length of the camera. Since the pose of the robot's camera is given in robot coordinate system, the point **M** given in camera coordinate system can easily be expressed in the coordinate system of the follower robot as follows

$$\mathbf{M_R} = \left[{}^{R}x_{\mathbf{M}}, {}^{R}y_{\mathbf{M}}, {}^{R}z_{\mathbf{M}}\right]^T = \left[{}^{C}z_{\mathbf{M}} + \Delta l, -{}^{C}x_{\mathbf{M}}, -{}^{C}y_{\mathbf{M}}\right]^T \tag{13}$$

Then the position (x_l, y_l) of the leader robot with respect to the follower robot can be estimated by

$$\begin{aligned} x_l &= {}^{R}x_{\mathbf{M}} \\ y_l &= {}^{R}y_{\mathbf{M}} \end{aligned} \tag{14}$$

4.2 Formation Control with Bezier Trajectory

Defining a Bezier curve between the leader and follower robots is of key importance [25]. As shown in Fig. 15, one of such curves is defined by four points P_0–P_3. The endpoints P_0 and P_3 are determined by the positions of the follower and leader robots respectively, while the control points P_1 and P_2 are chosen along the lines defined by the orientations of the robots. This cubic Bezier curve is given by (15).

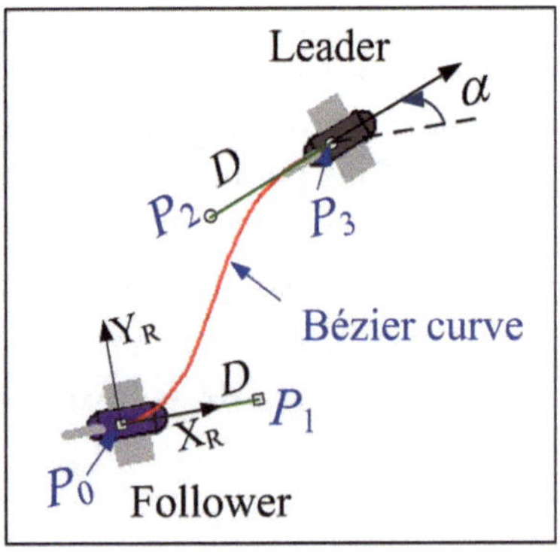

Fig. 15 Definition of Bezier trajectory between leader and follower robots

$$P(t)\begin{bmatrix} x(t) \\ y(t) \end{bmatrix} = (1-t)^3 P_0 + 3t(1-t)^2 P_1 + 3t^2(1-t)P_2 + t^3 P_3 \tag{15}$$

where $t \in [0, 1]$ is the curve parameter, and the four points P_0–P_3 are all defined in the 2-D coordinate system (X_R, Y_R) of the follower robot

$$P_0 = \begin{bmatrix} 0 \\ 0 \end{bmatrix}, \; P_1 = \begin{bmatrix} D \\ 0 \end{bmatrix}, \; P_2 = \begin{bmatrix} x_l - D\cos\alpha \\ y_l - D\sin\alpha \end{bmatrix}, \; P_3 = \begin{bmatrix} x_l \\ y_l \end{bmatrix} \tag{16}$$

The values of x_l, y_l, and α are obtained from the vision module described above. D is the distance between P_0 and P_1, also the distance between P_2 and P_3.

Then the expressions of $x(t)$ and $y(t)$ are the following:

$$\begin{cases} x(t) = (3D\cos\alpha - 2x_l + 3D)t^3 + 3(-D\cos\alpha + x_l - 2D)t^2 + 3Dt \\ y(t) = (3D\sin\alpha - 2y_l)t^3 + 3(-D\sin\alpha + y_l)t^2 \end{cases} \tag{17}$$

As mentioned above, the well-established Bezier trajectory states two interesting properties: (1) the curve passes through the two endpoints, and (2) the curve is tangent to the vectors $\overrightarrow{P_0P_1}$ and $\overrightarrow{P_2P_3}$ at the endpoints.

4.2.1 Orientation Control

To achieve a column formation, the angular velocity of the follower ω_f must correspond to the curvature of the Bezier trajectory at $P_0(t = 0)$ each time step.

According to the curvature formula of curve parametric equation, on the Bezier curve, the curvature at any point is calculated by

$$\kappa(t) = \frac{|x'(t)y''(t) - y'(t)x''(t)|}{[x'(t)^2 + y'(t)^2]^{3/2}} \tag{18}$$

The first-derivative and second derivative of Eq. (17) are substituted into (18), let $t = 0$, then the curvature of this curve at P_0 is computed by

$$\kappa(0) = \frac{2(y_l - D\sin\alpha)}{3D^2} \tag{19}$$

So the desired angular velocity of the follower can be calculated by

$$\omega_f = \frac{V_f}{R} = V_f\kappa(0) = \frac{2V_f(y_l - D\sin\alpha)}{3D^2} \tag{20}$$

where the linear velocity of the follower robotic fish V_f can be obtained using fuzzy controller described later.

Equation (20) accords with the nonholonomic constraint. In this method, although the value of D can be set arbitrarily, it does have an effect to define the control points which should be spaced up to a value of D proportional to the distance between the follower and leader robots for invariance to scale. Here for both pairs of endpoint and control point, we set $D = u|P_3 - P_0| = u\sqrt{x_l^2 + y_l^2}$. The scale factor u needs to be regulated, so the optimization problem of Bezier curve's curvature is presented.

Considering the nonholonomic properties and the inherent kinematic constraints of the robotic fish, it needs smooth and stable trajectory, which indicates that the evaluation criterion can be chosen as minimizing the square sum of curvature variation by searching appropriate scale factor u. The evaluation function is set as

$$f(u) = \int \left(\frac{\mathrm{d}\kappa(s)}{\mathrm{d}s}\right)^2 \mathrm{d}s \tag{21}$$

where s is the arc length.

Hence, the problem is transformed to be the expression of nonlinear constrained optimization problem as follows

$$\begin{aligned} &\min f(u) \quad u \in R \\ &\text{s.t.} \quad 0 < u < 1 \end{aligned} \tag{22}$$

Combining the penalty function and the original objective function [26], a new objective function without any constrained conditions is obtained

$$F(u, M_k) = f(u) + M_k(\min\{0, u\})^2 + M_k(\min\{0, 1 - u\})^2 \tag{23}$$

Algorithm 1 Penalty function methods for optimization problem with inequality constraints

Require: $u_{0,0}, \varepsilon > 0, M_0 > 0, c \in [4, 10], k = 0$ {ε is permissible error}

repeat
 $k = k + 1$
 $M_k = cM_{k-1}$
 $F(u, M_k) = f(u) + M_k(\min\{0, u\})^2 + M_k(\min\{0, 1 - u\})^2$
 if $k > 0$ **then**
  $u_{k,0} = u^*_{k-1}$ {$u_{k,0}$ is the starting point for the minimization of $F(u, M_k)$}
 end if
 compute u^*_k {This step involves a complete unconstrained minimization by one-dimensional search}
until $k > 0$ *and* $|u_k - u_{k-1}| < \varepsilon$

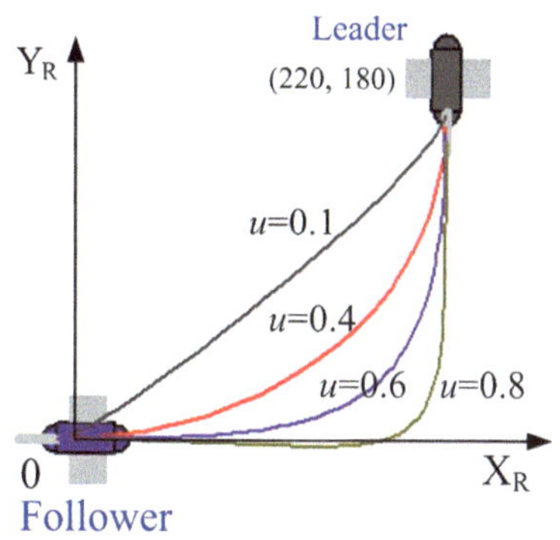

Fig. 16 Bezier curves with different values of the scale factor u

where M_k is penalty factor, $M_k > 0$, $M_1 < M_2 < \cdots < M_k < M_{k+1} < \cdots$, and $\lim_{k \to \infty} M_k = +\infty$. According to the practical experience, we choose $M_{k+1} = cM_k$, $c \in [4, 10]$.

Since the constrained optimization problem (22) can be changed into the unconstrained optimization problem (23), one-dimensional search is used to solve the optimization problem without any constraints. Algorithm 1 gives details of the penalty function approach [27]. Finally, the optimal solution $u_k^* \approx 0.4$ when the iteration is stopped. Therefore we choose $D = 0.4\sqrt{x_l^2 + y_l^2}$.

Simulation results are given to confirm the validity of the optimization algorithm. The Bezier curves with different values of the scale factor u are obtained for a same pair of leader and follower robots, and their curvatures have been optimized and non-optimized, as shown in Fig. 16. This indicates that the optimization process can make the robotic fish perform a smooth and stable trajectory.

In our experiments, the angular velocity of the follower robot is obtained from the optimal Bezier trajectory and updated at each image processing. When the new position and orientation angle of the leader are estimated, the optimized Bezier points are re-defined and the optimal curve is re-drawn, then the angular velocity is recomputed accordingly.

4.2.2 Speed Control

To keep formations, the following robotic fish adjusts its speed and tries to maintain a desired distance to its leader. In order to provide a scientific mechanism for reasoning and decision making with uncertain and imprecise information, fuzzy logic controller (FLC) is adopted to adjust the linear velocity of the follower robotic fish V_f. The leader robotic fish is assumed to follow an arbitrary trajectory. The linear velocity of the leader robotic fish V_l is known and can be regarded as an exogenous input of the follower by communications. Let Dist be the distance between the leader and the follower robotic fish, which is obtained by $\sqrt{x_l^2 + y_l^2}$. Let D_{set} denote the desired distance between them.

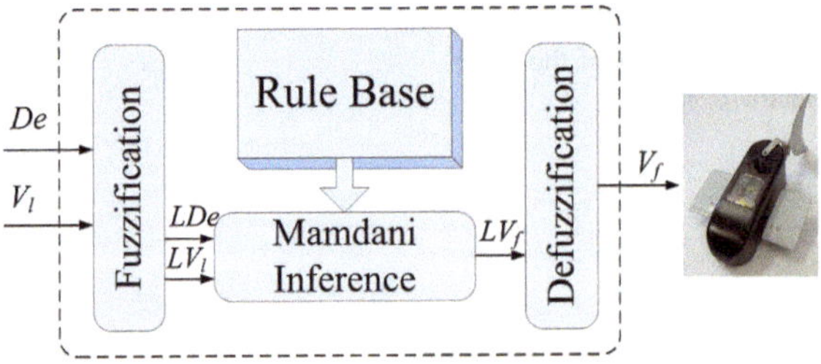

Fig. 17 Fuzzy logic controller

To let the follower robotic fish swim in the expected position relative to its leader, the linear velocity of the follower robotic fish V_f can be computed according to the values including the distance error D_e between the real distance to its leader Dist and the desired distance D_{set}, and the linear velocity of its leader robot V_l, through a set of fuzzy logic rules. Here $V > 0$ means forward swimming of the robotic fish while $V < 0$ means backward swimming. The inputs of the fuzzy logic rules are D_e and V_l, and the output is V_f. The block diagram of the fuzzy controller is shown in Fig. 17.

The next step in FLC design is to represent the variables by linguistic terms. Let LD_e, LV_l, and LV_f denote fuzzy variable sets associated with linguistic variables D_e, V_l, and V_f respectively. Firstly, LD_e and LV_l are represented by the linguistic fuzzy sets $\{NB, NS, ZE, PS, PB\}$, abbreviated from *negative big*, *negative small*, *zero*, *positive small*, *positive big*, respectively, with the membership functions shown in Fig. 18a. LV_f is represented by $\{NB, NS, ZE, PS, PB\}$, abbreviated from *negative big*, *negative small*, *zero*, *positive small*, *positive big*, respectively, with the membership function shown in Fig. 18b. The parameters in the membership functions can be derived and tuned through the experiments.

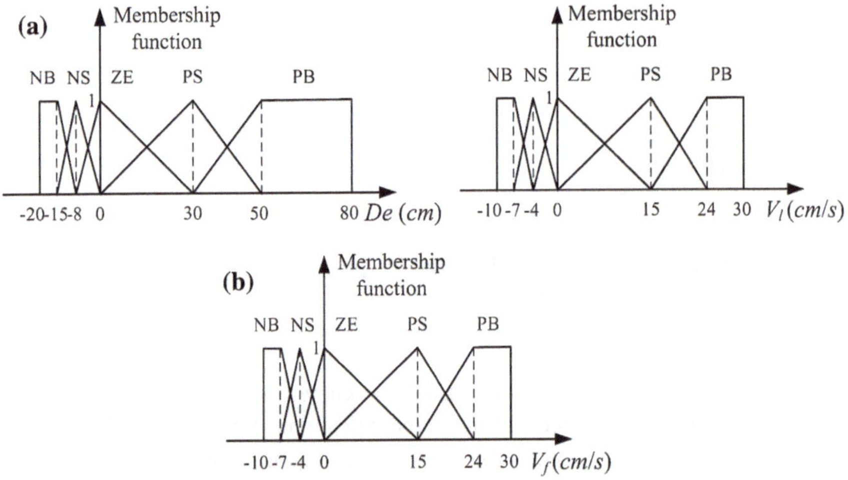

Fig. 18 Membership functions. **a** Membership functions of D_e, V_l. **b** Membership function of V_f

Table 1 Rule table for speed control of the robotic fish

V_f		V_l				
		NB	NS	ZE	PS	PB
D_e	NB	NB	NB	NB	NS	ZE
	NS	NB	NB	NS	ZE	PS
	ZE	NB	NS	ZE	PS	PB
	PS	NS	ZE	PS	PB	PB
	PB	ZE	PS	PB	PB	PB

The following step of developing a FLC is to specify the fuzzy rules. Based on the experimental experience, a fuzzy rule table which involves the fuzzy rules of the inference engine is designed. The size of rule table is completely dependent on the number of input fuzzy sets of the system. IF-THEN rule is adopted and a 2-D (5×5) rule table shown in Table 1 is built in our fuzzy control. V_f can be derived using the intuitive rule sets in this table. The fuzzy rules are given as:

If D_e is $LD_e(i)$ and V_l is $LV_l(j)$, then V_f is $b(ij)$

$(i = 1, 2, \ldots, 5$ and $j = 1, 2, \ldots, 5)$

where i and j denote the ith rule of fuzzy set LD_e and the jth rule of fuzzy set LV_l respectively, $b(ij)$ represents the point of minimum fuzziness in the consequent part of the rules.

To make all work together, an inference mechanism that generates the output signal is needed. Mamdani-type inference is employed here for speed control. The activation of the $i \times j$th rule triggered by an input containing D_e and V_l, V_f is then calculated by Mamdani inference with min for intersection and max for union.

At the defuzzification step, as shown in (24), the final linear velocity of the follower robotic fish V_f is obtained using the center-of-gravity (Centroid) defuzzification method.

$$V_{f\_\text{crisp}} = \frac{\sum_{i,j} \eta(ij) b(ij)}{\sum_{i,j} \eta(ij)} \tag{24}$$

where $\eta(ij)$ denotes the firing strength of a fuzzy control rule given by $\eta(ij) = \mu_{LD_e(i)} \wedge \mu_{LV_l(j)}$, $\mu_{LD_e(i)}$ and $\mu_{LV_l(j)}$ represent the membership functions of the linguistic values $LD_e(i)$ and $LV_l(j)$, respectively.

4.2.3 Extension to Other Formations

The establishing of Bezier trajectory between leader and follower robots can achieve a column formation directly. To keep specified geometrical formations, the following robotic fish tries to maintain a desired distance and desired angle relative to its leader. When all the pairs of leader and follower robots are in the expected positions, the desired formations are established. Forming arbitrary shapes can be realized by the introduction of virtual leaders. These virtual leaders do not

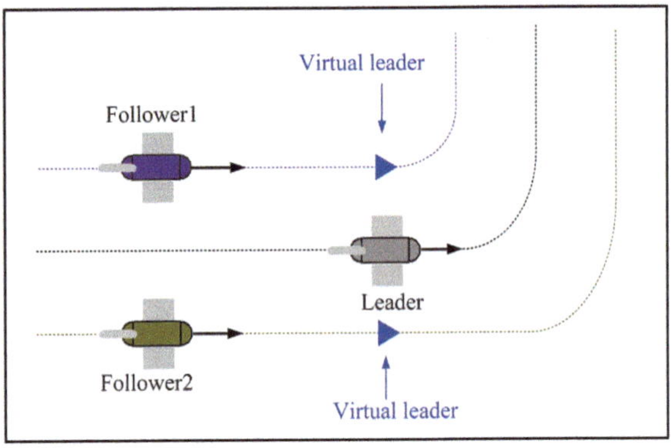

Fig. 19 Triangular formation with virtual leaders

correspond to the exact position of the leader robots, but with an added displacement. For instance, Fig. 19 illustrates a triangular formation where the reference point for the follower robots is not the leader robot but a point with a given offset on its left and right side respectively. Since the position and the orientation angle of the leader robot can be estimated, it is easy to obtain that information of the virtual leaders.

4.3 Simulation and Experimental Results

In what following, the effectiveness of the proposed approach is confirmed via simulations and some experimental results.

4.3.1 Simulations

A simulation platform for multiple robotic fish formation control is designed to study the formation control strategies for multiple robotic fish. The simulations were done using Visual C++.

Column Formation As shown in Fig. 20, the three robotic fish performed a column formation. Starting from the lower-left corner, the leader robotic fish (fish A) moves along a pre-programmed path, consisting of a straight line, two consecutive approximate 90° turns, and another straight line. Fish B and C start near the fish A, and each of them follows its preceding one with Bezier curve to generate a column formation. The trajectories in the figure illustrated that the tracking behaviors were quite smooth in all of the leader-follower pairs. Though the trajectories are not exactly coincident, simulated results confirm the validity of the approach.

Triangular Formation Figure 21 shows the simulation results of a three-robotic-fish team forming equilateral triangle. Initially these robotic fish start in

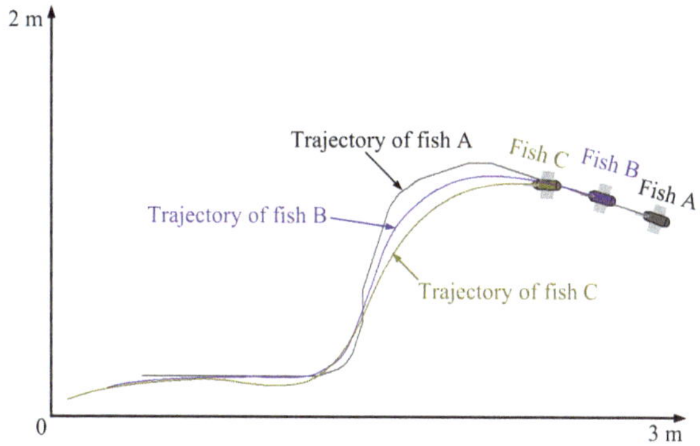

Fig. 20 Simulation of a 3-robotic-fish column formation

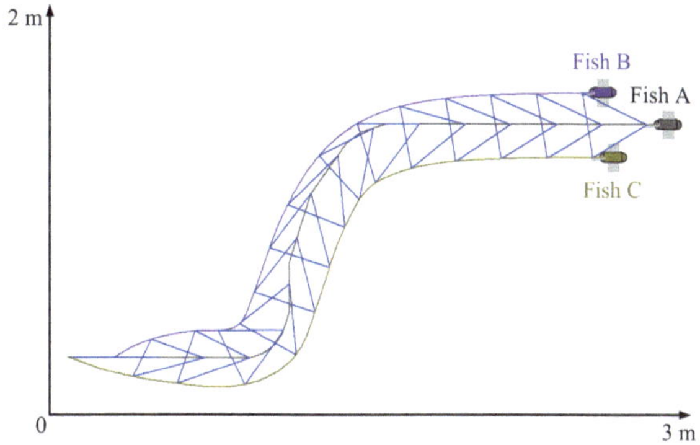

Fig. 21 Simulation of a 3-robotic-fish equilateral triangular formation

a line, then move to an equilateral triangle, and keep this formation during two turns of the leader fish A.

Switch between Formations The switching results between different geometric formations are also given via simulations. Figure 22 shows trajectories of a simulation run where three robotic fish were required to establish a triangular formation, then switched from the triangle to a column passing through the channel, and back again to a triangle after passing.

4.3.2 Experimental Results

Experiments with the autonomous robotic fish were carried out in an indoor swimming tank with the size of 3 m × 2 m and with still water of 0.25 m in depth. In our experiment, we tested the performance in formation control of three autonomous robotic fish during passing a channel which located at the center of the swimming tank, as shown in Fig. 23a. Fish A was marked with a color pattern consisting of a red rectangle and a golden rectangle, while fish B was specified with a color pattern of pink and green rectangles.

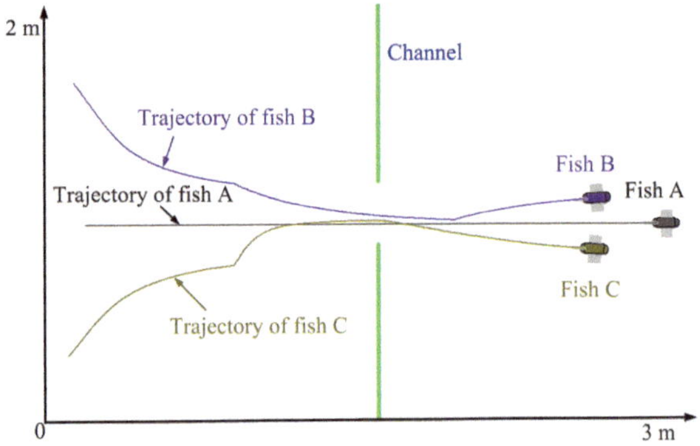

Fig. 22 Three robotic fish switch from a triangle to a column and then back to a triangle during passing a channel

The three robotic fish, starting from one side of the swimming tank, aimed to converge to a triangular formation, then switched from the triangle to a column passing through the channel, and back again to a triangle after passing. Initially, fish A was designated as the leader of both fish B and C. They first performed a triangular formation. Then, when fish A passed the channel, the three robotic fish were ordered to switch to a column formation so that they could pass through the channel orderly. As a result, fish C broke the relation with fish A and took fish B as the new leader. As the view field of the camera was limited, fish C turned to search for its new leader fish B. Finally, after each fish passed the channel successfully, the robotic fish were ordered again to re-establish the triangular formation. Figure 23 shows the scenarios of the experiment.

Although the experiments are influenced by the uncertainties and complexity of the underwater environment, e.g., waves produced by fish may cause uncertainty and inaccuracy to precise position control of them, the experimental results are still successful and promising.

5 Cooperative Transportation

This Section presents a cooperative underwater box-pushing scenario, in which three autonomous robotic fish that sense, plan and act on its own move an elongated box from some initial location to a goal location. With the onboard monocular camera, the robotic fish can estimate the pose of the object in the swimming tank. Considering the complexity of the underwater environment and the limited capability of a single robotic fish, the authors address the task by decomposing it into three subtasks and assigning them to capable robotic fish. With one robotic fish observing the box at the goal location and two robotic fish pushing the left and right ends of the box, the box can be moved gradually towards the goal location. The subtask consists of a series of behaviors, each designed to fulfil one step of the subtask. The robotic fish coordinate through explicit communications and distribute

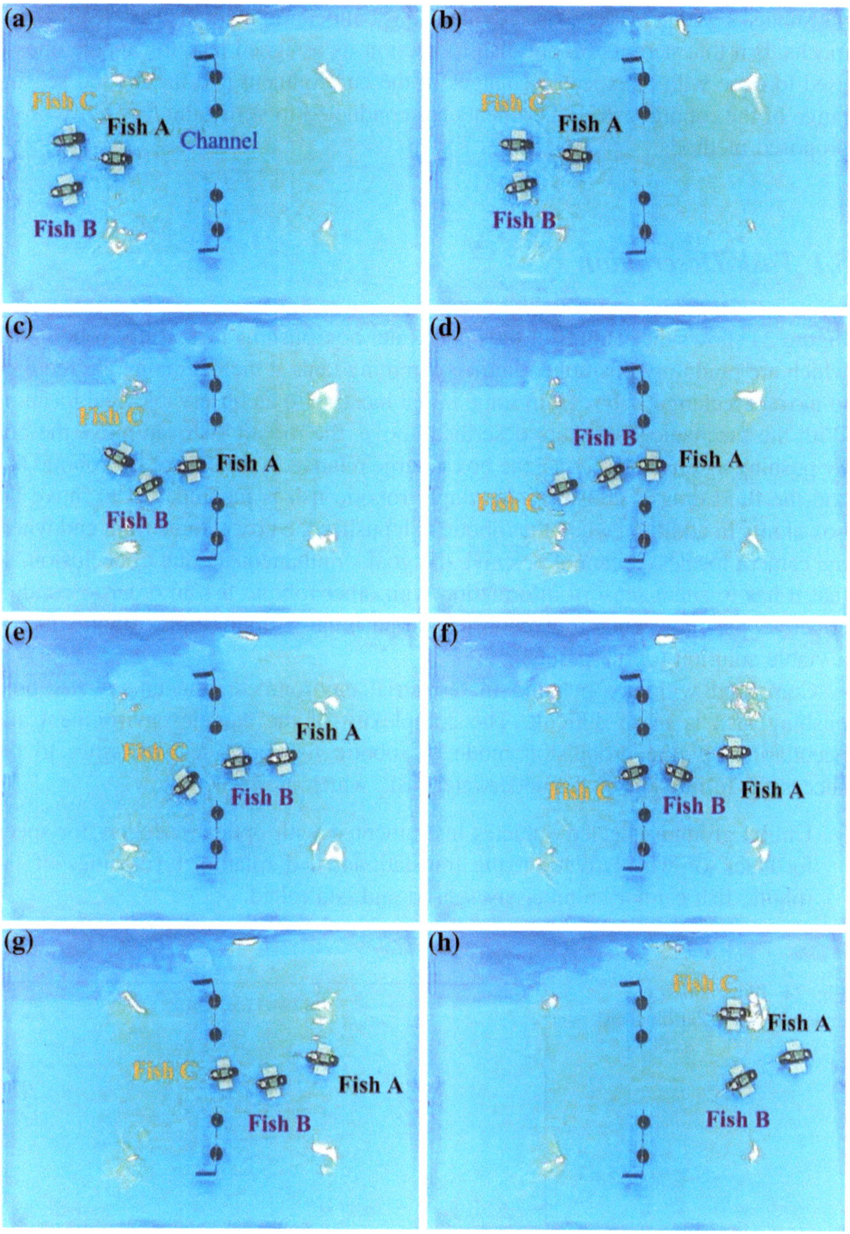

Fig. 23 Scenarios: Three robotic fish switch from a triangle to a column and then back to a triangle during passing a channel. **a** 0 s, **b** 4 s, **c** 7 s, **d** 12 s, **e** 15 s, **f** 17 s, **g** 19 s, **h** 24 s

the subtasks with a market-based dynamic task allocation method. Task reallocation mechanism that permits robotic fish to auction its assigned task to capable ones is used to cope with unexpected changes in the environment and the limited sensing range of the robotic fish. Experiments are conducted to verify the feasibility of the proposed methods.

5.1 Task Description

Figure 24 shows the settings of the underwater box-pushing task. Three robotic fish which are randomly positioned in the swimming tank at the beginning are required to move a rectangular box from some initial location to an observable goal location. With the mechanical structure described above, the robotic fish can move the box by pushing against it. Because the box is large relative to the size of the robotic fish and the fluid drag is considerable, single robotic fish is incompetent to move the box alone. In addition, when the robotic fish pushes the box with its front end where the camera locates, it cannot perceive the goal simultaneously due to occlusion, so that it has to share sensory information with other robotic fish in order to conduct effective pushing. Therefore coordinated pushing with multiple robotic fish is a viable solution to this problem.

Compared with box-pushing in terrestrial environment, the underwater box-pushing task is more difficult. The complexity of the aquatic environment and peculiarities of the propulsion mode of robotic fish pose several issues to the successful fulfillment of the underwater task, which are listed below:

- Unlike ground wheeled vehicles instrumented with optical encoders for speed feedback of wheel rotation, the translational and rotational velocities of the robotic fish cannot be precisely sensed and controlled.

Fig. 24 Illustration of underwater box-pushing task

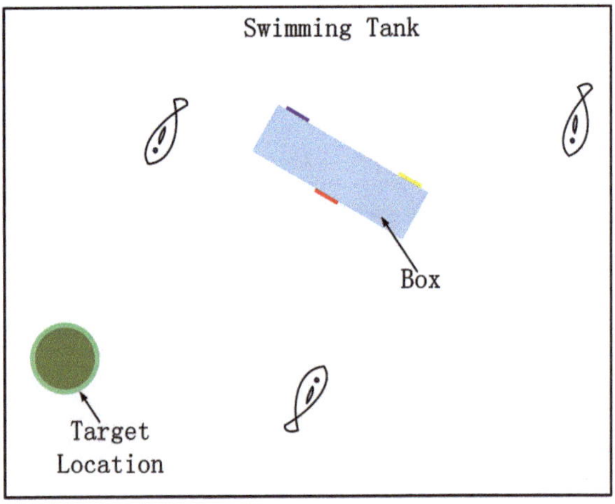

- The underwater image is plagued by several factors including distance-dependent visibility, ambient light, scattering and absorption, which make it difficult to perceive the box and the target accurately.
- Desired position and orientation of the box can hardly be reached with pushing actions due to the apparent effect of inertial drift in underwater environment.
- Waves occur when the robotic fish flaps to swim. The motion of the robotic fish and the box will be mutually affected through the coupling of waves, which further complicates the problem.

5.2 Vision Processing and Object Pose Estimation

Each object within the swimming task is marked with specified colors. The goal of vision processing is to identify the colored object of interest using the monocular camera and to estimate the distance and bearing of each object with respect to the robotic fish. Given the dynamic nature of the task and the full autonomy of the robotic fish, the vision algorithms should be both robust and efficient, consuming only a fraction of the CPU resources and leaving the remainder of computing capability for robot cognition. The vision algorithm consists of the following steps which are performed in a consecutive order on each frame:

- Thresholding: This step is to map each pixel in the raw YCbCr image into a color class label based on a threshold rectangular in the Cb and Cr chrominance dimensions. The Y components are disregarded due to their dramatic fluctuation to illuminance variations. A 2-D lookup table is constructed with predefined threshold values for fast classification.
- Blob formation: In this step, neighboring pixels belonging to the same color class are grouped together and merged into a single structure called blob. The segmented pixels are firstly run-length encoded forming horizontal color strips, and then runs that are 4-connected by one or more pixels are merged into a blob.
- Extracting blob information: For each blob, the following statistics are calculated: centroid, bounding box and area. Blobs of the same color are then sorted by area so that the largest blobs with area bigger than a threshold value can be identified as valid object.

The color fiducials of the objects are designed to be of specific shape and with unique color. The goal location is specified with a post wrapped with a green strip. The box is attached with a red square fiducial in the middle of one side and purple and yellow fiducials on two ends of the other side. The robot coordinate system $X_R Y_R Z_R$ has its origin at the focal point of the camera, its X_R-axis pointing forward, its Y_R-axis pointing through the left-hand side and its Z_R-axis pointing upward. The robotic fish swims in the horizontal plane and the vertical distance from the camera to the center of fiducials remains constant. Given the size of the fiducials and the information of corresponding projected color blobs, the pose of the

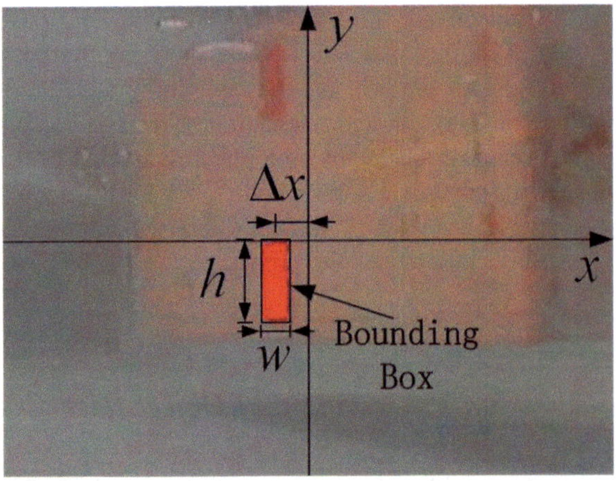

Fig. 25 The projected *color* blob of the *red square* fiducial on the box

objects measured in the robot coordinate system can be estimated on the basis of a pinhole camera model [28].

The projected image of a square fiducial will appear in the image plane with a projection distortion as shown in Fig. 25.

The height of the bounding box is h, the width is w and the offset from the center of the bounding box to the center of image in the x direction is Δx. The side length of the square fiducial is L. The posture estimates of the object, as illustrated in Fig. 26, can be calculated as:

$$d = f\frac{L}{h} \tag{25}$$

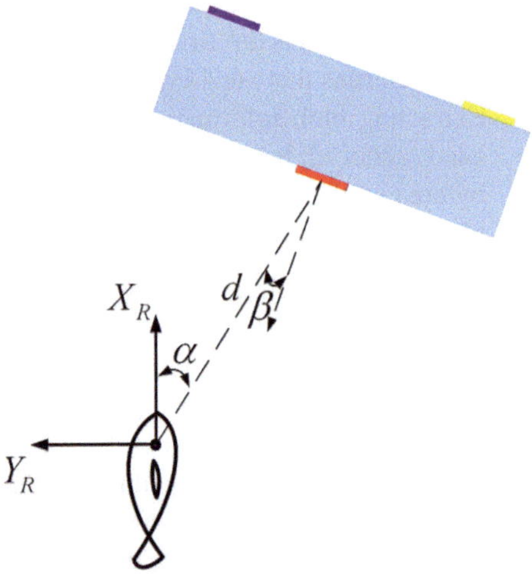

Fig. 26 Illustration of parameters in object pose estimation

$$\alpha = \Delta x \frac{\gamma}{X_{\text{res}}} \tag{26}$$

$$\beta = \begin{cases} \arccos(\frac{w}{h}), & \text{if the color blob is higher on the left side;} \\ -\arccos(\frac{w}{h}), & \text{otherwise.} \end{cases} \tag{27}$$

where d is the distance between the robotic fish and the object, f denotes the focal length of the camera, α represents the angle between the heading direction and the direction to object, γ and X_{res} are camera parameters representing the horizontal filed of view and the horizontal resolution of the camera respectively, and β is the angle of incidence to the object.

5.3 Decomposition of Underwater Box-Pushing Task

The division of labor mechanism has been widely used by humans and animals to address complex tasks. By breaking the task up into smaller pieces and assigning jobs to capable team members, the performance of the task can be maximized. Considering the requirements and settings of the underwater box-pushing task, we propose here a multirobot cooperative underwater transportation system in which two robotic fish are responsible for pushing the box whereas the third robotic fish acts as an environment-embedded sensor for pose perception of the box. With this division of labor approach, the overall box-pushing task can be decomposed by hand into the following three subtasks: Push-Left, Push-Right and Observe. Each subtask can be executed completely by a single robotic fish. The robotic fish that carry out the Push-Left and Push-Right subtasks can see the yellow and purple fudicials on the box respectively, while the robotic fish performing the Observe subtask is positioned at the goal location and can see the red fiducial on the opposite side of the box. The observing robotic fish calculates the pose of box and communicates with the pushing robotic fish to ensure that the box is moved towards the goal location. The scenario of coordinated underwater box-pushing with three robotic fish is illustrated in Fig. 27.

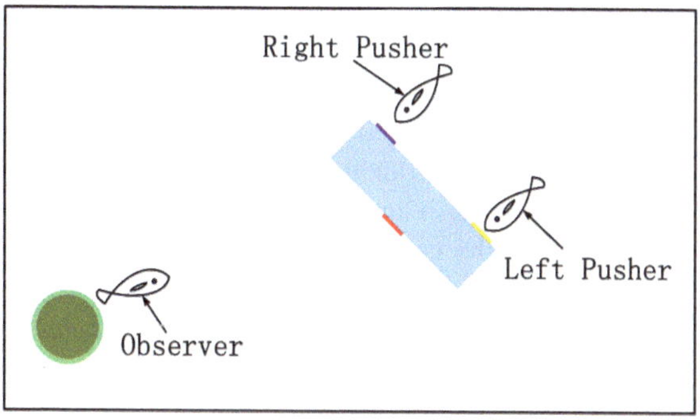

Fig. 27 Scenario of coordinated box-pushing with three robotic fish

The subtask can be achieved with a set of related behaviors, each designed to execute one stage of the subtask. The behaviors are executed in sequence and the transition from one behavior to another is triggered by real-time perception of the robotic fish. To reduce the oscillations at the anterior part of the robotic fish caused by the flapping movements of the tail fin, a hybrid swimming pattern is utilized for each behavior. This swimming pattern, which has been experimentally validated to produce minimum oscillations at the head, uses synchronized pectoral fins for thrust generation and tail fin as a rudder.

The Push-Left and Push-Right subtasks are designed in the same way and we will take the Push-Left subtask as an example to illustrate their implementation details. The robotic fish assigned with the Push-Left subtask has the corresponding end of the box in its FOV. To perform the task, the robotic fish must swim towards the left end of the box until it has contact with the box and then starts pushing the box with its head. The Push-Left subtask includes the following two primitive behaviors:

- Approach-Box-Left-End: This behavior consists in attracting the robotic fish to the left end of the box. The flapping frequency of the pectoral fins, which determines the translational speed of the robotic fish, varies with distance to the target. Outside a controlled zone, the robotic fish swims at maximum speed. When the robotic fish enters the controlled zone, the flapping frequency decreases linearly from maximum to zero. Within the dead zone, the robotic fish stops flapping pectoral fins and slowly drifts to the box in order to realize soft contact with the box. This behavior terminates 0.5 s after the robotic fish enters the dead zone. Although it is rather difficult to achieve zero speed when the robotic fish docks at the box, fierce collision can be avoided with the above speed control method. The swimming direction of the robotic fish is controlled with the angular offset of tail fin. A simple proportional controller is used to regulate the swimming direction of the robotic fish. The motion of the robotic fish is governed by the following equations:

$$f_i = \begin{cases} f_p^{\max} & \text{if } d_l > C_l \\ \frac{(d_l - D_l) f_p^{\max}}{C_l - D_l} & \text{if } D_l < d_l \le C_l \\ 0 & \text{if } d_l \le D_l \end{cases} \quad i = 1, 2 \tag{28}$$

$$\bar{\theta}_3 = K_l \alpha_l \tag{29}$$

  where $f_p^{\max}$ is the maximum flapping frequency of the pectoral fins, d_l is the distance from the robotic fish to the fiducial on the left end of the box, K_l is the gain of the proportional controller, α_l is the angle between the heading direction of the robotic fish and the direction to the left end of the box, C_l and D_l specify the radii of the controlled and dead zone respectively. Figure 28 illustrates the parameters used in this behavior.
- Push-Box-Left-End: The robotic fish executes this behavior to exert pushing force on the left end of the box so that the pose of the box can be changed. To

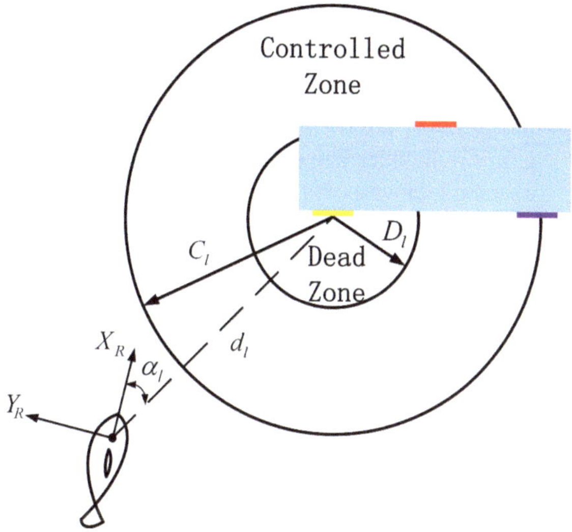

Fig. 28 Illustration of parameters in approach-box-left-end behavior

push the box in an effective manner, the robotic fish flaps its pectoral fins with maximum frequency. As the box moves, the robotic fish regulates the angular offset of the tail fin with a proportional controller to keep touch with the left end of the box with its head. The robotic fish is controlled by:

$$f_i = f_p^{\max}, \quad i = 1, 2 \tag{30}$$

$$\bar{\theta}_3 = K_l \alpha_l \tag{31}$$

The function of the Observe subtask is to direct the pushing robotic fish based on visual perception at the goal location. The observing robotic fish first swims to the proximity of the goal, swirls to search the box and then starts guiding the pushing robotic fish. The following primitive behaviors are linearly combined to realize the Observe subtask:

- Approach-Goal: This behavior enables the robotic fish to swim towards the goal location. The robotic fish swims with maximum speed until it gets reasonably close to the goal and switches to the next behavior. A proportional controller is used to control the angular offset of the tail fin. The equations that determines the motion of the robotic fish are:

$$f_i = f_p^{\max}, \text{ for } d_g > C_g, \quad i = 1, 2 \tag{32}$$

$$\bar{\theta}_3 = K_l \alpha_l \tag{33}$$

where d_g is the distance from the robotic fish to the goal and C_g is a threshold distance for this behavior.

- Search-Box: With this behavior, the robotic fish swirls to find the box. While flapping its pectoral fins with maximum frequency, the tail fin is biased $\frac{\pi}{3}$ to the left or right side. The effected motion with this behavior is turning clockwise or counterclockwise with a small radius. The mathematical formulation of this behavior is:

$$f_i = f_p^{\max}, \quad i = 1, 2 \tag{34}$$

$$\bar{\theta}_3 = \frac{\pi}{3} \text{ or } -\frac{\pi}{3} \tag{35}$$

- Monitor: The robotic fish executes this behavior when it has found the box. Although the robotic fish doesn't move with this behavior, a series of complex operations are performed. The robotic fish continuously estimates the pose of the box, and then directs the other robotic fish to push the box towards itself with a coordination protocol. The implementation details of the coordination methods are described in the following subsection.

With the above task decomposition scheme, the underwater box-pushing scenario can be executed concurrently with the three robotic fish. Figure 29 shows the decomposition scheme of the underwater box-pushing task.

The robotic fish that is not performing any of the subtasks executes Safe-Wander behavior, which enables random motion without colliding with obstacles, i.e., the box or other robotic fish. Both the flapping frequency of the pectoral fins and

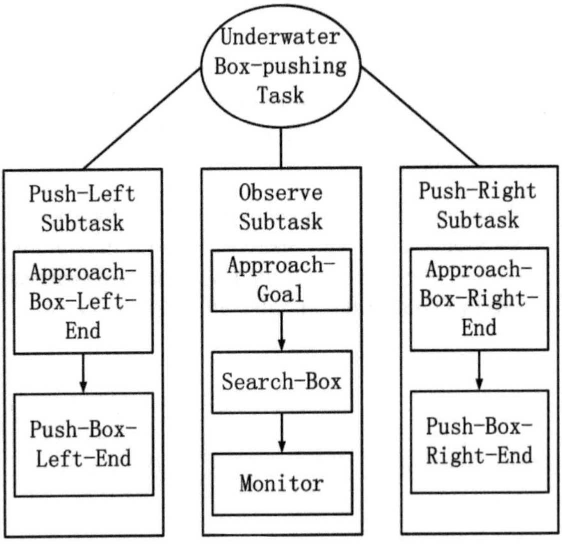

Fig. 29 Decomposition scheme of the underwater box-pushing task

angular offset of the tail fin are randomly generated if there is no obstacle in the FOV or the obstacle is far away, otherwise the robotic fish turns to avoid the obstacle. The control parameters of this behavior are determined as:

$$f_i = f_p^{\mathrm{rdm}}, \quad i = 1, 2 \tag{36}$$

$$\bar{\theta}_3 = \begin{cases} \phi_t^{\mathrm{rdm}} & \text{if no obstacle in FOV or } d_o > C_o \\ \frac{\pi}{3} & \text{if } d_o \leq C_o \text{ and } \alpha_o < 0 \\ -\frac{\pi}{3} & \text{if } d_o \leq C_o \text{ and } \alpha_o \geq 0 \end{cases} \tag{37}$$

where f_p^{rdm} and ϕ_t^{rdm} are random values that are periodically generated, d_o denotes the distance from the robotic fish to the obstacle, C_o is a threshold distance, and α_o is the angle between the heading direction of the robotic fish and the robot-to-obstacle direction.

5.4 Dynamic Task Allocation

The successful fulfillment of the underwater box-pushing task requires to determine which robotic fish should execute which subtask, also known as the task allocation problem in multirobot systems. Given the dynamic nature of the underwater environment and the motion uncertainties of the box, the assignment of robotic fish to subtask is a dynamic process and needs to be continuously adjusted to improve overall system performance. Dynamic task allocation among the robotic fish is achieved through deliberate communications and negotiations. After introducing the communication infrastructure of the team of robotic fish, the market-based dynamic task allocation method for the underwater box-pushing task will be presented.

The robotic fish negotiate with each other through explicit communications using the onboard serial RF communication hardware. Messages are broadcast at baud rate of 19,200 bits per second to all of the robotic fish. Each robotic fish has a unique ID number that identifies itself in the communication network. To guarantee collision-free access to the radio channel, a timed token-passing protocol is employed. There exists a token traveling among the robotic fish in a circular fashion and each robotic fish can transmit messages only when it possesses the token. The basic idea is to assign each robotic fish a time budget, which is the maximum time the robotic fish is permitted to speak every time it receives the token. In case the robotic fish holding the token malfunctions and fails to forward the token, a timer on the next robotic fish expires and a new token can be created to recover from the system failure.

In recent years, task allocation has been a hot research topic in the field of multirobot systems [29–31]. Among a number of task allocation models and philosophies, market-based approaches have gained considerable popularity as flexible and efficient mechanisms for distributed task allocation [32, 33]. Inspired by the original Contract Net Protocol of Smith [34], market-based approaches use auction mechanisms for task allocation. In these approaches, tasks available to be allocated are put up for auction and candidate robots submit bids that are their costs or utility estimates associated with completing the tasks. Once all bids have been received or a prespecified deadline has passed, the auctioneer evaluates all the submitted bids and awards the robot with the highest bid a contract to execute the tasks. In the underwater box-pushing scenario, tasks are allocated via a sequence of single-item sealed-bid auctions, in which tasks are auctioned one at a time and the bidder submits its bid honestly without knowing the others' bid.

At the start, all robotic fish execute Safe-Wander behavior and a host computer, acting on behalf of the human supervisor, announces the Observe subtask. The robotic fish receive the information about the task being auctioned and then respond with a bid. The value of the bid is estimated by the robotic fish based on its perceived path cost to the goal location, which is calculated as:

$$U = \begin{cases} \frac{k_1}{d_g} + \frac{k_2}{|\alpha_g|}, & \text{if the goal is in FOV;} \\ 0, & \text{otherwise.} \end{cases} \tag{38}$$

where α_g represents the angle between the heading direction of the robotic fish and the robot-to-goal direction, k_1 and k_2 are positive constant parameters. Once the host computer receives all bids, it informs the robotic fish with the highest bid to perform the Observe subtask. If the goal is not observable to all robotic fish, i.e., all robotic fish submit zero-valued bids, the auctioneer will continuously broadcasts the Observe subtask until it is allocated.

Once the robotic fish is assigned with the Observe subtask, it will consecutively execute Approach-Goal behavior, Search-Box behavior, and finally Monitor behavior. With Monitor behavior, the robotic fish periodically auctions Push-Left and Push-Right subtasks according to the pose of the box. The other two robotic fish bid the pushing subtasks with their position estimates with respect to the left and right ends of the box. The angle of incidence β_b to the box is used to decide which subtask should be auctioned. If β_b is greater than a threshold value β_b^{thr}, only the Push-Left subtask is auctioned; when $\beta_b < -\beta_b^{\text{thr}}$, only Push-Right subtask is announced; otherwise both subtasks are auctioned. The contract of a pushing task is considered overdue by the observing robotic fish when the orientation of the box is changed so significantly that new pushing strategy should be employed. The robotic fish cannot hold two or more tasks at the same time, therefore only robotic fish with no assigned task can bid newly-announced tasks.

Since the task environment is dynamic and complex, the robotic fish may fail to achieve the task it has committed to. For example, when the robotic fish assigned with Observe subtask executes Approach-Goal behavior, other robotic fish may appear in its FOV and occlude the goal. A pushing robotic fish may also lose sight of the box due to the unpredictable motion of the box caused by waves or the pushing of other robotic fish. To cope with these situations, we allow for re-auctioning task when the robotic fish is no longer competent for its assigned task. In the auction algorithm of task reallocation, the auctioneer itself can also bid the task in case it will become suitable for the task again. The task re-allocation mechanism is especially useful in the underwater box-pushing task, since the environment is highly dynamic and the sensing capability of robotic fish is very limited.

5.5 Experiments

The underwater box-pushing experiments with three robotic fish are conducted in an indoor swimming tank with size of 3000 mm × 2000 mm and with still water of 300 mm in depth. The dimension of the elongated box is 850 mm × 300 mm × 150 mm (length × width × height). The scenarios within the swimming tank are captured with an overhead camera and recorded for offline analysis. A host computer, which connects with an RF communication module, is used to assign the Observe subtask to the robotic fish at the beginning. The box is initially positioned in such a posture that the robotic fish at the goal location can see the side attached with red fiducial.

Typical experiment scenarios of the underwater box-pushing task are shown in Fig. 30.

The goal is located at the bottom-left corner of the swimming tank. Figure 30a shows the initial scenario in which three robotic fish that are randomly positioned in the swimming tank start by executing Safe-Wander behavior and the host computer auctions the Observe subtask. At 2.0 s in Fig. 30b, the robotic fish on the left sees the goal and gets the Observe subtask. As shown in Fig. 30c, the robotic fish assigned with Observe subtask swims towards the goal location while the other two robotic fish still execute Safe-Wander behavior. After the robotic fish on the left arrives at the goal location, it turns around to face the box and then starts auctioning Push-Left and Push-Right subtasks (see Fig. 30d). The other two robotic fish bid the pushing subtasks and move the box by pushing against the fiducials that correspond to their allocated tasks. In Fig. 30e–g, the box floats gradually towards the observing robotic fish. The orientation of the box is adjusted by repeated pushing on its left and right ends. At 44.0 s, the box is successfully moved to the goal location, as shown in Fig. 30h.

The performance of the experiments can be measured by the time the three robotic fish take to move the box from the same initial location to the goal location. A total of 30 runs of the experiment were conducted with seven failures and uneven consuming time for successful trials (see Fig. 31 for statistic results). The failures are mainly caused by the significant rotation of the box to a posture in which the robotic fish at the goal location cannot see it. Because the box is constantly drifting under the influence of waves, the robotic fish assigned with the pushing task may be far away from the box and cannot reach the task location before the box disappears from the FOV of the observing robotic fish. The thrust force of the robotic fish is quite limited and cannot counteract the fast rotation of the box. The observing

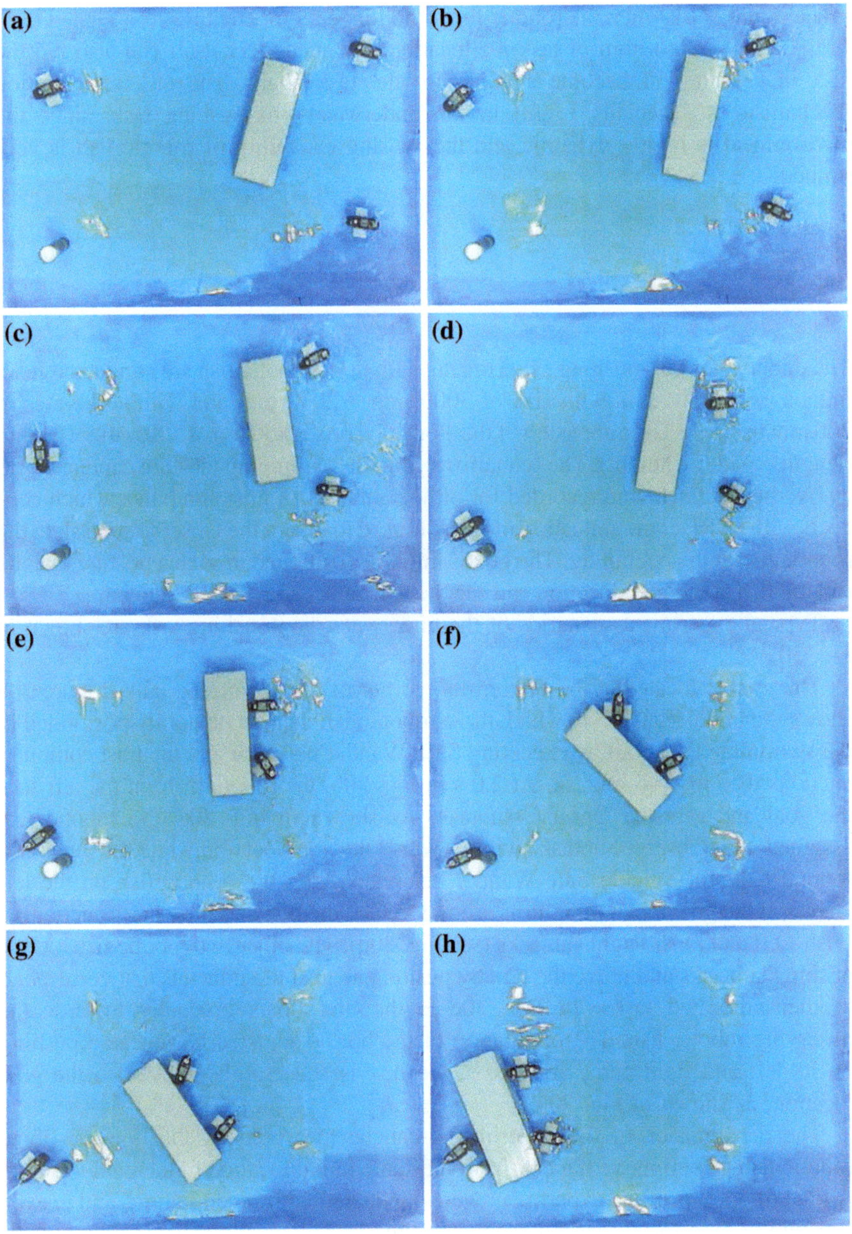

Fig. 30 Experiment scenarios of underwater box-pushing. **a** 0.0 s, **b** 2.0 s, **c** 8.0 s, **d** 16.0 s, **e** 20.0 s, **f** 28.0 s, **g** 36.0 s, **h** 44.0 s

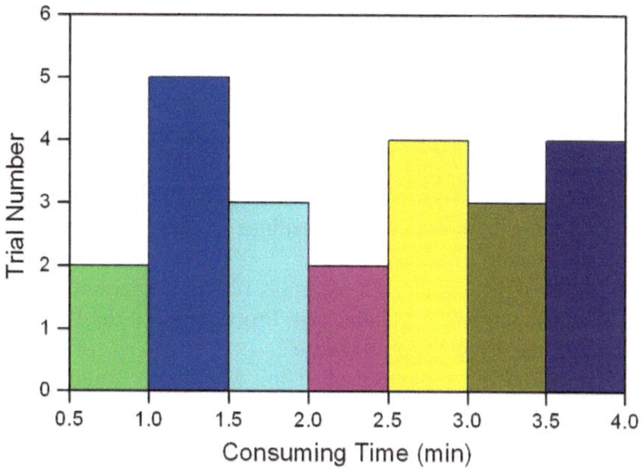

Fig. 31 Statistics of consuming time for successful trials

robotic fish itself is also drifting, which may results in a situation that the box is moved to the observing robotic fish but far away from the goal location. Trials that take long time to complete usually involve reallocations of the Observe subtask. If the robotic fish executing Monitor behavior loses sight of the box, the Observe subtask will be re-auctioned which corresponds to a new start of the mission. Since the robotic fish not performing any subtask executes Safe-Wander behavior that generates motion with random speed and orientation, the time the robotic fish takes to see the task location and undertake the subtask is random. Moreover, the poor quality of the underwater image occasionally causes erroneous pose estimate of the box and induces undesired pushing on the box. Therefore coordinated box-pushing in the highly dynamic and complex underwater environment without any global position information about the robotic fish and the box is a particularly challenging problem.

6 Conclusion

A multiple autonomous robot system with fish-like robots with built-in visual sensor was developed in this chapter. With elaborately designed control and cooperation strategies, three cooperation tasks, namely target tracking and collision avoidance, formation control, and cooperative transportation, were successfully accomplished with multiple robotic fish in the aquatic environment. Compared with cooperation tasks carried out in the terrestrial environment, underwater tasks are much more difficult due to the strong hydrodynamic disturbances, degraded visual data and the reduced ability of motion control. The robotic fish and cooperation methods presented in this study can help to achieve more complex tasks in the real world, such as oceanographic exploration, multi-target tracking and underwater salvage.

References

1. Mataric MJ, Sukhatme GS (2001) Task allocation and coordination of multiple robots for planetary exploration. In: Proceedings of the international conferences on advanced robotics, pp 61–70
2. Kolling A, Carpin S (2008) Multirobot surveillance: an improved algorithm for the GRAPH-CLEAR problem. In: Proceedings of the IEEE international conferences on robotics and automation, pp 2360–2365
3. Koes M, Nourbakhsh I, Sycara K (2006) Constraint optimization coordination architecture for search and rescue robotics. In: Proceedings of the IEEE international conferences on robotics and automation, pp 3977–3982
4. Amigoni F, Gasparini S, Gini M (2006) Building segment-based maps without pose information. Proc IEEE 94:1340–1359
5. Pereira GAS, Campos MFM, Kumar V (2004) Decentralized algorithms for multi-robot manipulation via caging. Int J Robot Res 23:783–795
6. Bandyopadhyay PR (2005) Trends in biorobotic autonomous undersea vehicles. IEEE J Ocean Eng 30:109–139
7. Colgate JE, Lynch KM (2004) Mechanics and control of swimming: a review. IEEE J Ocean Eng 29:660–673
8. Fish FE, Lauder GV (2006) Passive and active flow control by swimming fishes and mammals. Annu Rev Fluid Mech 38:193–224
9. Lauder GV, Tytell ED (2006) Hydrodynamics of undulatory propulsion. Fish Physiol 23:425–468
10. Hu Y, Wang L, Zhao W, Wang Q, Zhang L (2007) Modular design and motion control of reconfigurable robotic fish. In: Proceedings of international conferences on decision and control, pp 5156–5161
11. Triantafyllou MS, Triantafyllou GS (1995) An efficient swimming machine. Sci Am 272:64–70
12. Morgansen KA, Triplett BI, Klein DJ (2007) Geometric methods for modeling and control of free-swimming fin-actuated underwater vehicles. IEEE Trans Robot 23:1184–1199
13. Gerkey BP, Mataric MJ (2002) Pusher-watcher: an approach to fault-tolerant tightly-coupled robot coordination. In: Proceedings of the IEEE international conferences on robotics and automation, pp 464–469
14. Takahashi H, Nishi H, Ohnishi K (2004) Autonomous decentralized control for formation of multiple mobile robots considering ability of robot. IEEE Trans Ind Electron 51:1272–1279
15. Defoort M, Floquet T, Kököy A, Perruquetti W (2008) Sliding-mode formation control for cooperative autonomous mobile robots. IEEE Trans Ind Electron 55:3944–3953
16. Nouyan S, Groß R, Bonani M, Mondada F, Dorigo M (2009) Teamwork in self-organized robot colonies. IEEE Trans Evol Comput 13:695–711
17. Lauder GV, Madden PGA (2006) Learning from fish: kinematics and experimental hydrodynamics for roboticists. Int J Autom Comput 3(4):325–335
18. Lenser S, Bruce J, Veloso M (2001) CMPack: a complete software system for autonomous legged soccer robots. In: Proceedings of the 5th international conferences autonomous agents, pp 204–211
19. Bruce J, Balch T, Veloso M (2000) Fast and inexpensive color image segmentation for interactive robots. In: Proceedings of the IEEE/RSJ international conference on intelligent robots and systems, pp 2061–2066
20. Fasola J, Veloso M (2006) Real-time object detection using segmented and grayscale images. In: Proceedings of the IEEE international conferences on robotics and automation, pp 4088–4093
21. Bandlow T, Klupsch M, Hanek R, Schmitt T (1999) Fast image segmentation, object recognition and localization in a RoboCup scenario. In: RoboCup-99: Robot Soccer World Cup III, pp 174–185

22. Dudek G, Jenkin M, Prahacs C, Hogue A, Sattar J, Giguere P, German A, Liu H, Saunderson S, Ripsman A, Simhon S, Torres LA, Milios E, Zhang P, Rekletis I (2005) A visually guided swimming robot. In: Proceedings of the IEEE/RSJ international conferences on intelligent robots and systems, pp 3604–3609
23. Bruce J (2000) Realtime machine vision perception and prediction. Undergraduate thesis, School of Computer Science, Carnegie Mellon University, Pittsburgh
24. Balch T, Arkin RC (1998) Behavior-based formation control for multirobot teams. IEEE Trans Robot Autom 14(6):926–939
25. Prautzsch H, Boehm W, Paluszny M (2002) Bezier and B-Spline techniques. Springer, New York
26. Tang HW, Qin XZ (2004) Practical optimization method, 3rd edn. Da Lian University of Technology Press, Dalian
27. Scales LE (1985) Introduction to non-linear optimization. Springer, New York
28. Sonka M, Hlavac V, Boyle R (1999) Image processing, analysis, and machine vision, 2nd edn. PWS Publishing, Pacific Grove
29. Fua C, Ge SS (2005) COBOS: cooperative back-off adaptive scheme for multi-robot task allocation. IEEE Trans Robot 21:1168–1178
30. Parker LE, Tang F (2006) Building multirobot coalitions through automated task solution synthesis. Proc IEEE 94:1289–1305
31. Chen J, Sun D (2007) Resource constrained multirobot task allocation based on leader-follower coalition methodology. Int J Robot Res. doi:10.1177/0278364910396552
32. Dias MB, Zlot R, Kalra N, Stentz A (2006) Market-based multirobot coordination: a survey and analysis. Proc IEEE 94:1257–1270
33. Gerkey BP, Mataric MJ (2002) Sold!: auction methods for multirobot coordination. IEEE Trans Robot Autom 18:758–768
34. Smith RG (1980) The contract net protocol: high-level communication and control in a distributed problem solver. IEEE Trans Comput C-29:1104–1113

Fish–Robot Interactions: Robot Fish in Animal Behavioral Studies

Sachit Butail, Nicole Abaid, Simone Macrì and Maurizio Porfiri

Abstract In this chapter, we focus on the use of robotic fish in animal behavior studies. Specifically, we describe the design and control of a low-cost robot along with accompanying enabling technologies for use in animal experiments. The robotic fish appearance and movement are inspired by the zebrafish animal model. The robot is capable of autonomous underwater operation. Two behavioral studies demonstrate the use of the robotic fish to test hypotheses on zebrafish social behavior. In the first study exploring zebrafish preference in a binary choice test, we find that the robot is able to elicit attraction in both individuals and small shoals when the other alternative is an empty compartment. At the same time, between conspecifics and the robot, zebrafish prefer the former, highlighting design choices that need further improvement. The second study describes the interaction between the robot and shoals of zebrafish in a free-swimming environment. The robot swims autonomously along predefined circular trajectories at three different speeds, corresponding to increasing tail-beat frequency. The robot is found to modulate zebrafish shoal cohesion, confirming expectations from the preference study result.

S. Butail · S. Macrì · M. Porfiri (✉)
Department of Mechanical and Aerospace Engineering,
New York University Polytechnic School of Engineering, Brooklyn, NY, USA
e-mail: mporfiri@nyu.edu

S. Butail
e-mail: sachitb@gmail.com

S. Macrì
e-mail: simone.macri@iss.it

N. Abaid
Department of Biomedical Engineering and Mechanics, Virginia Polytechnic Institute and State University, Blacksburg, VA, USA
e-mail: nabaid@vt.edu

S. Macrì
Section of Behavioral Neuroscience, Department of Cell Biology and Neuroscience, Istituto Superiore di Sanità, Rome, Italy

S. Butail
Indraprastha Institute of Information Technology Delhi (IIITD), New Delhi, India

© Springer-Verlag Berlin Heidelberg 2015

R. Du et al. (eds.), *Robot Fish*, Springer Tracts in Mechanical Engineering,
DOI 10.1007/978-3-662-46870-8_12

In summary, the robotic fish platform described in this chapter provides a viable and fully controllable three-dimensional interactive tool for animal behavior experiments.

1 Introduction

Animals possess high interindividual differences in their behavioral response to the same environment, even within the same species [1]. Obtaining a consistent response in behavioral studies where live stimuli are used as independent variables is therefore a challenging task. In this respect, robots constitute a valid tool for testing hypotheses that would otherwise require extensive animal training and use [2, 3]. As controllable machines can be made to look [4], sound [5], or even smell [6] like animals, robots can be assigned a set of repeatable behaviors to elicit consistent response from subjects [7], thus contributing to our understanding of animal behavior [8, 9]. Furthermore, with frequent field deployments and greater degree of autonomy [10], robots hold the promise of assisting behavioral studies in the wild.

Fish-like robots have been used to gain insight into quorum sensing [4], leadership [11], swimming hydrodynamics [12], and the effects of psychotropic drugs on the regulation of emotions [7, 13]. Because fish in a majority of these studies use vision as their primary sensory modality, the robot's morphology plays an important role in obtaining a consistent response. Studies in [4, 11] have shown that it is possible to regulate fish behavior with a life-sized rigid replica that is maneuvered inside a tank with a mobile magnetic base. At the same time, investigations using a bioinspired robotic fish with undulating body parts that mimic fish locomotion have established that body movement plays an important role in fish perception of their robotic analogs [12, 14, 15]. The bioinspired robotic fish used in these studies has the swimming mechanism onboard, making it a viable alternative for autonomous operation [15].

In this chapter, we describe the design of a low-cost, modular, bioinspired robotic fish platform including the accompanying enabling technologies that are used to quantify animal behavior and response. The original design of the bioinspired robotic fish is inspired by zebrafish, a model organism frequently used in neurobehavioral, developmental, and preclinical research [16–18]. Since its inception, several versions of the robotic fish platform have been used in a range of experimental studies to investigate spatial preference, boldness and shyness, anxiety-related response, hydrodynamic implications of swimming, information flow during social interactions, effect of color morphs on courtship, and collective behavior in fish [7, 12–15, 19–30]. Here, we summarize two of those studies. The first study focuses on the preference of zebrafish individuals and shoals for an anchored version of the robotic fish [20], and the second study focuses on the response of small shoals to an autonomous version of the robotic fish [15].

2 Customizable Robotic Platform for Lab Fish Studies

In this section, we describe the hardware components of the robotic platform and the enabling technologies used in behavioral studies [15, 31]. We designed the robotic platform for low-cost assembly, customization, and ease of implementation. The robot has the actuation and control mechanism onboard and can be controlled to perform specific maneuvers, similar to other prototypes [32–40]. Fish–robot interactions and fish behavior are quantified using standard methods of data collection and data assimilation that require minimal user training.

2.1 Hardware

The robotic fish body is modeled in a computer aided design (CAD) software (Fig. 1). We selected the body size to (a) match the aspect ratio of a zebrafish, and (b) to contain sufficient room for housing the electronics needed for autonomous operation. The robot itself consists of two parts, a body and a movable tail. A flexible caudal fin attached to the tail provides the necessary propulsion for swimming underwater. The length, width, and height of the robot body are 15.4 cm, 4.8 cm, and 2.6 cm, respectively. This size permits inserting a servomotor in the tail section; the body section can hold a microcontroller for controlling the servomotor, a transceiver to send and receive the control signals, and a rechargeable battery that can hold charge for up to an hour of regular operation.

The total cost to manufacture a single robot including the electronics is less than 100 USD [41]. The robotic fish is fabricated in a 3D prototyping machine using ABS

Fig. 1 The robotic fish is designed to match the zebrafish color pattern, aspect ratio, and caudal fin shape. The robot has onboard electronics and motor that allow it to swim autonomously

plastic. A single-cell rechargeable lithium polymer battery, an Arduino Pro mini microcontroller (Sparkfun Electronics, Boulder, Colorado, USA), and an nRF2401A transceiver chip (Nordic Semiconductor, Oslo, Norway) are all assembled outside the body and can be replaced if needed. A Hitec HS-55 servomotor (Hitec RCD 155 USA Inc., Poway, California, USA) in the tail section actuates the body–tail joint and propels the robot in a carangiform/subcarangiform movement. The robot is remotely controlled using an Arduino Duemilanove microcontroller (Sparkfun Electronics, Boulder, Colorado, USA) that interfaces with a computer and an nRF2401A transceiver chip. The onboard microcontroller is used to control the tail-beat frequency, amplitude, and offset, which in turn controls the robot speed and turning rate [31]. The robot can be controlled in real time to perform interactive experiments. In [14], for example, the platform is augmented to allow the robotic fish to beat its tail in response to fish position. This is achieved by tracking the fish position in real time to control the tail-beat frequency of the anchored robot.

The robotic platform is highly customizable. Besides its use in laboratory experiments, we have extensively used the robotic fish in outreach programs that encourage the involvement of K-12 students in science, technology, engineering, and mathematics (STEM) fields [42–45]. For example, students in an outreach activity have been tasked with designing the geometry of the caudal fin to maximize the swimming thrust based on observations of animal morphology at the New York Aquarium [42–44]. The modular design also permits disassembly and changing of individual parts on-the-go making the platform ideal for testing hardware performance [41]. The caudal fin can be attached and removed easily to test the effect of different shapes and sizes on swimming performance [42]. The body can be painted with nontoxic pigments to match the color pattern of a zebrafish. Finally, to make the robot accessible for users in a wide age-group, such as those who are likely to participate in a public event, we custom designed a smart phone application to replace the remote control unit [44].

2.2 Enabling Technologies

Animal behavior studies often entail the continuous observation of live animals over a considerable amount of time [46]. Quite often, this task is performed by human observers and is prone to bias and fatigue, whereby an experimenter may involuntarily score a behavior that is not present. In this respect, a data assimilation workflow that allows automatic quantification of fish behavior would overcome all these limitations. Enabling technologies for such a workflow include a video multitarget tracking system and behavioral analysis scripts that can calculate behavioral measures directly from video data [47–49]. Running in real time, these tools offer the capability to control the robot to perform specific maneuvers [15, 27] as well as to respond to fish behavior [14].

While human-assisted behavioral phenotyping was adopted in the first study discussed in this chapter, a multitarget tracking software was developed in

MATLAB (R2011a, Mathworks, Natick, Massachusetts, USA) for the second study. The input to the tracking system was video from an overhead camera view. The output was two-dimensional position and velocity estimates of each fish (and the robot, if present) in the tank at each frame. The tracking algorithm included a measurement extraction procedure where, in each frame, individual fish were segmented as blobs after background subtraction. An optimal filtering algorithm called a Kalman filter was used to estimate the position and velocity of each fish. A global optimal assignment algorithm [50] was used to preserve fish identities in scenarios where the fish swam close to each other. In a recent version of the tracking system [51], fish occlusions are resolved automatically as follows: a normal distribution of fish size in pixels is created and updated at each frame until five hundred points are available. Once the distribution is constructed, each blob on the frame is checked to ensure that it is less than two standard deviations of the average size. If the size of the blob is larger, an expectation–maximization (EM) algorithm is used to split the large blob into individual blobs. In particular, the EM algorithm optimally fits multiple Gaussian distributions to the occluded blob so that individual fish shapes are approximated as two-dimensional ellipses [52]. The tracking system consists of a graphical user interface, also developed in MATLAB, which is used to manually verify and repair fish trajectories. In the event of an unresolved occlusion, missed detection, or a false detection, a user can switch, add, and delete tracks projected on the video.

Trajectory data from the tracking system is stored in the form of text files that can be further processed for behavioral analysis. For example, the following select behavioral measures (Fig. 2) can be automatically computed from the tracking system output:

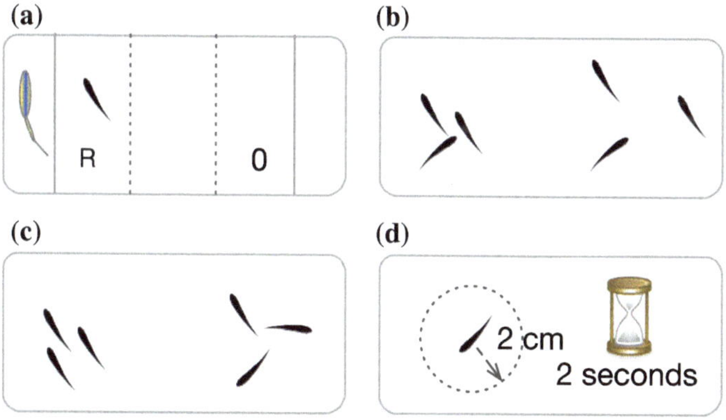

Fig. 2 Trajectory data is used to classify fish behavior into **a** preference for the robot (R) compared to the empty side (0) in a tripartitioned tank; **b** cohesion, where *left* group is more cohesive than the one on the *right*; **c** polarization, where the group on the *left* is more polarized than the one on the *right*; and **d** freezing, which is the percentage of experimental time when the fish stays within a radius of 2 cm for 2 s or more

Preference: the experimental tank is virtually divided into three parts and the time spent by the fish in each part is recorded. Preference for a given choice is then computed as the ratio of the time spent near one end of the tank and the total time spent by the subject in the two parts near the tank ends. For a shoal, preference is computed for each fish individually and averaged over the experimental time.

Cohesion: the degree of cohesion of zebrafish shoals is computed using individual fish positions in terms of the average nearest neighbor distance (ANND). Given the two-dimensional position of fish i at time k as $\mathbf{r}_i[k]$, the ANND at k is

$$\mathrm{ANND}[k] = \frac{1}{N}\sum_{i=1}^{N} \min_{j\in\{1,\ldots,N\},j\neq i}\left(\|\mathbf{r}_i[k] - \mathbf{r}_j[k]\|\right), \tag{1}$$

where N is the total number of fish and $\|\cdot\|$ denotes the standard Euclidean norm. Another measure of cohesion is the average pairwise distance (APD), which is computed by averaging the distances between all pairs of shoal members within the focal group.

Polarization: the degree of group coordination is calculated using fish velocity in terms of polarization that quantifies the degree of alignment in fish motion. Given the two-dimensional velocity $\mathbf{v}_i$ of fish i at time k, polarization is computed as

$$\mathrm{Pol}[k] = \frac{1}{N}\left\|\sum_{i=1}^{N}\hat{\mathbf{v}}_i[k]\right\|, \tag{2}$$

where $\hat{\mathbf{v}}_i[k] = \frac{\mathbf{v}_i[k]}{\|\mathbf{v}_i[k]\|}$ is the direction of motion. Polarization varies between 0 and 1, with a value of 1 corresponding to all fish moving in the same direction and close to 0 if the fish move in randomly distributed directions.

Freezing: fish stress is measured directly from trajectory data in terms of the amount of time spent freezing during each trial. A fish is considered freezing if it spends two continuous seconds within a ball of radius 2 cm [14].

3 Fish Behavioral Studies

3.1 Zebrafish Animal Model

Behavioral research rests upon our understanding of model organisms that share similarities with mammals at developmental, genetic, and behavioral levels [53]. Among such organisms, zebrafish (*Danio rerio*) is rapidly emerging as a valid animal model [54–59] due to its elevated degree of homology with human genes, ability to rapidly absorb psychoactive compounds with minimal invasiveness, short intergeneration time, and high stocking densities [16].

Adult zebrafish are typically 3–4 cm in size and are characterized by a striped color pattern on their body (Fig. 1), hence the name "zebrafish" [60]. They exhibit strong shoaling behavior that has been associated with improved foraging efficiency and predator detection [61]. Zebrafish are especially useful in robotics-based research due to their propensity to form groups based dominantly on vision [16, 17, 62], a sensory modality that can be preferentially exploited by designing the robot morphology to match that of the fish shape and color pattern.

3.2 Preference Experiments

The classical preference test utilizes an experimental setup where a fish is observed as it swims between two different choices presented on either end of an experimental tank [63–65]. In our case, this setup serves to test the hypothesis that a robotic fish, which is inspired by zebrafish in its shape, color pattern, and motility, will attract single individuals as well as shoals of zebrafish. The robotic fish is anchored to one side of the tank and actuated externally. The size of the robotic fish allows for housing all the electronics necessary for self-propulsion, toward further implementation beyond choice tests. This permits using the results of this experiment to inform future studies. For example, protocols for free-swimming and field experiments that require the robot to be autonomous and interact with the fish without any tethering can be designed on the basis of the average distance of the fish from the robot quantified from preference studies. Similarly, the color pattern and preferential frequency in one study can be used as a reference condition in the next [15, 21].

The zebrafish used in this study were acquired from local pet stores (Petland Discounts, Brooklyn and New World Aquarium, Manhattan, New York City, USA) and acclimatized for at least two weeks in the housing facility at New York University Polytechnic School of Engineering before use in the experiments. The housing tanks were maintained at 26 ± 1 °C temperature and 7.2 pH. Lighting was controlled according to a 12 h light/day circadian rhythm [66] and the stocking density was maintained at less than 1 fish per liter at all times.

The experimental setup consisted of a 74 cm × 30 cm × 30 cm glass tank with the longest side partitioned into three distinct regions using a perforated transparent plexiglass (Fig. 3). The central region where the experimental subjects were present was 54 cm long and the choice regions on either side of the tank were 10 cm long each. The plexiglass partition permitted physical separation between the subject and the stimuli while preserving visual communication. The tank surface was lit by two 50 W fluorescent lamps on either side to ensure a uniform and diffused lighting. Fish behavior was recorded with a high definition video camera (Canon, Vixia HG20, Japan) mounted approximately 150 cm above the test tank. The test setup was isolated from external disturbances using black curtains on all four sides of the tank.

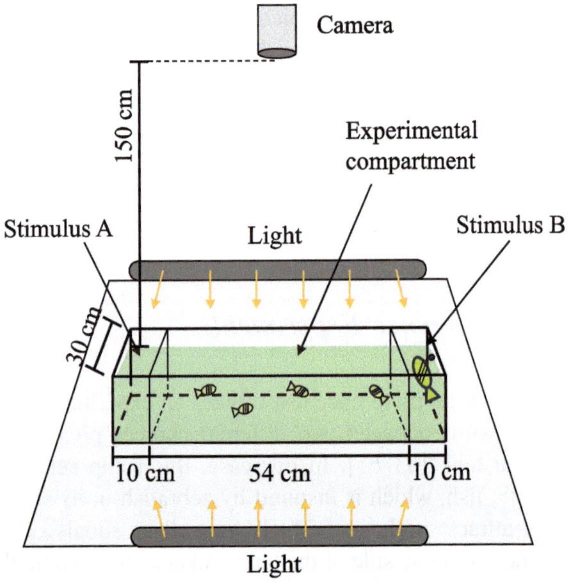

Fig. 3 Schematic of the preference test experimental apparatus. The two choices are the robotic fish and an empty compartment (© IOP Publishing. Reproduced by permission of IOP Publishing and [19]. All rights reserved)

The experimental procedure consisted of sixteen experimental conditions[1] that tested different combinations of choices with single individuals and small shoals of zebrafish (Fig. 4). In particular, the choices available for the subjects in the central compartment were Robot (R), one fish (1), ten fish (10), static replica that had no tail actuation (SR), and empty compartment (0). To investigate the effect of noise due to servomotor, the absence of visual cues, one-way visual feedback, and physical separation, the setup was modified so that experiments were performed without lighting (Dark), using one-way glass (Glass), and without the transparent plexiglass (Free). Ten trials were conducted for each condition, where each trial consisted of a 10 min habituation period followed by a 5 min experimental time during which the fish were observed every thirty seconds. The robotic fish, wherever present, was anchored to one side of the tank at an angle of 45° with tail beating at 2.3 Hz frequency at 3 cm amplitude (except in the case of SR when tail-beat was absent). The tail-beat frequency and amplitude were selected on the basis of engineering constraints and biological observations. Experimental trials for each condition were distributed uniformly between 10 am and 7 pm to ensure no bias existed because of the time of the day. In this first exploratory study involving large fish populations, fish were sometimes tested more than once in a randomized protocol.

Fish behavior was quantified in terms of preference for a stimulus, APD and Extended ANND (EANND). Given two choices A and B, and n_A and n_B the number of instances spent by a subject near the stimulus A and B, the preference for A was

[1]The experimental procedure was approved by Polytechnic Institute of New York University (now New York University Polytechnic School of Engineering) Animal Welfare Oversight Committee AWOC-2011-101.

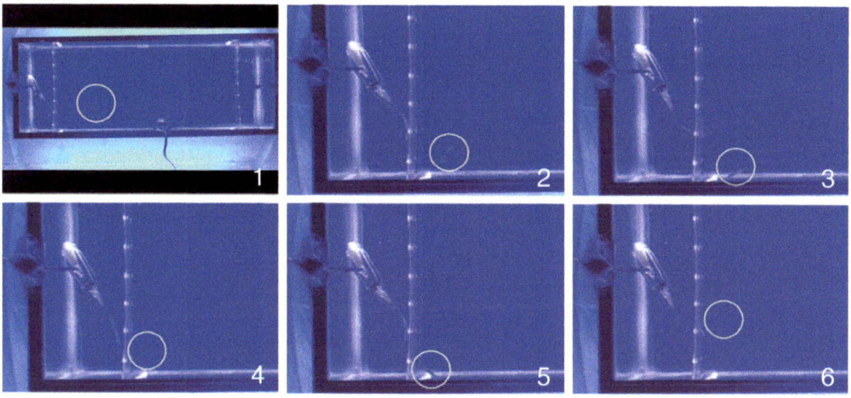

Fig. 4 Successive snapshots of a sample experimental video. Six frames from an experimental video one-second apart show the robotic fish and the live zebrafish. The first frame is at full resolution and the following frames zoomed in for better visibility. The fish is *circled* in each frame

computed as $n_A/(n_A + n_B)$. Extended ANND was computed in the same manner as ANND in (1) but including the robotic fish as part of the group and only considering positions along the length of the tank. Fish preference was compared to no preference within a condition using chi-square statistical tests and fish APD (also computed along the length of the tank) and EANND were compared using one-way ANOVA [67]. Post hoc comparisons, wherever significance was found, were performed using Fisher's protected least significant difference tests. Significance level was set to $p < 0.05$.

The results from statistical comparisons are summarized in Table 1 and Fig. 5. These results show that both individuals and small shoals tend to join larger shoals when given a choice between a shoal of ten conspecifics and one. Comparison between an empty compartment and robotic fish preference indicate that both individuals and small shoals of zebrafish tend to prefer the bioinspired robotic fish, and that this preference is independent of the existence of a physical barrier between them. At the same time, when given a choice between the robotic fish and live conspecifics, the subjects preferred the live fish, indicating that the robotic fish is not perceived as a conspecific. Control conditions show that the noise from the servomotor in the dark has a negative effect on fish preference whereby they spend significantly more time near the empty compartment when unable to see the robotic fish. The presence of holes or visual feedback did not have a significant effect on fish preference. Finally, robot body movement played an important role as shown by the increased preference for the robot moving its tail over a static replica.

3.3 Free-Swimming Experiments

The results from preference experiments demonstrate that zebrafish individuals and shoals preferred the robotic fish to an empty compartment. The robotic fish design

Table 1 Synopsis of the main results

Conditions	Dist. from stimulus (cm)	Center (%)	Preference (%)	χ^2 ($df = 1$)	p	APD (cm)	EANND (cm)
Single fish							
1 v 0	18.4	24.6	72.0	117.2	0	–	–
10 v 0	16.0	25.6	79.6	208.8	0	–	–
10 v 1	17.1	20.4	71.6	119.3	0	–	–
R v 0	22.6	40.9	63.7	35.4	0	–	–
R v 1	29.6	33.4	40.6	19.0	0	–	–
R v 0 (Free)	30.1	20.1	62.3	38.8	0	–	–
0 v 0	26.3	37.3	53.4	2.4	0.080	–	–
R v 0 (dark)	31.2	41.8	38.1	26.4	0	–	–
R v SR	23.5	43.3	56.6	7.9	0.003	–	–
1 v 0 (dark)	27.1	44.7	51.2	0.3	0.702	–	–
R v 1 (glass)	31.9	41.4	35.0	42.1	0	–	–
Shoal of four fish							
1 v 0	30.5	35.8	61.1	101.0	0	10.5	3.9
10 v 0	15.4	28.8	83.8	1041.9	0	10.5	3.8
10 v 1	17.2	26.4	77.1	692.8	0	12.3	3.3
R v 0	24.2	46.7	60.5	74.9	0	9.5	3.7
R v 1	31.0	43.1	34.5	175.4	0	14.1	3.4

Distance from stimulus is the mean position from the first stimulus in the condition. The length of the central compartment is 54 cm in all conditions except those that are performed without partitions (Free) where it is 74 cm. Center denotes the average percent of occurrences over all trials in the condition in which the subjects were found in the central compartment. Preference gives the mean percent preference for the side of the apparatus containing the first stimulus. Chi-square values compare the preference for either stimulus. $p < 10^{-5}$ is shown as 0 (© IOP Publishing. Reproduced by permission of IOP Publishing and [20]. All rights reserved)

permits autonomous operation using onboard electronics. In a second study, we allow the robotic fish to swim autonomously with the help of the online tracking system to test the hypothesis that robot fish spatial movement will modulate the behavior of small shoals of zebrafish.

Zebrafish used in this study were acquired from an online aquarium (LiveAquaria.com, Rhinelander, Wisconsin, USA) and maintained in 37.8 l tanks at a stocking density of at most 1 fish per two liters. The fish were acclimatized for at least 10 days to the new environment before starting the experiments. Fish tanks were lit under a 12 h light/day circadian rhythm [66]. Water temperature and pH in the holding and experimental tanks were maintained at 27 ± 1 °C and 7.2, respectively.

The experimental setup comprised a large square water tank of 120 cm side length and 20 cm high mounted on an aluminum frame (Fig. 6). The water level was maintained at 10 cm during the experiments. A Web camera (Life Cam,

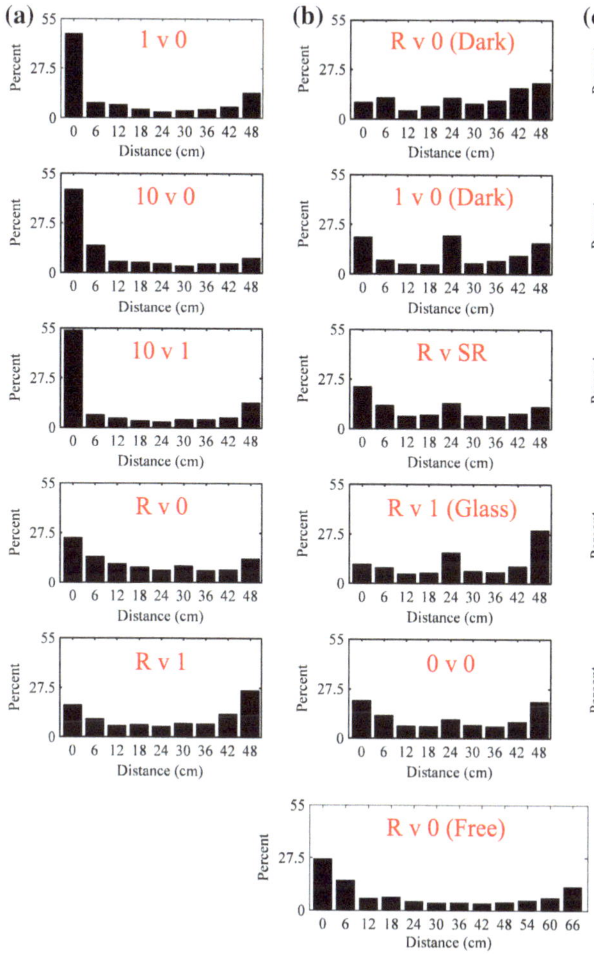

Fig. 5 Graphical synopsis of preference data: histograms of position data points for zebrafish position frequency for each tested pair of conditions reported in Table 1. Column **a** presents main experiments on single individuals, column **b** presents control experiments on individuals, and column **c** presents main experiments on shoals. In the condition labels, R is the bioinspired robot and SR is the static replica. In addition, Free refers to the free-swimming scenario, Dark to the experiments in the dark, and Glass to the use of one-way glass partitions. Note that the region available for fish to swim in the free-swimming case is larger than all other conditions (© IOP Publishing. Reproduced by permission of IOP Publishing and [20]. All rights reserved.)

Microsoft Corp., Seattle, Washington, USA) was mounted 150 cm above the water surface to film an overhead view of the tank. The tank was lit by diffused light from four 150 W fluorescent tubes mounted 100 cm above the water surface. The multitarget tracking and control algorithm was run on a 2.5 GHz Pentium dual core desktop computer with 3 GB memory. The experimental setup was isolated using dark curtains on all sides of the tank.

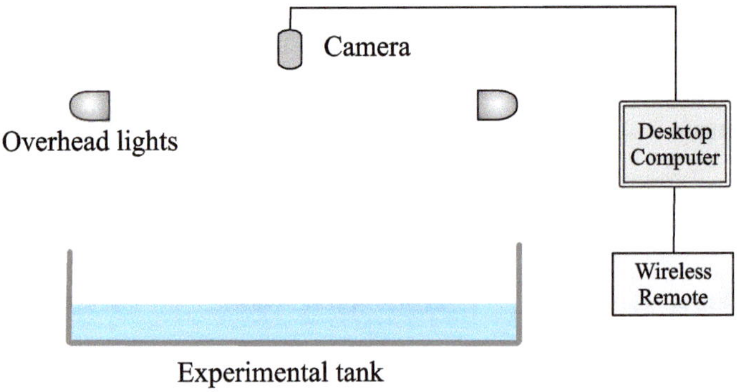

Fig. 6 Schematic of the free-swimming test experimental apparatus. The experimental apparatus consisted of a square shallow tank and an overhead camera for real-time tracking (Reproduced with permission from [15])

The control algorithm consisted of a Proportional Integral Derivative (PID) controller to maneuver the robotic fish in fixed-size circular trajectories in the presence of groups of zebrafish (Fig. 7). The control signal was sent to the robot via the transceiver every three-fifths of a second to follow a set of sixteen waypoints, $\mathbf{w}_s$, $s = 1, \ldots, 16$, spaced equally on a 40 cm circle centered in the tank in a clockwise motion. In particular, the control input to attain the desired direction of motion $\hat{\mathbf{v}}_R^d[k]$ was computed using estimates of robot position $\mathbf{r}_R[k]$ and velocity $\mathbf{v}_R[k]$ at each k as

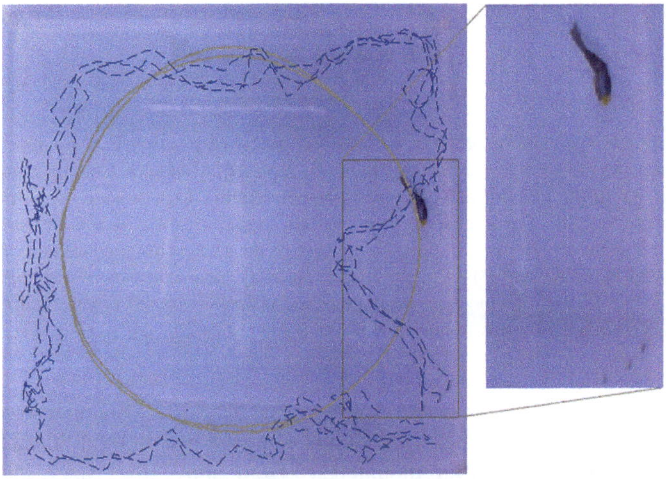

Fig. 7 Overhead view of robotic fish swimming autonomously in the presence of a zebrafish shoal. Individual trajectories of robotic fish and zebrafish are also shown as tan *solid lines* and *blue dashed lines*, respectively

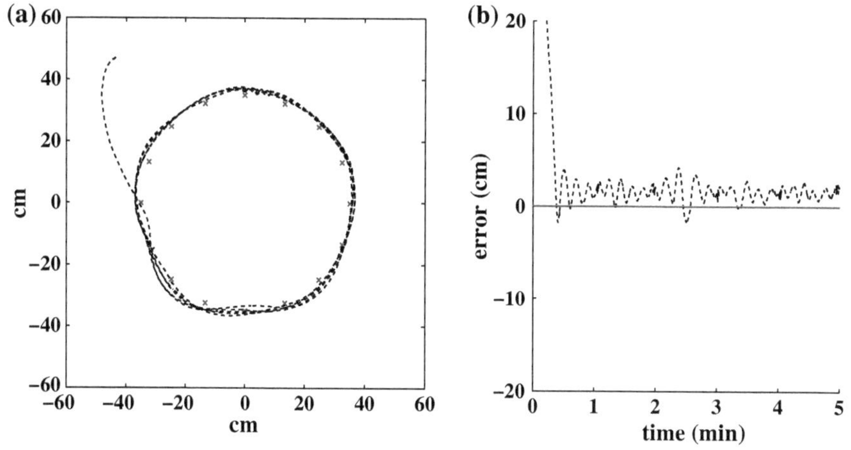

Fig. 8 Robot trajectory with reference to waypoints (**a**) and the error (**b**) (Reproduced with permission from [15])

$$\hat{\mathbf{v}}_R^d[k] = \frac{\mathbf{w}_s[k] - \mathbf{r}_R[k]}{||\mathbf{w}_s[k] - \mathbf{r}_R[k]||},$$

where $\mathbf{w}_s[k]$ is the waypoint-to-reach at the current time-step. At frame k', when the robot was within a threshold distance (15 cm) of the current waypoint-to-reach, the value of the current waypoint was updated. The control input was computed as a function of the error $e[k] = \sin(\theta[k])$, where $\theta(k) = \arg(\hat{\mathbf{v}}_R[k] - \mathbf{v}_R^d[k])$ is the angle between the robot direction of motion and the desired direction of motion. The resulting PID control was

$$u[k] = K_p e[k] + K_i \sum_{l=k'}^{k} e[l]\Delta t + K_d \frac{\Delta e[k]}{\Delta t},$$

where K_p, K_i, and K_d are the proportional, integral, and derivative control gains and $\Delta e[k] = e[k] - e[k-1]$. The control gains were tuned so that the robot followed the circle closely for over five minutes in the test trials. Figure 8 shows the robot trajectory in comparison to the waypoints on the tank region.

We tested four experimental conditions[2] covering a range of swimming speeds (0, 2, 3, and 4 cm/s) corresponding to different tail-beat frequencies (0, 1, 2, and 3 Hz) of the robotic fish as it swam in fixed-size circles within the tank. The tail-beat frequencies corresponded to variations about the 2 Hz value used earlier in the

[2]Experiments followed protocol numbers AWOC-2012-101 and AWOC 2013-103 that were approved by the Animal Welfare Oversight Committee of the Polytechnic Institute of New York University (now New York University Polytechnic School of Engineering).

preference tests [20]. The remote control unit was programmed to receive control parameters namely tail-beat frequency, tail-beat amplitude, and tail-section offset via a universal serial bus (USB), which were then transmitted wirelessly to the robot. In our experiments, we kept the tail-beat amplitude constant at 20°. The heading of the robot was controlled by varying the tail-section offset from a trim value of zero degrees when the robot body was in line with the tail section. Additional experiments where the fish were observed without a robot in the tank (No Robot), and where the robotic fish was held stationary in a place with its tail beating at 2 Hz (Fixed) were tested to control for the presence of robot in the tank and its body movement. Eight trials were performed for each condition with three experimentally naive fish used per trial.

Fish response to the robot was quantified in terms of average speed, ANND, and polarization. Fish interaction with the robot was quantified using average and minimum distance to the robot, and relative group speed. Freezing behavior was measured in terms of percentage of the total experimental time. All comparisons were made using one-way ANOVA statistical tests and post hoc comparisons were made using Tukey-HSD tests [67]. One-way ANOVA was used to compare the experimental conditions with the control conditions.

The results of the experiment are summarized in Fig. 9. Statistical comparisons of group behavior show that group cohesion (ANND) varied significantly with robot speed ($p = 0.0126$). The maximum and minimum values of ANND at 14.87 and 3.6 cm were recorded when the robot swam at 3 and 4 cm/s, respectively. Post hoc comparisons with control conditions did not show a significant difference due to the presence of the robot (No robot and 0 Hz, $p = 0.099$) and the presence of tail-beat movement only (Fixed and 2 Hz, $p = 0.243$). Group coordination as measured using polarization failed to reach statistical significance ($p = 0.123$). As with cohesion, the presence of the robot (No robot and 0 Hz, $p = 0.172$) and tail-beat movement (Fixed and 2 Hz, $p = 0.740$) did not have an effect on polarization. Although group speed was not affected by the robot speed ($p = 0.151$), the presence of a robot produced a significant effect (No robot and 0 Hz, $p < 0.01$). Finally, the time spent freezing was not significantly affected by robot speed ($p = 0.171$), robot presence (No robot and 0 Hz, $p = 0.091$), or due to tail-beat movement (Fixed and 2 Hz, $p = 0.642$).

Fish–robot interactions measured in terms of average ($p = 0.067$) and minimum ($p = 0.093$) distance to the robot were seemingly affected by the robot speed but failed to reach statistical significance. While the average distance to the robot stayed more than 45 cm, the 2 Hz condition saw the largest difference of 9.5 cm between average and minimum distance. Relative group speed varied significantly ($p = 0.0154$) with robot speed. Post hoc comparisons show that the fish tend to match their speed with the robot closely at 2 Hz (3 cm/s) and that this value of relative speed is significantly different from when the robot was stationary or swimming at 1 Hz (2 cm/s).

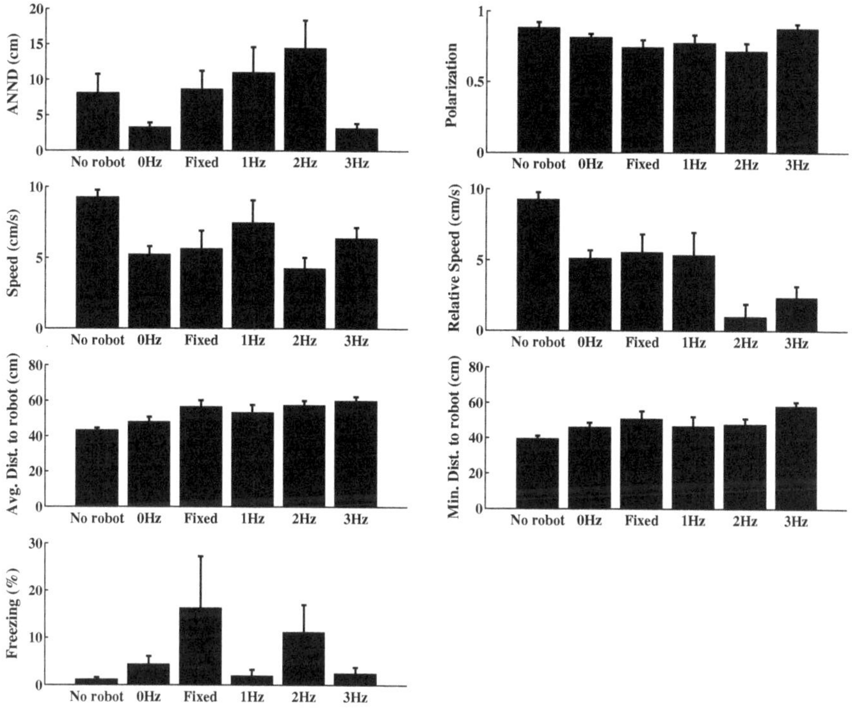

Fig. 9 Group cohesion measured using ANND and group relative speed varied significantly with robot speed (Reproduced with permission from [15])

4 Discussion and Conclusion

The two experimental studies described here demonstrate the capability of using a bioinspired robotic fish to modulate live zebrafish behavior. Results from the first study show that both individual fish and small shoals display a robust attraction toward the robotic fish when given a choice between the robot and an empty compartment. This preference is lost when the choice is between the robot and live zebrafish, showing that the former is not perceived as a conspecific. This same response is confirmed in the second study where the ANND of fish shoals was found to be considerably smaller than their distance to the robot.

The robotic fish in both studies present competing cues that on one hand attract the zebrafish through its color pattern, body movement, and aspect ratio [21]; on the other hand it repels them with its servomotor noise [20]. At the same time, preference due to the possibility of the robot being inspected as a novel object is remote because of the 10 min habituation time [68] that provides ample opportunity for the fish to come close. Fish shoals in the free-swimming test maintain a larger distance from the robot as compared to those in the preference test. The closest analog in the preference test is perhaps the R v 0 (free) condition that allows direct physical contact, where the fish demonstrate significant preference for the robot. Since in the free-swimming test the robot is additionally covering a large experimental region, the presence of relatively large fish–robot distances suggests that the robot spatial movement is unlikely to constitute an attractive stimulus. While we cannot dismiss the possibility that the robot is perceived as a predator in the free-swimming environment, we do not observe a significantly larger freezing response in the presence of the robot. Compared to the preference test setup, a free-swimming environment with a mobile robot presents a complex interplay of cues that require further studies investigating the perception of robot motion by live zebrafish. In this respect, we have conducted experiments to study the effect of number of robots, their speed, and their configuration on zebrafish [27].

The collective behavior of the shoals is modulated by the robotic fish in both the studies. In the preference test, the APD of approximately three body lengths and an EANND of approximately one body length indicates that the preference of the shoal toward the robot is likely a result of one-to-one interaction and not individual preference, which would otherwise result in a ceiling effect. In the free-swimming scenario, though the shoals maintain a relatively larger distance from the robot, the fish exhibit maximum disparity between minimum and average distance to the robot when the robot's tail is beating at 2 Hz. This is also the frequency at which the fish match their speed closely to the robot, and the shoal is least cohesive, suggesting that visual cues at this combination of speed and body movement are relevant in shaping fish–robot interactions. More importantly, this combination of body and spatial movement of the robotic fish, where the fish tend to explore the unconstrained free-swimming environment, shows that both types of motion differentially modulate zebrafish behavior.

In summary, an ethorobotics approach as described in this chapter presents an important direction for the design of a robotic fish—one that bears direct relevance to how robots may aid biology and in turn benefit their own design. The modular design used here makes the robotic fish easy to customize; the actuation mechanism adds a natural undulating movement that is shown to affect fish preference. The robotic platform is complemented with enabling technologies that allow controlling the robotic fish to perform specific maneuvers in a free-swimming environment, and opens the possibility to actively interact with the fish [14]. Finally, open problems include mitigating the repelling cues such as servomotor noise by using the alternative propulsion techniques such as ionic polymer metal composites and piezoelectric materials [29, 30, 69–73].

Acknowledgments This research was supported by National Science Foundation under grants CMMI- 0745753, DGE-0741714, CMMI-1129820, and CMMI-1433670. The authors would also like to thank Giovanni Polverino, Vladislav Kopman, and Tiziana Bartolini who have contributed to the research efforts summarized in this chapter.

References

1. Boissy A (1995) Fear and fearfulness in animals. Q Rev Biol 70(2):165–191
2. Krause J, Winfield AFT, Deneubourg J (2011) Interactive robots in experimental biology. Trends Ecol Evol 26(7):369–375
3. Rossi C, Coral W, Barrientos A (2012) Robotic fish to lead the school. In: Palstra AP, Planas JV (eds) Swimming physiology of fish. Springer, New York, pp 407–421
4. Ward AJW, Sumpter DJT, Couzin ID, Hart PJB, Krause J (2008) Quorum decision-making facilitates information transfer in fish shoals. Proc Natl Acad Sci USA 105(19):6948–6953
5. Partan SR, Larco CP, Owens MJ (2009) Wild tree squirrels respond with multisensory enhancement to conspecific robot alarm behaviour. Anim Behav 77(5):1127–1135
6. Halloy J, Sempo G, Caprari G, Rivault C, Asadpour M, Tâche F, Saïd I, Durier V, Canonge S, Amé JM, Detrain C, Correll N, Martinoli A, Mondada F, Siegwart R, Deneubourg JL (2007) Social integration of robots into groups of cockroaches to control self-organized choices. Science 318(5853):1155–1158
7. Spinello C, Macrì S, Porfiri M (2013) Acute ethanol administration affects zebrafish preference for a biologically-inspired robot. Alcohol 47(5):391–398
8. Todd DJ (1993) Mobile robots-the lessons from nature. Robot Biol Syst Towards New Bionics 102(1):193–206
9. Webb B (2002) Can robots make good models of biological behaviour? Behav Brain Sci 24(06):1033–1050
10. Tokekar P, Branson E, Vander Hook J, Isler V (2013) Tracking aquatic invaders: autonomous robots for monitoring invasive fish. Robot Autom Mag 20(3):33–41
11. Faria JJ, Dyer JRG, Clément RO, Couzin ID, Holt N, Ward AJW, Waters D, Krause J (2010) A novel method for investigating the collective behaviour of fish: introducing 'Robofish'. Behav Ecol Sociobiol 64(8):1–8
12. Marras S, Porfiri M (2012) Fish and robots swimming together: attraction towards the robot demands biomimetic locomotion. J R Soc Interface 9(73):1856–1868
13. Cianca V, Bartolini T, Porfiri M, Macrì S (2013) A robotics-based behavioral paradigm to measure anxiety-related responses in zebrafish. PLoS ONE 8(7):e69661
14. Kopman V, Laut J, Polverino G, Porfiri M (2013) Closed-loop control of zebrafish response using a bioinspired robotic-fish in a preference test. J R Soc Interface 10(78):20120540
15. Butail S, Bartolini T, Porfiri M (2013) Collective response of zebrafish shoals to a free-swimming robotic fish. PLoS ONE 8(10):e76123
16. Miklósi A, Andrew RJ (2006) The zebrafish as a model for behavioral studies. Zebrafish 3(2):227–234
17. Gerlai R (2010) Zebrafish antipredatory responses: a future for translational research? Behav Brain Res 207(2):223–231
18. Kalueff AV, Stewart AM (eds) (2012) Zebrafish protocols for neurobehavioral research. Humana Press, New York
19. Ladu F, Bartolini T, Panitz S, Chiarotti F, Butail S, Macrì S & Porfiri M (2015) Live predators, robots, and computer-animated images elicit differential avoidance responses in zebrafish, Zebrafish. doi: 10.1089/zeb.2014.1041
20. Polverino G, Abaid N, Kopman V, Macrì S, Porfiri M (2012) Zebrafish response to robotic fish: preference experiments on isolated individuals and small shoals. Bioinspir Biomim 7(3):036019
21. Abaid N, Bartolini T, Macrì S, Porfiri M (2012) Zebrafish responds differentially to a robotic fish of varying aspect ratio, tail beat frequency, noise, and color. Behav Brain Res 233 (2):545–553
22. Polverino G, Porfiri M (2013) Mosquitofish (*Gambusia affinis*) responds differentially to a robotic fish of varying swimming depth and aspect ratio. Behav Brain Res 250(1):133–138
23. Polverino G, Porfiri M (2013) Zebrafish (*Danio rerio*) behavioural response to bioinspired robotic fish and mosquitofish (*Gambusia affinis*). Bioinspir Biomim 8(4):044001

24. Polverino G, Phamduy P, Porfiri M (2013) Fish and robots swimming together in a water tunnel: robot color and tail-beat frequency influence fish behavior. PLoS ONE 8(10):e77589
25. Abaid N, Marras S, Fitzgibbons C, Porfiri M (2013) Modulation of risk-taking behaviour in golden shiners (*Notemigonus crysoleucas*) using robotic fish. Behav Process 100:9–12
26. Butail S, Ladu F, Spinello D, Porfiri M (2014) Information flow in animal-robot interactions. Entropy (Spec Iss Inf Dyn Syst Complex Syst) 16(3):1315–1330
27. Butail S, Polverino G, Phamduy P, Del Sette F, Porfiri M (2014) Influence of robotic shoal size, configuration, and activity on zebrafish behavior in a free-swimming environment. Behav Brain Res 275:269–280
28. Phamduy P, Polverino G, Fuller RC, Porfiri M (2014) Fish and robot dancing together: bluefin killifish females respond differently to the courtship of a robot with varying color morphs. Bioinspir Biomim 9(3):036021
29. Aureli M, Porfiri M (2010) Coordination of self-propelled particles through external leadership. Europhys Lett 92(4):40004
30. Aureli M, Fiorilli F, Porfiri M (2012) Portraits of self-organization in fish schools interacting with robots. Phys D Nonlinear Phenom 241(9):908–920
31. Kopman V, Laut J, Acquaviva F, Rizzo A, Porfiri M (2014) Dynamic modeling of a robotic fish propelled by a compliant tail. IEEE J Ocean Eng PP(99):1–13
32. Strefling PC, Helium AM, Mukherjee R (2012) Modeling, simulation, and performance of a synergistically propelled ichthyoid. IEEE/ASME Trans Mechatron 17(1):36–45
33. Low KH, Chong CW (2010) Parametric study of the swimming performance of a fish robot propelled by a flexible caudal fin. Bioinspir Biomim 5(4):046002
34. Barrett DS, Triantafyllou MS, Yue DKP, Grosenbaugh MA, Wolfgang MJ (1999) Drag reduction in fish-like locomotion. J Fluid Mech 392:183–212
35. Shao J, Wang L, Yu J (2008) Development of an artificial fish-like robot and its application in cooperative transportation. Control Eng Pract 16(5):569–584
36. Malec M, Morawski M, Zajac J (2010) Fish-like swimming prototype of mobile underwater robot. J Autom Mobile Robot Intell Syst 4:25–30
37. Liu J, Hu H (2010) Biological inspiration: from carangiform fish to multi-joint robotic fish. J Bionic Eng 7(1):35–48
38. Morgansen KA, Triplett BI, Klein DJ (2007) Geometric methods for modeling and control of free-swimming fin-actuated underwater vehicles. IEEE Trans Robot 23(6):1184–1199
39. Guo J (2006) A waypoint-tracking controller for a biomimetic autonomous underwater vehicle. Ocean Eng 33(17):2369–2380
40. Alvarado PV y, Youcef-Toumi K (2006) Design of machines with compliant bodies for biomimetic locomotion in liquid environments. J Dyn Syst Meas Control 128(1):3–13
41. Kopman V, Porfiri M (2012) Design, modeling, and characterization of a miniature robotic-fish for research and education in biomimetics and bioinspiration. IEEE/ASME Trans Mechatron 18(2):471–483
42. Abaid N, Kopman V, Porfiri M (2013) An attraction toward engineering careers: the story of a Brooklyn outreach program on biomimetics, underwater robotics, and marine science for K-12 students. IEEE Robot Autom Mag 20(2):31–39
43. Laut J, Bartolini T, Porfiri M (2014) Bioinspiring an interest in STEM. IEEE Trans Educ 58(1):48–55
44. Abaid N, Bernhardt J, Frank JA, Kapila V, Kimani D, Porfiri M (2013) Controlling a robotic fish with a smart phone. Mechatronics 23(5):491–496
45. Abaid N, Yuvienco C, Kapil V, Iskander M (2011) Mechatronics mania at the inaugural USA science and engineering festival. IEEE Control Syst Mag 31(5):105–124
46. Blaser R, Gerlai R (2006) Behavioral phenotyping in zebrafish: comparison of three behavioral quantification methods. Behav Res Methods 38(3):456–469
47. Pérez-Escudero A, Vicente-Page J, Hinz RC, Arganda S, de Polavieja GG (2014) idTracker: tracking individuals in a group by automatic identification of unmarked animals. Nat Methods 11:743–748

48. Delcourt J, Denoël M, Ylieff M, Poncin P (2013) Video multitracking of fish behaviour: a synthesis and future perspectives. Fish Fish 14:186–204
49. Dell AI, Bender JA, Branson K, Couzin ID, de Polavieja GG, Noldus LP, et al. (2014) Automated image-based tracking and its application in ecology. Trends Ecol Evol 29:417–428
50. Kuhn HW (1955) The Hungarian method for the assignment problem. Nav Res Logist Q 2:83–97
51. Ladu F, Butail S, Macrì S, Porfiri M (2014) Sociality modulates the effects of ethanol in zebrafish. Alcohol Clin Exp Res 38(7):2096–2104
52. Carson C, Belongie S (2002) Blobworld: image segmentation using expectation-maximization and its application to image querying. IEEE Trans Pattern Anal Mach Intell 24(8):1026–1038
53. Lieschke GJ, Currie PD (2007) Animal models of human disease: zebrafish swim into view. Nat Rev Genet 8(5):353–367
54. Grunwald DJ, Eisen JS (2002) Headwaters of the zebrafish—emergence of a new model vertebrate. Nat Rev Genet 3(9):717–724
55. Mathur P, Guo S (2010) Use of zebrafish as a model to understand mechanisms of addiction and complex neurobehavioral phenotypes. Neurobiol Dis 40(1):66–72
56. Gerlai R (2012) Using zebrafish to unravel the genetics of complex brain disorders. In: Current topics in behavioral neurosciences, behavioral neurogenetic, vol 12. Springer, Berlin, pp 3–24
57. Goldsmith JR, Jobin C (2012) Think small: zebrafish as a model system of human pathology. J Biomed Biotechnol 2012:817341
58. Kalueff AV, Steward AM, Gerlai R (2014) Zebrafish as an emerging model for studying complex brain disorders. Trends Pharmacol Sci 35(2):63–75
59. Stewart AM, Nguyen M, Wong K, Poudel MK, Kalueff AV (2014) Developing zebrafish models of autism spectrum disorder (ASD). Prog Neuro-Psychopharmacol Biol Psychiatry 50:27–36
60. Laale HW (1977) The biology and use of zebrafish, *Brachydanio rerio* in fisheries research. A literature review. J Fish Biol 10(2):121–173
61. Buske C, Gerlai R (2011) Shoaling develops with age in Zebrafish (*Danio rerio*). Prog. Neuro-Psychopharmacol Biol Psychiatry 35(6):1409–1415
62. Saverino C, Gerlai R (2008) The social zebrafish: behavioral responses to conspecific, heterospecific, and computer animated fish. Behav Brain Res 191(1):77–87
63. Pritchard VL, Lawrence J, Butlin RK, Krause J (2001) Shoal choice in zebrafish, *Danio rerio*: the influence of shoal size and activity. Anim Behav 62(6):1085–1088
64. Bass SLS, Gerlai R (2008) Zebrafish (*Danio rerio*) responds differentially to stimulus fish: the effects of sympatric and allopatric predators and harmless fish. Behav Brain Res 186(1):107–117
65. Snekser JL, Ruhl N (2010) The influence of sex and phenotype on shoaling decisions in zebrafish. Int J Comp Psychol 23(1):70–81
66. Cahill G (2002) Clock mechanisms in zebrafish. Cell Tissue Res 309(1):27–34
67. McDonald JH (2009) Handbook of biological statistics. Sparky House Publishing, Baltimore
68. Wong K, Elegante M, Bartels B, Elkhayat S, Tien D, Roy S, Goodspeed J, Suciu C, Tan J, Grimes C, Chung A, Rosenberg M, Gaikwad S, Denmark A, Jackson A, Kadri F, Chung KM, Stewart A, Gilder T, Beeson E, Zapolsky I, Wu N, Cachat J, Kalueff AV (2010) Analyzing habituation responses to novelty in zebrafish (*Danio rerio*). Behav Brain Res 208(2):450–457
69. Guo S, Fukuda T, Asaka K (2003) A new type of fish-like underwater microrobot. IEEE/ASME Trans Mechatron 8(1):136–141
70. Aureli M, Kopman V, Porfiri M (2010) Free-locomotion of underwater vehicles actuated by ionic polymer metal composites. IEEE/ASME Trans Mechatron 15(4):603–614
71. Chen Z, Shatara S, Tan X (2010) Modeling of biomimetic robotic fish propelled by an ionic polymer–metal composite caudal fin. IEEE/ASME Trans Mechatron 15(3):448–459
72. Chen Z, Um TI, Bart-Smith H (2011) A novel fabrication of ionic polymer–metal composite membrane actuator capable of 3-dimensional kinematic motions. Sens Actuators A Phys 168(1):131–139
73. Cen L, Erturk A (2013) Bio-inspired aquatic robotics by untethered piezohydroelastic actuation. Bioinspir Biomim 8(1):016006

MIX
Papier aus verantwortungsvollen Quellen
Paper from responsible sources
FSC® C105338

FSC
www.fsc.org

If you have any concerns about our products,
you can contact us on
ProductSafety@springernature.com

In case Publisher is established outside the EU,
the EU authorized representative is:
Springer Nature Customer Service Center GmbH
Europaplatz 3, 69115 Heidelberg, Germany

Printed by Libri Plureos GmbH
in Hamburg, Germany